METHODS IN MOLECULAR BIOLOGY

Series Editor
John M. Walker
School of Life Sciences
University of Hertfordshire
Hatfield, Hertfordshire, AL10 9AB, UK

For further volumes:
http://www.springer.com/series/7651

Listeria monocytogenes

Methods and Protocols

Edited by

Kieran Jordan

Teagasc Food Research Centre, Fermoy, Cork, Ireland

Edward M. Fox

Food Microbiology and Safety Group, Animal, Food and Health Sciences, CSIRO, Werribee, VIC, Australia

Martin Wagner

University of Veterinary Medicine Vienna, Vienna, Austria

Humana Press

Editors
Kieran Jordan
Teagasc Food Research Centre
Fermoy, Cork, Ireland

Edward M. Fox
Food Microbiology and Safety Group
Animal, Food and Health Sciences, CSIRO
Werribee, VIC, Australia

Martin Wagner
University of Veterinary Medicine Vienna
Vienna, Austria

ISSN 1064-3745   ISSN 1940-6029 (electronic)
ISBN 978-1-4939-0702-1   ISBN 978-1-4939-0703-8 (eBook)
DOI 10.1007/978-1-4939-0703-8
Springer New York Heidelberg Dordrecht London

Library of Congress Control Number: 2014937660

Printed on acid-free paper

Humana Press is a brand of Springer
Springer is part of Springer Science+Business Media (www.springer.com)

Preface

From its first description by Murray et al. in 1926 (referred to as *Bacterium monocytogenes*), *Listeria monocytogenes* has frequently been associated with infection of humans and animals [1, 2]. The dual lifestyle of *L. monocytogenes*, from environmental saprophyte to pathogen, has sparked interest in scientists across a range of fields and has advanced our understanding of the biology of the bacterium [3]. The evolution of this understanding has been characterized by many notable milestones. Studies on the ecology of *L. monocytogenes* illustrated its ubiquitous nature, with a large range of environments harboring the organism, including soil, plant material, water, and wastewater, to carriage by animals and humans, often asymptomatically [4, 5]. Although current knowledge suggests that cases of human listeriosis are almost exclusively through foodborne infection, this critical transmission vector became clear only during the 1980s, largely the result of a series of high-profile disease outbreaks, perhaps most notable of which was the Canadian outbreak of 1981, linked to contaminated coleslaw [6]. With many foodborne outbreaks recorded globally every year since then, some of which have been amongst the most severe of any attributed to a bacterial pathogen [7, 8], *L. monocytogenes* has been a driving force in the development of current disease surveillance and control strategies. This includes global surveillance networks such as PulseNet, which allows international comparison of different strains of *L. monocytogenes*. Along with these advances in the epidemiology of the organisms, other strides were being made in the understanding of the pathogenesis of the organism, including its intracellular nature and how this contributed to crossing three key barriers—the intestinal barrier, the blood–brain barrier, and also the fetoplacental barrier, perhaps most characteristic of this pathogen [9, 10, 11]. The knowledge of this intricate mode of infection has led to the recent reformation of the interaction between *L. monocytogenes* and humans, which has seen the agent of one of the most severe bacterial diseases of humans being used in the fight against cancer, one of the leading causes of human mortality [12, 13].

This long journey in the understanding of *L. monocytogenes* has been achieved through a vast array of research covering a wide range of scientific areas, including, in recent years, molecular methodologies. These achievements have often been made through innovative strategies devised to address many different questions regarding the biology of the organism, from pathogenicity and virulence to characterization and tracking sources, and are characterized by the development of many scientific methodologies.

Methods in Molecular Biology is a series of books that presents a step-by-step protocol approach to experimentation. Each protocol opens with an introductory overview, a list of the materials and reagents needed to complete the experiment, and is then followed by a detailed procedure supported with a notes section offering tips and tricks of the trade as well as troubleshooting advice. The protocols are comprehensive and reliable.

As *Listeria monocytogenes* continues to be a major threat to public health, this book in the series is a timely addition. It brings together protocols and methodologies that are used in research to gain a better understanding of *Listeria* at a molecular level. The topics covered include sampling in order to isolate *Listeria*, methods for their identification and characterization, methods for gene manipulation, and finally methods for the control of the organism. The book will contribute towards the harmonization of the methods used and will therefore benefit all those interested in Listeria research.

Fermoy, Ireland   *Kieran Jordan*
Werribee, VIC, Australia   *Edward M. Fox*
Vienna, Austria   *Martin Wagner*

References

1. Murray EGD, Webb RA, Swann MBR (1926) A disease of rabbits characterized by large mononuclear leucocytosis, caused by a hitherto undescribed bacillus, *Bacterium monocytogenes* (n. sp.). J Pathol Bacteriol 29:407–439

2. Gray ML, Killinger AH (1966) *Listeria monocytogenes* and listeric infections. Bacteriol Rev 30:309–382

3. Freitag NE, Port GC, Miner MD (2009) *Listeria monocytogenes*—from saprophyte to intracellular pathogen. Nat Rev Microbiol 7:623

4. McCarthy SA (1990) *Listeria* in the environment. In: Miller AJ, Smith JL, Somkuti GA (eds) Foodborne listeriosis. Society for Industrial Microbiology. Elsevier Science Publishing, Inc., New York, pp 25–29

5. Grif K, Patscheider G, Dierich MP, Allerberger F (2003) Incidence of fecal carriage of *Listeria monocytogenes* in three healthy volunteers: a one-year prospective stool survey. Eur J Clin Microbiol Infect Dis 22:16–20

6. Schlech WF III, Lavigne PM, Bortolussi RA, Alien AC, Haldane EV, Wort AJ, Hightower AW, Johnson SE, King SH, Nicholls ES, Broome CV (1983) Epidemic listeriosis-evidence for transmission by food. N Engl J Med 308:203–206

7. Linnan MJ, Mascola L, Lou XD, Goulet V, May S, Salminen C, Hird DW, Yonekura ML, Hayes P, Weaver R, Audurier A, Plikaytis BD, Fannin SL, Kleks A, Broome CV (1988) Epidemic listeriosis associated with Mexican-style cheese. N Engl J Med 319:823–828

8. Bille J (1990) Epidemiology of human listeriosis in Europe with special reference to the Swiss outbreak. In: Miller AJ, Smith JL, Somkuti GA (eds) Foodborne listeriosis. Society for Industrial Microbiology. Elsevier Science Publishing, Inc., New York, pp 71–74

9. Chakraborty T, Goebel W (1988) Recent developments in the study of virulence in *Listeria monocytogenes*. Curr Top Microbiol Immunol 138:41–48

10. Lecuit M (2005) Understanding how Listeria monocytogenes targets and crosses host barriers. Clin Microbiol Infect 11:430–436

11. Seveau S, Pizarro-Cerda J, Cossart P (2007) Molecular mechanisms exploited by *Listeria monocytogenes* during host cell invasion. Microbes Infect 9:1167–1175

12. Rothman J, Paterson Y (2013) Live-attenuated *Listeria*-based immunotherapy. Expert Rev Vaccines 12:493–504

13. Le DT, Dubenksy TW Jr, Brockstedt DG (2012) Clinical development of *Listeria monocytogenes*-based immunotherapies. Semin Oncol 39:311–322

Contents

Contributors

GONÇALO ALMEIDA • *Laboratório Associado, Escola Superior de Biotecnologia, Universidade Católica Portuguesa/Porto, Porto, Portugal*

CORINNE AMAR • *Gastrointestinal Bacteria Reference Unit, Public Health England, London, UK*

REHA ONUR AZIZOGLU • *Department of Food, Bioprocessing and Nutrition Sciences, North Carolina State University, Raleigh, NC, USA*

MOHAMMED BAHEY-EL-DIN • *Department of Pharmaceutical Microbiology, Faculty of Pharmacy, Alexandria University, Alexandria, Egypt*

ANDREI SORIN BOLOCAN • *Faculty of Food Science and Engineering, Dunarea de Jos University of Galati, Galati, Romania*

AIDAN CASEY • *Teagasc Food Research Centre, Fermoy, Cork, Ireland*

PAUL D. COTTER • *Teagasc Food Research Centre, Fermoy, Cork, Ireland*

PASCALE COSSART • *Institut Pasteur, Unité des Interactions Bactéries Cellules, Paris, France*

MARION DALMASSO • *Teagasc Food Research Centre, Fermoy, Cork, Ireland*

KAREN DALY • *Department of Microbiology, University College Cork, Cork, Ireland*

DRISS ELHANAFI • *Biomanufacturing Training and Education Center, North Carolina State University, Raleigh, NC, USA*

MARCEL R. EUGSTER • *Institute of Food, Nutrition and Health, ETH Zurich, Zurich, Switzerland*

VÂNIA FERREIRA • *Laboratório Associado, Escola Superior de Biotecnologia, Universidade Católica Portuguesa/Porto, Porto, Portugal*

EDWARD M. FOX • *Food Microbiology and Safety Group, Animal, Food and Health Sciences, CSIRO, Werribee, VIC, Australia*

CORMAC G.M. GAHAN • *Department of Microbiology and School of Pharmacy, University College Cork, Cork, Ireland; Alimentary Pharmabiotic Centre, University College Cork, Cork, Ireland*

LISA GORSKI • *Produce Safety and Microbiology Research Unit, United States Department of Agriculture, Agricultural Research Service, Albany, CA, USA*

MANSEL GRIFFITHS • *Canadian Research Institute for Food Safety, University of Guelph, Guelph, ON, Canada*

CAITRIONA M. GUINANE • *Teagasc Food Research Centre, Fermoy, Cork, Ireland*

MICHEL HÉBRAUD • *INRA, Clermont-Ferrand Research Centre, UR454 Microbiology, Saint-Genès Champanelle, France*

MARTA HERNÁNDEZ • *Instituto Tecnológico Agrario (ITACyL), Valladolid, Spain*

COLIN HILL • *Department of Microbiology, University College Cork, Cork, Ireland*

KIERAN JORDAN • *Teagasc Food Research Centre, Fermoy, Cork, Ireland*

SOPHIA KATHARIOU • *Department of Food, Bioprocessing and Nutrition Sciences and Biomanufacturing Training and Education Center, North Carolina State University, Raleigh, NC, USA*

ANDREAS KÜHBACHER • *Institut Pasteur, Unité des Interactions Bactéries Cellules, Paris, France*

MARK L. LAWRENCE • *College of Veterinary Medicine, Mississippi State University, Starkville, MS, USA*

MARTIN J. LOESSNER • *Institute of Food, Nutrition and Health, ETH Zurich, Zurich, Switzerland*

RUI MAGALHÃES • *Centro de Biotecnologia e Química Fina, Laboratório Associado, Escola Superior de Biotecnologia, Universidade Católica Portuguesa/Porto, Porto, Portugal*

STAVROS G. MANIOS • *Agricultural University of Athens, Athens, Greece*

MAZIN MATLOOB • *Canadian Research Institute for Food Safety, University of Guelph, Guelph, ON, Canada*

ROBERTA MAZZA • *Sez. di Ispezione degli Alimenti di Origine Animale, Dipartimento di Medicina Veterinaria, Università degli Studi di Sassari, Sassari, Italia*

RINA MAZZETTE • *Sez. di Ispezione degli Alimenti di Origine Animale, Dipartimento di Medicina Veterinaria, Università degli Studi di Sassari, Sassari, Italia*

OLIVIA MCAULIFFE • *Teagasc Food Research Centre, Fermoy, Cork, Ireland*

CRISTINA MENA • *Centro de Biotecnologia e Química Fina, Laboratório Associado, Escola Superior de Biotecnologia, Universidade Católica Portuguesa/Porto, Porto, Portugal*

ANCA IOANA NICOLAU • *Faculty of Food Science and Engineering, Dunarea de Jos University of Galati, Galati, Romania*

CONOR P. O'BYRNE • *Bacterial Stress Response Group, Microbiology, School of Natural Sciences, National University of Ireland, Galway, Ireland*

JAVIER PIZARRO-CERDÁ • *Institut Pasteur, Unité des Interactions Bactéries Cellules, Paris, France*

SWETHA REDDY • *College of Veterinary Medicine, Mississippi State University, Starkville, MS, USA*

DAVID RODRÍGUEZ-LÁZARO • *Instituto Tecnológico Agrario (ITACyL), Valladolid, Spain; Microbiology Section, Faculty of Sciences, University of Burgos, Burgos, Spain*

IRENE RÜCKERL • *Institute of Milk Hygiene, Milk Technology and Food Science, Department of Veterinary Public Health and Food Science, University of Veterinary Medicine, Vienna, Austria*

KATHRIN RYCHLI • *University of Veterinary Medicine Vienna, Vienna, Austria*

MATHIAS SCHMELCHER • *Institute of Food, Nutrition and Health, ETH Zurich, Zurich, Switzerland*

STEPHAN SCHMITZ-ESSER • *Institute for Milk Hygiene, University of Veterinary Medicine Vienna, Vienna, Austria*

JOANA SILVA • *Laboratório Associado, Escola Superior de Biotecnologia, Universidade Católica Portuguesa/Porto, Porto, Portugal*

PANAGIOTIS N. SKANDAMIS • *Agricultural University of Athens, Athens, Greece*

BEATRIX STESSL • *University of Veterinary Medicine Vienna, Vienna, Austria*

PAULA TEIXEIRA • *Centro de Biotecnologia e Química Fina, Laboratório Associado, Escola Superior de Biotecnologia, Universidade Católica Portuguesa/Porto, Porto, Portugal*

MARTA UTRATNA • *Bacterial Stress Response Group, Microbiology, School of Natural Sciences, National University of Ireland, Galway, Ireland*

MARTIN WAGNER • *University of Veterinary Medicine Vienna, Vienna, Austria*

Part I

Detection, Quantification, and Confirmation

Chapter 1

Sampling the Processing Environment for *Listeria*

Anca Ioana Nicolau and Andrei Sorin Bolocan

Abstract

This chapter describes in detail the procedures used when sampling for *Listeria* in food processing environments. Sampling of food contact surfaces, non-food contact surfaces, and liquids such as drain effluents are addressed. Sponge stick swabs are considered advantageous for surface sampling and tips regarding their application are given. Liquids are collected using sterile dippers and the procedure for their correct use is described. Advice on places to sample, the best time for sampling and the frequency of sampling are also given. Such details help hygienists/microbiologists to be successful in their attempts to isolate strains of *Listeria*, even if such bacteria are well attached to surfaces or located in niches that are difficult to reach.

Key words *Listeria monocytogenes*, Persistence, Food processing environment, Harborage site, Swab, Surface sampling, Biofilm, Liquid sampling

1 Introduction

Some strains of *Listeria monocytogenes* persist in food processing environments for extended periods of time, sometimes longer than 10 years, and many of the persistent strains are responsible for outbreaks of listeriosis [1, 2]. Awareness of the presence of *L. monocytogenes* can help prevent, or at least minimize product contamination, especially when the final product is a ready-to-eat food. Correct sampling is of paramount importance in controlling *L. monocytogenes* in processing environments and in isolating strains that can then be studied further.

L. monocytogenes persistence can be partially explained by the organisms' capacity to adhere to the materials commonly used in the food industry and to form thick complex biofilms that are more difficult to remove than adherent single cells [3]. Also, *L. monocytogenes* cells are significantly more resistant to disinfection in biofilms than their free-living counterparts [4]. Studies demonstrated that flagellum-mediated motility is critical for *L. monocytogenes* biofilm formation on abiotic surfaces [5]. *L. monocytogenes* has four

Kieran Jordan et al. (eds.), *Listeria monocytogenes: Methods and Protocols*, Methods in Molecular Biology, vol. 1157, DOI 10.1007/978-1-4939-0703-8_1, © Springer Science+Business Media New York 2014

to six peritrichous flagella per cell and their formation is temperature dependent, temperatures below 30 °C, as are encountered in dairy and meat products companies, favoring formation [6].

Good growth conditions for *L. monocytogenes* can be found in so-called harborage sites or niches (i.e., shelters due to unhygienic design of equipment and premises or unhygienic or damaged surfaces), where food and moisture accumulate. A minimum initial cell load is necessary for bacteria to persist in such sites [7].

Sponges as well as swabs remain the primary device choice for microorganism detection on surfaces because of their simplicity, affordability, and ability to access a diversity of areas within a food processing facility [8]. While swabs are indicated for geometrically abnormal spaces, sponges are used in a similar way and are advantageous for sampling larger surface areas in order to increase the likelihood of capturing pathogens in low numbers such as *L. monocytogenes* [9]. The representatives of the European Union Reference Laboratory for *L. monocytogenes* also consider that wipe sampling methods (using swabs, sponges, pads, or cloths/tissues) are the only ones appropriate to use for *L. monocytogenes* detection [10] due to the possibility offered by these devices to scrub the surfaces and to detach the biofilms formed on them.

A wipe sampling method collects the microorganisms on a moisturized wiping device as a result of its zigzag movement in two perpendicular directions on the chosen surface, usually outlined by a sterile area template. The wiping device is then returned to the plastic bag that has kept it sterile before usage. This is done in order to protect the device against contamination during transportation to the testing laboratory for analysis.

Besides surface samples, liquid samples from drainage effluents, standing water, melt water from thawed processing ice, and vacuum or drip pan condensate should be also collected to be able to have the full picture of *Listeria* harborage sites in food processing environments. Liquid samples of 100 mL are collected using sterile dippers and transported to the laboratory for analysis. The results obtained for surfaces and liquids are used to monitor *Listeria* presence within a processing environment and to obtain isolates for further characterization. In addition, once the contamination status of a processing facility is known, corrective action to protect consumers can be taken.

2   Materials

The list below contains those materials that are necessary to collect samples from food processing environments. Samples from surfaces, including food-contact surfaces (FCS) and non-food-contact surfaces (NFCS), and liquid samples, including floor and drain effluents, are in view. The field materials for sampling the processing environment are as follows:

1. Wipe sampling devices:
   (a) Stick swabs, sterilized sticks tipped with cotton or synthetic material free of microbicidal substances, individually contained in a sterilized tube (*see* **Note 1**).
   (b) Sponge sticks, free of microbicidal substances, sterilized and individually packed in a sterilized plastic bag (*see* **Note 2**).
2. Diluents: sterile 1 % peptone solution or 1/4 strength Ringer's solution, distributed in tubes or bottles (*see* **Note 3**).
3. Neutralizing solution (*see* **Note 4**).
4. Templates for outlining the sampling area (*see* **Note 5**).
5. Disposable sterile gloves (optional).
6. Cool box, with pre-frozen refrigerant packs and cardboard or foam separator for sample transportation to the laboratory.
7. Sterilized absorbant paper, capable of absorbing stagnating water (i.e., pools of water on the floors) (*see* **Note 6**).
8. Sterilized dippers to hold 100 mL individually contained in a sterilized bag (*see* **Note 7**).
9. Basket to hold the sample bags.
10. Scissors (sterile).
11. Marking pen.
12. Plastic bags for trash.
13. Disinfectant (*see* **Note 8**).
14. Hand sanitizer.
15. Butane pencil torch (*see* **Note 9**).
16. Gas refiller for butane pencil torch.
17. Lighter.

3 Methods

Only correct sampling will provide reliable and accurate data on the microbiological status of different surfaces and fluids. The methods presented below can help technicians/operators to adhere to good laboratory practices and scientifically proven techniques, in order to generate consistent microbiological data.

3.1 Surface Sampling

The method most commonly used for the detection of *L. monocytogenes* presence in a food processing environment is based on surface wiping and uses a sponge-type swab (*see* **Note 10**).

To sample surfaces using sponge-type swabs, an operator (*see* **Note 11**) should follow the steps below:

1. Wash hands with soap and disinfect them with a hand disinfectant or use a hand sanitizer (*see* **Note 12**).

2. Put on a pair of sterile gloves.

3. Take the template (if used) from its protective package and place it on the surface to be sampled. Use templates outlining an area of 90 cm².

4. Take a sample bag containing the sponge stick and label it using a permanent marker (*see* **Note 13**).

5. Keep the bag horizontally and open it by pulling the tabs on either side of the top of the bag to create enough space for removing the sponge.

6. Push the sponge from the outside with one hand and pull the stick with your other hand to leave the handle protruding from the opening. Grasp the handle behind the thumb-stop and remove the sponge stick from the bag (*see* **Note 14**).

7. Moisten the sponge, if you are using a dry sponge (*see* **Note 15**).

8. Wipe the sampling area outlined by the template with the sponge (*see* **Note 16**).

9. Return the sponge to the pouch. Do not insert the sponge any further than the thumb-stop marked on the stick.

10. Pinch the sponge from outside of the bag with thumb and forefinger, then bend the handle back and forth to break-off the sponge at the score mark below the sponge edge. Discard the stick to the trash bag.

11. Expel as much air you can from the bag by pressing it with the hands.

12. Fold down the top of the pouch three times and use the wire tabs to secure the bag.

13. Put the bag in the cool box (*see* **Note 17**).

14. Discard the gloves in the trash bag.

15. Sanitize the wiped area immediately after sampling using the butan pencil torch on nonflammable surfaces and disinfectant on flammable surfaces (*see* **Note 18**). If using tissues/wipes for disinfection, discard the used ones in the trash bag.

16. Discard the trash bag in appropriate trash containers. If such containers are not available at the food processing premises, transport them to the laboratory, but place them in a different container than that used for samples.

In food establishments, known from previous sampling to be uncontaminated food processing environments, composite sampling may be applied in order to reduce the sampling costs. To obtain a composite sample, the same sponge or swab is used to wipe up to five different surfaces (*see* **Note 19**).

To sample surfaces using cotton bud swabs, an operator should follow the same steps mentioned for sponge sticks, but using templates that outline a smaller area (25 cm²) and adapting the procedure taking into consideration that swabs came in containers with caps and the stick is not removed after sampling.

For each series of surface samples taken, it is useful to include a negative control sample. This is obtained by removing the sponge or swab from the bag or container using sterile gloves and then reintroducing it into the bag or container. The negative control sample should be coded in such way that it is not recognized as such during samples processing in the laboratory. Samples are immediately transported to the laboratory under refrigeration and analyzed within 24 h. For a video on sampling surfaces, watch http://www.youtube.com/watch?v=tXXkSHbL8DE (Accessed September 26, 2013).

Report presence or absence of *L. monocytogenes* at the sampling location indicating the size of the sampled area (e.g., present per 900 cm²) (*see* **Note 20**).

3.2 Collecting Liquid Samples

To collect liquid samples an operator should follow the steps below:

1. Wash hands with soap and disinfect them with a hand disinfectant or use a hand sanitizer (*see* **Note 12**).

2. Put on a pair of sterile gloves.

3. Take the dipper out of its protective bag by grasping the handle.

4. Label the dipper using a permanent marker (*see* **Note 13**).

5. Open the lid being careful to handle the inside.

6. Immerse the dipper in the liquid that have to be collected, keep it immersed as long as is necessary to fill it, then gently take it out from the liquid.

7. Close the lid tightly.

8. Absorb the external liquid with a tissue and discard the tissue in the trash bag.

9. Return the dipper to its sterile bag. Hold it in your hand from the exterior of the bag and take out the handle. Discard the handle in the trash bag.

10. Put the dipper in the cool box.

11. Discard the gloves in the trash bag.

As for surface samples, liquid samples are also immediately transported to the laboratory, kept under refrigeration and analyzed within 24 h.

3.3   Sampling Sites

Sample sites within a processing environment have to be chosen to find contamination, if *Listeria* is present. As *Listeria*-free environments are difficult to maintain, it is considered normal to occasionally find *Listeria* in a food processing environment. The choice of sampling sites must be justified and documented in the food safety program. Sites where *L. monocytogenes* is likely to establish and multiply are presented in Table 1. These sites are classified in two priority zones. Zone 1 sites are those that are typically contaminated with *L. monocytogenes*, while Zone 2 sites are those that can harbor *Listeria* cells in a processing environment. Samples should be taken from surfaces situated both in Zone 1 and Zone 2.

3.4   Sampling Time and Frequency

A minimum of five environmental sites should be sampled monthly for *Listeria* by any food business operator and should be analyzed throughout the year regardless the production volume. Over time, samples should cover all important work surfaces. It is necessary to alternate sampling sites each week in order to get an entire set within 1–3 months. Processing environment sampling should be done either during production or immediately after production, but before cleaning and disinfection. Additional sampling could be done after cleaning and disinfection, in order to determine the efficiency of the cleaning program.

Finding positive samples of *L. monocytogenes*, when sampling a processing environment, provides an early warning of its potential presence in the products obtained within the tested environment. The history of sampling results and trends in contamination with *Listeria*, the production amount, the type of products, the facility layout, and the product flow must be taken into consideration when establishing the monitoring frequency. Many food processors find useful to base the monitoring frequency applied for different areas on Criticality Indexes (*see* **Note 21**) or Zones of Risk System (*see* **Note 22**).

Whenever there is a positive environmental sample for *Listeria*, the food processor should increase the frequency of environmental testing to weekly and continue to test until the environmental testing program achieves negative results for 3 consecutive weeks.

Table 1
Sites for checking on *L. monocytogenes* presence [11]

Priority zones	Examples of sampling sites
Zone 1	Equipment that comes into contact with cooked product (e.g., slicers, dicers, hoppers), spiral freezers, conveyors, tables, benches, cutting boards
Zone 2	Floors, walls, ceilings, drain outlets, pools of water (e.g., on the floors of a manufacturing area or cold room), condensate from refrigeration evaporators (cold rooms), chiller doors, switches, floor joints/crevices, underneath shelving, and work surfaces

The purpose of having an increased testing frequency is to monitor the effectiveness of the corrective action undertaken by the food operator. If positive results are obtained after corrective action, the corrective action should be revised and the revised action implemented.

4   Notes

1. Cotton bud-type swabs are appropriate for sampling small surfaces (max. 100 cm^2), where cracks and crevices exist, or confined spaces, usually the ones that could be described as niches for *Listeria* (fins on cooling units, motor housings, chain conveyor links, bearings on conveyors and inside hollow rollers, knife holders, screw holes). Care should be taken not to break the swab stick when wiping.

2. Sponge-type swabs are preferred to cloths, tissues, gauze pads, while sponge stick swabs are preferred to simple sponges because they allow taking a sample without directly handling the sponge. They should be used for sampling large surfaces (900–1,000 cm^2), either flat or non-flat, and are also effective in sampling drains, pipes, and places around equipments. They are made on cellulose and can be purchased in dry format or with neutralizing broth (*see* **Note 4**) or lecithin and with and without gloves (Fig. 1).

Fig. 1 Sponge stick swab

3. As sampling is indicated to be performed during or at the end of processing, simple diluents as peptone water (1‰) or Ringer solutions (1/4 strength) can be successfully used to moisten the wiping devices. Phosphate-buffered diluents are not recommended due to their negative impact on the culturability of stressed cells. Fraser broth or half Fraser broth should not be used in place of a diluent since they could favor growth of *L. monocytogenes* in the processing site. It is indicated to have a tube with diluent for each surface sample that have to be taken in the processing environment. These will not be necessary, if sponges in wet formats are used.

4. When the presence of a residual disinfectant is expected on a surface, it is strongly recommended to use a neutralizer instead of a simple diluent. When there is no residual disinfectant, use of neutralizing broth is not recommended [10]. As neutralizing agent it is used very often the Dey-Engley (D/E) neutralizing broth, which is commercially available or prepared according to the formula: tryptone (5 g), yeast extract (2.5 g), dextrose (10.0 g), sodium thioglycollate (1.0 g), sodium thiosulfate (6.0 g), sodium bisulfite (2.5 g), Polysorbate 80 (5.0 g), soy bean lecithin (7.0 g), bromocresol purple (0.02 g), distilled water (1.0 L), pH 7.6 ± 0.2 at 25 °C.

5. *L. monocytogenes* cells are not evenly distributed on a surface and comparisons of results from large and small areas is not possible. To avoid this situation, templates that outline the sampling area in order to sample the same area each time have to be used. Consistency related to the size of the sampled area will help environmental officers/inspectors to monitor trends regarding *L. monocytogenes* presence in the processing environment over time. Flexible stencils made on PTFE, Teflon, or silicon are more desirable to those made on stainless steel especially when the sampling area is large. Fixing stencils with adhesive strips during the wiping process is not recommended because residues could be left behind when the strips are peeled. Stencils with edges specially designed to be held in place by hand during sampling are preferred. Used templates should be sterilized before disposal or reuse.

6. Sterilized absorbing paper should be used whenever the area to be sampled is too wet (i.e., pools of water on the floor) to remove liquid in excess. It should be gently applied on the wet surface. Several absorbing paper may be used, if necessary.

7. Commercially available dippers are transparent polystyrene or shatterproof polypropylene (PP) recipients with different capacities, which are individually wrapped to ensure complete sterility. They have closures made of PP and handles that can be easily removed, enabling the sample to be packed easily in the cool-box (Fig. 2).

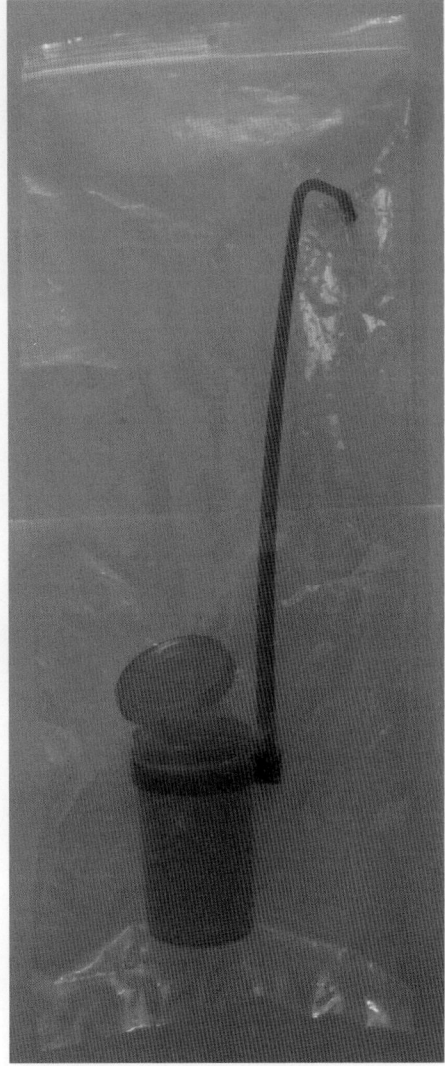

Fig. 2 Dipper that can be used for liquid samples

8. A 70 % alcohol solution, which will be applied on surfaces using paper tissues, or single-use quaternary/alcohol-based wipes are particularly useful for this purpose.

9. A pencil torch (Fig. 3) is a specialized tool that emits a flame to heat nonflammable items. It is cordless, which gives the user the possibility to move freely. It is useful to have such a tool for sterilizing the scissor and the area templates when reusing them and for sterilizing the sampling area after taking the sample.

10. Sponges are indicated to sample flat surfaces as tables, benches, boards, trays, floors, conveyors. Sponges can also

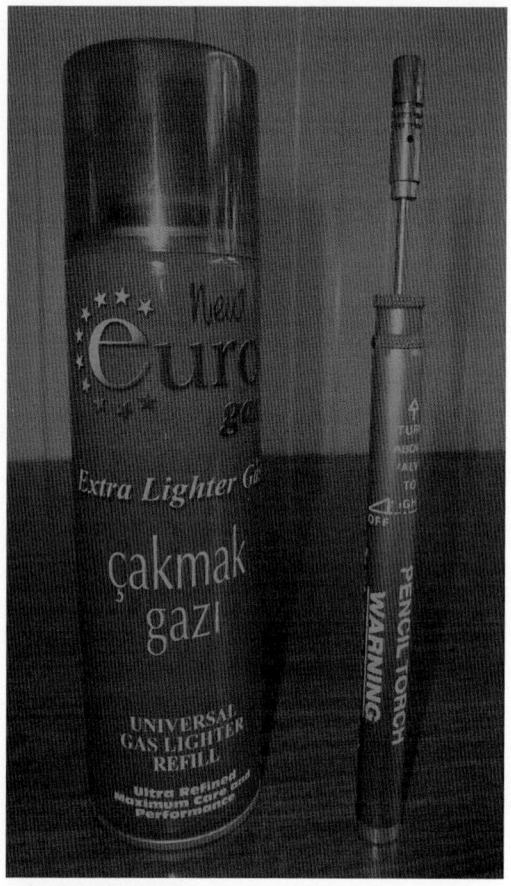

Fig. 3 Pencil torch and gas refiller

be used on slicers, dicers, packing machines, and other processing equipment.

11. Sampling for *Listeria* and sample analysis must be conducted by persons with appropriate microbiological training. If possible, it is important to have the same employee conducting the testing (using the same analytical method) on a regular basis to ensure consistency of the procedures. The operators have to wear the working equipment that is authorized for the premises where sampling is performed.

12. Hand disinfection is recommended even if the sampler decides to wear gloves during sampling. Hand disinfection performed as part of the procedure used for hand washing and disinfection of personnel entering into a food processing area is satisfactory for taking the first surface sample. Between samples, the sampler could use a hand sanitizer.

13. A label should contain information regarding the company name, date, time, location of the sampling area. A predetermined numeric or alphanumeric coding system is recommended.

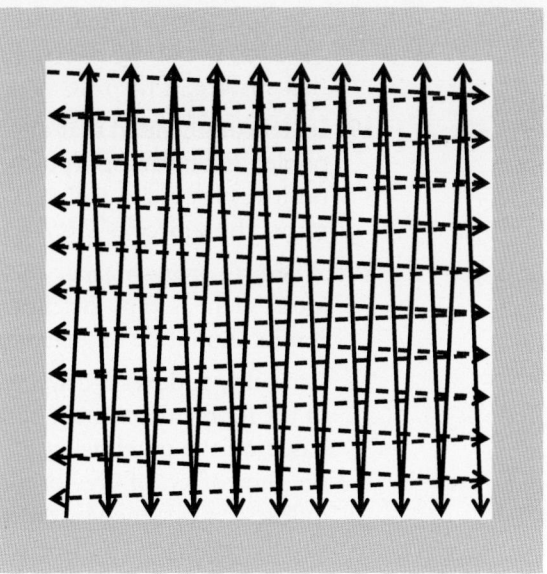

Fig. 4 Wiping procedure for surface sampling

14. The inside of the bag should not be touched. This is the reason for wearing gloves when the operator is not experienced with the technique (gloves are sterile and, if the operator accidentally touches the inside of the bag, he/she will not contaminate it). An experienced operator will be skilled to work in such way to not touch the inside of the bag, so he/she can decide not to wear gloves. Do not touch the stick below the thumb stop.

15. If you are not using a premoistened sponge and the surface to be sampled is dry, use sterile peptone water to moisten the sponge. If the surface to be sampled is already wet, such as a drip tray or a conveyor, it is best to rehydrate the sponge using the moisture on the surface being tested.

16. Move the sponge up and down to describing about ten Vs along the area outlined by the template. Apply a reasonable force to be able to remove particles of dust and organic material containing bacteria (to be successful in collecting surface samples wiping should be firm). When finish the lines, turn the sponge and continue to wipe the surface in the same way but from left to right (Fig. 4).

17. The cool-box should be prepared for sample storage before taking the samples. To do this, introduce the pre-frozen gel packs into the box and place a cardboard on top of gel packs and then put the samples. Cover the samples with a foam plug or a cardboard and send the boxes to the lab.

18. Disinfection of the sampled area will return it to hygienic condition, including removing any small amounts of residual

liquid left by the premoistened sponge, and will eliminate any question about the contamination of the products that touched that surface, if the tests are positive for *Listeria*.

19. It is recommended that only samples from the same zone to be pooled for testing purposes. If composite samples prove to be positive, each site will subsequently be tested individually to reveal the source of *Listeria*.

20. When *L. monocytogenes* is detected on unmeasured surface areas such as pipe interiors, nozzles, valves, or gaskets, the results should be reported specifying the entire sampling site.

21. Criticality Indexes is a system that assigns an index to each area of the food environment based on risk. The assessment of risk is based on the potential impact that associated hazards may have on the safety of the products being manufactured. Higher indexes are obtained by dirtier activities, areas where dirty activities are performed in close relative proximity to clean areas, areas which are often wet or have high levels of staff activity.

22. Zones of Risk or Zoning System is the physical or visual division of a food production factory into subareas, leading to the segregation of different activities with different hygiene levels.

References

1. Kathariou S (2002) *Listeria monocytogenes* virulence and pathogenicity, a food safety perspective. J Food Prot 65:1811–1829
2. Tompkin RB (2002) Control of *Listeria monocytogenes* in the food-processing environment. J Food Prot 65:709–725
3. Møretrø T, Langsrud S (2004) *Listeria monocytogenes*: biofilm formation and persistence in food-processing environments. Biofilms 1:107–121
4. Pan Y, Breidt F Jr, Kathariou S (2006) Resistance of *Listeria monocytogenes* biofilms to sanitizing agents in a simulated food processing environment. Appl Environ Microbiol 72:7711
5. Lemon KP, Higgins DE, Kolter R (2007) Flagellar motility is critical for *Listeria monocytogenes* biofilm formation. J Bacteriol 189:4418–4424
6. Peel M, Donachie W, Shaw A (1988) Temperature-dependent expression of flagella of *Listeria monocytogenes* studied by electron microscopy, SDS-PAGE and western blotting. J Gen Microbiol 134:2171–2178
7. Carpentier B, Cerf O (2011) Persistence of *Listeria monocytogenes* in food industry equip-
ment and premises. Int J Food Microbiol 145:1–8
8. Foschino R, Picozzi C, Civardi A, Bandini M, Faroldi P (2003) Comparison of surface sampling methods and cleanability assessment of stainless steel surfaces subjected or not to shot peening. J Food Eng 60:375–381
9. Lindblad M (2007) Microbiological sampling of swine carcasses: a comparison of data obtained by swabbing with medical gauze and data collected routinely by excision at Swedish abattoirs. Int J Food Microbiol 118:180–185
10. Carpentier B, Barre L (2012) Guidelines on sampling the food processing area and equipment for the detection of *Listeria monocytogenes*, Version 3–20/08/2012, EURL for *Listeria monocytogenes*. Maisons-Alfort Laboratory for Food Safety, ANSES, France
11. NSW Food Authority (2008) Listeria Management Program (NSW/FA/FI034/0809). http://www.foodauthority.nsw.gov.au/_Documents/industry_pdf/listeria-management-program.pdf. Accessed May 2013

Chapter 2

Traditional Methods for Isolation of *Listeria monocytogenes*

Rui Magalhães, Cristina Mena, Vânia Ferreira, Gonçalo Almeida, Joana Silva, and Paula Teixeira

Abstract

Conventional methods for the detection of *Listeria monocytogenes* in foods and environmental samples relies on selective pre-enrichment, enrichment, and plating. This is followed by confirmation of suspected colonies by testing a limited number of biochemical markers.

Key words Culture methods, Enrichment, Detection, Enumeration, Confirmation, Selective media, ISO standards, Most Probable Number

1 Introduction

Detection and identification of *Listeria monocytogenes* in food and environmental samples traditionally involve culture methods based on selective pre-enrichment, enrichment, and plating. This is followed by confirmation of suspected colonies using colony morphology, sugar fermentation pattern, and hemolytic properties (Fig. 1). *L. monocytogenes* is a non-spore forming, catalase-positive, Gram-positive rod-shaped bacterium that shows hemolytic activity on blood agar.

On this basis, several methods were developed worldwide for the detection and/or enumeration of this pathogen. FDA-BAM [1], USDA [2] methods, and ISO 11290 standards [3, 4] are probably the most commonly used reference methods. The criteria of the EU Regulation 20073/2005 [5] define ISO 11290-1 [3] and ISO 11290-2 [4] as the reference methods for detection and enumeration, respectively, of *L. monocytogenes*. Negative results can be confirmed in 3–4 days, the time for a positive result is usually 5–7 days from sample collection.

It is well known that microorganisms in foods are often injured so that they become sensitive to the presence of selective agents

Kieran Jordan et al. (eds.), *Listeria monocytogenes: Methods and Protocols*, Methods in Molecular Biology, vol. 1157,
DOI 10.1007/978-1-4939-0703-8_2, © Springer Science+Business Media New York 2014

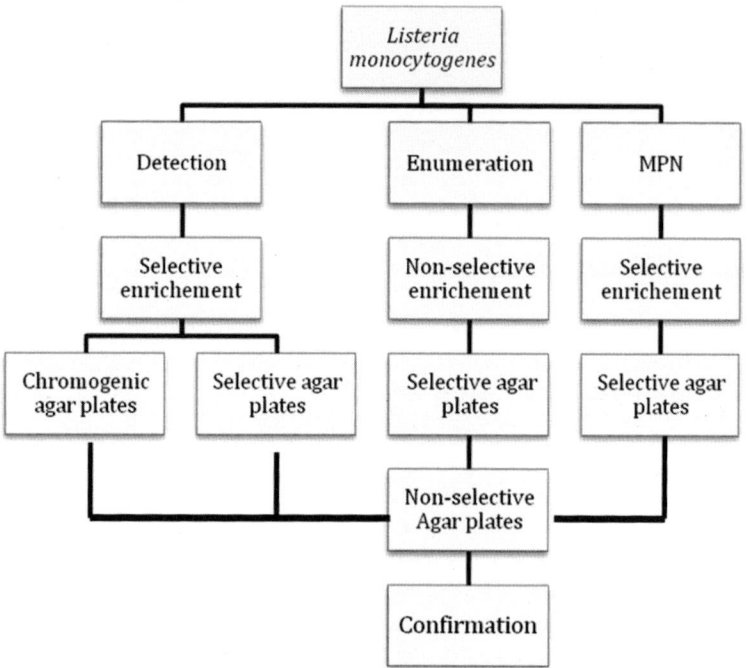

Fig. 1 Conventional approaches for the detection and enumeration of *Listeria monocytogenes*

present in media recommended for their isolation [6, 7]. In order to overcome this limitation, recovery of stressed cells is promoted by a pre-enrichment step in a non-selective broth prior to the selective enrichment and isolation on selective/differential agar media. Most conventional selective enrichment broths contain selective agents: nalidixic acid that inhibits growth of Gram-negative organisms; acriflavine that inhibits Gram-positive bacteria; cycloheximide that inhibits the growth of saprophytic fungi; and lithium chloride (LiCl) that inhibits enterococci. The selective agents commonly used in *L. monocytogenes* isolation media are acriflavine, LiCl, polymyxin B, and cephalosporins.

Detection of *L. monocytogenes* after enrichment is hindered by several factors namely the high population of competitive microflora, the low levels of the pathogen, and the interference of inhibitory food components. The higher growth rate of *L. innocua* in selective liquid media compared with *L. monocytogenes* can result in a high number of false-negative results on Polymyxin Acriflavine Lithium Chloride Ceftazidime Aesculin Mannitol (PALCAM) and Oxford (OXA) agars, the media initially recommended by ISO [8, 9]. Differentiation of colonies of *L. monocytogenes* from other non-pathogenic species of *Listeria* is not possible on these media—detection based on the hydrolysis of aesculin. Johansson [10] demonstrated that the selection of five colonies for confirmation from these media might not be sufficient if other *Listeria* species were present. In 2004, ISO modified the isolation as well as enumeration media for *L. monocytogenes*. The chromogenic medium Agar *Listeria* according to Ottaviani and Agosti (ALOA) was adopted as an obligatory selective and differential medium for the isolation of *Listeria* spp. and presumptive identification and enumeration of *L. monocytogenes* [3, 4]; detection based on the activity of the enzymes phosphatidylinositol phospholipase C and β-glucosidase. Lecithin present in the agar is hydrolyzed by phospholipase enzyme synthesized only by *L. monocytogenes* and *L. ivanovii* forming a white precipitation zone around the colony. β-Glucosidase cleaves the chromogenic substrate producing green-blue colonies [11]. In addition to ALOA, another selective medium at the choice of the laboratory (e.g., PALCAM or OXA) must be used [3, 4]. It is likely that the more selective/indicator media or methods that are used in the examination of a sample the more likely it is that the results obtained are representative of the true status of the pathogen in the sample.

In addition to the standard method for the enumeration of *L. monocytogenes* in food and environmental samples [4], the Most Probable Number (MPN) technique might be used to estimate the cell density in a test sample; it is particularly useful when low numbers of organisms are present. Generally, three tenfold serial dilutions are used in either a three or five tubes MPN series. Based on positive results achieved, an MPN table is used to infer the cell numbers in the original sample [12].

2 Materials

2.1 Media Preparation

Before sample examination for *L. monocytogenes*, microbiological media and all materials coming into contact with it must be sterile. During any subsequent handling of the bacterial cultures, unwanted or contaminant organisms must be excluded employing aseptic techniques. Complete instructions for the preparation of culture media (namely quantity of powder per liter and sterilization conditions) are given on the label of each bottle. Appropriate precautions must be taken when preparing media that contain toxic agents, particularly antibiotics. They should be handled with care avoiding dispersion of powder which can give rise to allergic or other reactions in laboratory personnel.

1. Rinse all glassware with the distilled/deionized water and make sure that the vessels are clean and free from toxic chemicals which may be absorbed on the surface of the glass.

2. Use freshly prepared distilled water. Use warm (50 °C) water to hasten the solution of the powder.

3. Prepare the medium in a flask about twice the final volume of the medium to allow adequate mixing.

4. With a clean spatula accurately weigh the prescribed amount of medium powder, avoiding inhaling the powder and prolonged skin contact. Close the medium container as soon as possible.

5. Pour half the required volume of distilled water in the flask, then the weighed quantity of medium. Add a stir bar and stir for a few minutes. Pour the rest of the distilled water, washing the sides of the flask to remove any adherent powder.

6. Agar-free media will usually dissolve on gentle agitation. Media containing agar should be heated to dissolve the agar before autoclaving. The media should clarify near boiling (90–100 °C). Do not allow to boil over.

7. Prior to sterilization, after the medium has been cooled to 25 °C, the final pH of the prepared medium must be checked to guarantee that it conforms to the label specification.

8. Most culture media will require final sterilization in an autoclave. Broth media can be distributed into individual lab tubes in the desired amount prior to sterilization. Place dissolved, loosely capped media in the autoclave. If using dehydrated commercial media, follow carefully the manufacture instructions for media preparation, sterilization (time/temperature), and storage conditions.

9. Carefully remove from autoclave and allow cooling to 50–60 °C.

10. For agar culture media, open a sterile package of Petri-dishes preserving the bag for later storage. Mark the sides of the dishes to indicate the type of media and pour about 15–20 mL of the medium, using aseptic technique. When plates have solidified, invert, place in 37 °C incubator for 24–48 h to check for sterility. Store in labelled plastic bag at 4 °C. Pre-warm before using.

2.2 Selective Enrichment Broth Media

1. Buffered *Listeria* Enrichment Broth: Buffered *Listeria* Enrichment Broth (BLEB) base is used in the FDA/BAM recommendations for selective enrichment procedure for isolation of *L. monocytogenes*. The medium BLEB is a modification of the initial formula developed by Lovett et al. [13], by the addition of disodium phosphate, which results in an increased buffering capacity of the medium and improvement of the enrichment properties. Selective agents can be added after an initial 4 h period to facilitate resuscitation, repair, and growth of injured *Listeria* cells.

Composition: casein enzymic hydrolysate, 17.0 g/L; dextrose, 2.5 g/L; dipotassium hydrogen phosphate, 2.5 g/L; disodium phosphate, anhydrous, 9.6 g/L; monopotassium phosphate, anhydrous, 1.35 g/L; papaic digest of soyabean meal, 3.0 g/L; sodium chloride, 5.0 g/L; sodium pyruvate, 1.0 g/L; yeast extract, 6.0 g/L. Selective agents include: acriflavine hydrochloride, 10 mg/L; nalidixic acid, 40 mg/L; and cycloheximide, 50 mg/L.

Preparation: dissolve the base components or commercial dehydrated medium base in distilled water, by heating if necessary. Adjust the pH if necessary, so that after sterilization it is 7.3 at 25 °C. Sterilize in the autoclave for 15 min at 121 °C.

The following filter sterilized supplements are aseptically added to the basal media at 47 °C immediately prior to use: 10 mg/L acriflavine hydrochloride (0.5 % aqueous solution); 40 mg/L nalidixic acid sodium salt (0.5 % aqueous solution); 50 mg/L cycloheximide (1 % solution in 40 % ethanol).

Appearance of prepared medium: clear, medium amber with none to moderate precipitate.

2. University of Vermont Medium: University of Vermont Medium (UVM) *Listeria* selective enrichment broth is based on the formula described by Donnelly and Baigent [14], and it is the media recommended in the USDA-FSIS method for isolation of *L. monocytogenes*. UVMI broth has been recommended as a primary enrichment broth for recovery of heat-injured *L. monocytogenes*.

Composition: beef extract, 5.0 g/L; casein enzymic hydrolysate, 5.0 g/L; disodium hydrogen phosphate, 12.0 g/L; aesculin, 1.0 g/L; monopotassium hydrogen phosphate, 1.35 g/L; proteose peptone, 5.0 g/L; sodium chloride, 20.0 g/L; yeast extract, 5.0 g/L. Selective agents for UVMI include: nalidixic acid, 20 mg/L; acriflavine hydrochloride, 12 mg/L.

Preparation: dissolve the base components or commercial dehydrated medium base in the distilled water, by heating if necessary. Adjust the pH if necessary, so that after sterilization it is 7.4 at 25 °C. Sterilize in the autoclave for 15 min at 121 °C.

The following filter sterilized supplements are aseptically added to the basal media at 47 °C immediately prior to use: 12 mg/L acriflavine hydrochloride (0.5 % aqueous solution); 20 mg/L nalidixic acid sodium salt (0.5 % aqueous solution).

Appearance of prepared medium: medium amber colored, slightly opalescent solution with a bluish tinge.

3. Fraser broth: Fraser broth base is recommended by the ISO 11290-1 [3], for the selective enrichment and enumeration of *L. monocytogenes* and other *Listeria* species in food and environmental samples, based on the formula described by Fraser and Sperber [15]. The base formula of the medium already includes antibiotics, but it is necessary to add the ferric ammonium citrate supplement. Half Fraser broth is used as the primary enrichment broth in the ISO methodology and consists of a modification of Fraser broth which contains half of the

concentration of nalidixic acid and acriflavine hydrochloride to aid in the recovery of stressed cells.

Composition: meat peptone, 5.0 g/L; tryptone, 5.0 g/L; beef extract, 5.0 g/L; yeast extract, 5.0 g/L; sodium chloride, 20.0 g/L; disodium hydrogen phosphate dehydrated, 12.0 g/L; potassium dihydrogen phosphate, 1.35 g/L; aesculin, 1.0 g/L; lithium chloride, 3.0 g/L. Selective agents for Fraser broth and half Fraser broth include nalidixic acid, acriflavine hydrochloride, and ferric ammonium citrate at different concentrations. Nalidixic acid sodium salt solution may be added to the base before autoclaving.

Preparation: dissolve the base components or commercial dehydrated medium base in the distilled water, by heating if necessary. Adjust the pH if necessary, so that after sterilization it is 7.2 at 25 °C. Sterilize in the autoclave for 15 min at 121 °C.

For half Fraser broth preparation, the following filter sterilized supplements are aseptically added to the basal medium at 47 °C immediately prior to use: ferric ammonium citrate 500 mg/L (5 % aqueous solution); nalidixic acid sodium salt, 10 mg/L (1 % in 0.05 M sodium hydroxide solution); 12.5 mg/L acriflavine hydrochloride (0.25 % aqueous solution).

For Fraser broth preparation, the following filter sterilized supplements are aseptically added to the basal medium at 47 °C immediately prior to use: ferric ammonium citrate 500 mg/L (5 % aqueous solution); nalidixic acid sodium salt, 20 mg/L (1 % in 0.05 M sodium hydroxide solution); 25 mg/L acriflavine hydrochloride (0.25 % aqueous solution).

Appearance of prepared medium: Straw colored solution.

2.3 Isolation Selective Media

Selective isolation media can be divided into two categories: aesculin-containing media and chromogenic media. The characteristic of colonies of *Listeria* spp. and *L. monocytogenes* are summarized in Table 1.

1. Aesculin containing media: Aesculin offers differential properties to the media. It is hydrolyzed by β-D-glucosidase, resulting in the formation of 6,7-dihydroxycoumarin that reacts with the ferric ions. All colonies of *Listeria* spp. are greyish-green with brown-black surrounding halos.

2. Polymyxin Acriflavine Lithium Chloride Ceftazidime Aesculin Mannitol Agar: Polymyxin Acriflavine Lithium Chloride Ceftazidime Aesculin Mannitol Agar (PALCAM) is based on the formulation of van Netten et al. [16], who developed this medium, highly selective due to the presence of LiCl, ceftazidime, polymyxin B, and acriflavine. The double indicator system (aesculin and ferrous iron and mannitol and phenol red) allows the easy differential between *L. monocytogenes*, which does not ferment mannitol, from contaminants, such as enterococci and staphylococci.

Table 1
Characteristics of typical colonies of *Listeria* species and *L. monocytogenes* on isolation media

	Medium	Characteristics of *Listeria* spp. colonies	Characteristics of *L. monocytogenes* colonies
Based on the activity of phosphatidylinositol phospholipase C	ALOA	*L. ivanovii*: blue-green regular round colonies with halo Other *Listeria*: blue-green regular round colonies with or without halo	Blue-green colonies with an opaque halo
	BCM	*L. ivanovii*: turquoise convex colonies with turquoise halos Other *Listeria*: white convex colonies; 2.0 mm without precipitates or halos	Turquoise convex colonies with turquoise halos
	Rapid'*L.mono*	*L. ivanovii*: blue-green colonies with a yellow halo Other *Listeria*: white, with or without a yellow halo	Blue (pale blue, grey-blue to dark blue) colonies
	CHROMagar *Listeria*	*L. ivanovii*: blue with white halo Other *Listeria*: blue without halo	Blue with white halo
Based on the hydrolysis of aesculin	OXA/MOX	At 24 h black with black halos After 48 h remain black with a black halo, but with a sunken center	At 24 h olive-green with black halo After 48 h become darker with a hollow black center surrounded by black zones
	PALCAM	Grey-green with a black halo	Grey-green with a black zone

Composition: protease peptones, 23.0 g/L; starch, 1.0 g/L; sodium chloride, 5.0 g/L; yeast extract, 3.0 g/L; D-glucose, 0.5 g/L; D-mannitol, 10.0 g/L; aesculin, 0.8 g/L; ferric ammonium citrate, 0.5 g/L; phenol red, 0.08 g/L; lithium chloride, 15 g/L; agar, 9–18 g/L. PALCAM selective supplement includes: polymyxin B, 10 mg/L; acriflavine, 5 mg/L; ceftazidime, 20 mg/L.

Preparation: dissolve the base components or commercial dehydrated medium base in the distilled water, by boiling. Adjust the pH if necessary, so that after sterilization it is 7.2 at 25 °C. Sterilize in the autoclave for 15 min at 121 °C.

The following filter sterilized supplements are aseptically added to the basal medium at 47 °C immediately prior to use:

10 mg/L of polymyxin B sulfate solution (1 % aqueous solution), 5 mg/L of acriflavine hydrochloride solution (0.05 % aqueous solution), and 20 mg/L of sodium ceftazidime pentahydrate solution (0.1 % aqueous solution). Mix gently before pour the medium into sterile Petri-dishes.

Appearance of prepared medium: Red gel.

3. Oxford Agar: Oxford *Listeria* Agar (OXA) is prepared according to the formulation of Curtis et al. [17] and is a specified plating medium in the FDA/BAM isolation procedure. Selectivity is increased by adding various antimicrobial agents (acriflavine, colistin sulfate, cefotetan, cycloheximide, and fosfomycin) to the Oxford *Listeria* Agar base.

Composition: protease peptones, 23.0 g/L; starch, 1.0 g/L; sodium chloride, 5.0 g/L; aesculin, 1.0 g/L; ferric ammonium citrate, 0.5 g/L; lithium chloride, 15.0 g/L; agar, 15.0 g/L. Selective supplements: acriflavine, 5 mg/L; cefotetan, 2 mg/L; colistin sulfate, 20 mg/L; cycloheximide, 400 mg/L; fosfomycin, 10 mg/L.

Preparation: dissolve the base components or commercial dehydrated medium base in the distilled water by boiling. Adjust the pH if necessary, so that after sterilization it is 7.0 at 25 °C. Sterilize in the autoclave for 15 min at 121 °C.

Then, after cooling to 47 °C, and immediately before use, aseptically add 10 mL of a filtered sterilized supplement solution containing: 0.4 g of cycloheximide, 0.02 g of colistin sulfate, 0.005 g of acriflavine hydrochloride, 0.002 g of cefotetan, 0.01 g of fosfomycin (dissolved in 5 mL of distilled water and 5 mL of ethanol). Mix gently before pour the medium into sterile Petri dishes.

Appearance of prepared medium: pale green-colored gel.

4. Modified Oxford Agar: Modified Oxford Agar (MOX) is a modification of the Oxford Agar medium referred above. MOX is recommended for isolating and identifying *L. monocytogenes* from processed meat and poultry products, while OXA is recommended for isolating *Listeria* from enrichment broth cultures. The difference between the two media relies on the selective supplements that are added to the oxford agar base formula: colistin and moxalactam.

The supplement for MOX includes colistin sulfate, 10 mg/L; and moxalactam, 20 mg/L.

2.4 Chromogenic Media

Culture media utilizing virulence factors of pathogenic *Listeria* spp. for selectivity are an attractive alternative to the conventional methods due to a more rapid detection of pathogenic *Listeria* spp. These types of media are available commercially in powder or ready-to-use agar plates.

1. Agar Listeria according to Ottaviani and Agosti: Agar *Listeria* according to Ottaviani and Agosti (ALOA) is a selective and differential medium for the isolation of *Listeria* spp. from foodstuffs and other samples and for the identification of *L. monocytogenes*. The selectivity of the medium is due to LiCl and to the addition of antimicrobial selective mixture containing ceftazidime, polymyxin B, nalidixic acid, and cycloheximide. The differential activity is due to the presence in the medium of the chromogenic compound for the detection of β-glucosidase, common to all *Listeria* species. The specific differential activity is obtained by means of a substrate (L-α-phosphatidylinositol) for a phospholipase C enzyme that is present in *L. monocytogenes* and in some strains of *L. ivanovii*. The combination of both substrates permits the differentiation of *Listeria* spp., which grow with a green-blue color, from the colonies of *L. monocytogenes*, which grow with a green-blue color surrounded by an opaque halo. Occasionally, some non-*Listeria* spp. appear green-blue with a halo, so confirmation of suspect colonies is necessary.

2. CHROMOagar: CHROMOagar *Listeria* easily differentiates *L. monocytogenes* from other *Listeria* spp. Colonies of *L. monocytogenes* appear a blue color, regular with a white halo. Other microorganisms are blue, colorless, other color, or inhibited. Some strains of *L. ivanovii* may also give blue colonies with a white halo. Some strains of *Bacillus cereus* can also grow as blue colonies but can easily be distinguished as they are much larger with an irregular edge to the colony and very large white halo.

3. Rapid'*L.mono* agar: The principle of RAPID'*L.mono* chromogenic agar medium relies on the specific detection of the phosphatidylinositol phospholipase C activity of *L. monocytogenes* and the inability of this species to metabolize xylose. The addition of xylose to the medium allows for differentiation of *L. monocytogenes* that form characteristic blue, pale blue, grey-blue to dark blue colonies without a yellow halo from *L. ivanovii* that produces blue-green colonies with a distinct yellow halo. Other *Listeria* spp. produce white colonies with or without a yellow halo. The selective supplement inhibits the majority of interfering flora, including Gram-positive and Gram-negative bacteria, yeasts and moulds.

4. Biosynth Chromogenic Medium: The Biosynth Chromogenic Medium I (BCMI) is based on the activity of phosphatidylinositol phospholipase C. The medium contains a novel enzyme substrate 5-Bromo-4-chloro-3-indoxyl-myo-inositol-1-phosphate, which enzymatic cleavage by *L. monocytogenes* and *L. ivanovii* leads to turquoise colonies, easy to enumerate. Non-pathogenic *Listeria* spp. appear clearly distinguishable

as white colonies. The Biosynth Chromogenic Medium II (BCMII) additionally combines the cleavage of X-phos-Inositol in forming turquoise colonies with the production of a white precipitate surrounding the colonies due to lecithinase activity. The inhibition of contaminants is increased by the addition of antibiotics and LiCl.

2.5 Nonselective Media

1. Tryptic Soy Agar Yeast Extract: Tryptic Soy Agar supplemented with 0.6 % of Yeast Extract (TSAYE) is a general purpose plating medium used for the isolation, cultivation, and maintenance of *Listeria* spp., namely for purification of colonies isolated on selective media (e.g., OXA or PALCAM). TSAYE plates can be examined for typical colonies under an obliquely transmitted light—Henry illumination test. Using a powerful source of beamed white light, striking the bottom of the plate in a 45° angle *Listeria* spp. colonies appear blue-grey to blue color and a granular surface.

   Composition: tryptone, 17 g/L; soya peptone, 3 g/L; sodium chloride, 5 g/L; dipotassium phosphate, 2.5 g/L; glucose, 2.5 g/L; yeast extract, 6 g/L; agar, 15 g/L.

   Preparation: Dissolve the components or the commercial dehydrated medium by boiling. Adjust the final pH 7.3 at 25 °C. Autoclave for 15 min at 121 °C.

   Appearance of prepared medium: Prepared medium is trace to slight hazy and yellow beige color.

2. Carbohydrate utilization Broth: This medium is used to differentiate *Listeria* species based on carbohydrate fermentation. This is a carbohydrate-free medium with bromocresol purple as pH indicator. Specific carbohydrates are added to the basal medium, and when inoculated with an organism that has the capacity to ferment the carbohydrate present, acid is produced and the indicator changes the medium color from purple to yellow. If the carbohydrate is not fermented, the color will remain unchanged.

   Composition: enzymatic digest of animal tissues, 10 g/L; meat extract, 1 g/L; sodium chloride, 5 g/L; bromocresol purple.

   Preparation: Dissolve the components or the commercial dehydrated medium by heating if necessary. Adjust the final pH 6.8 at 25 °C. Dispense appropriate amounts of the medium into tubes. Autoclave for 15 min at 121 °C.

   Carbohydrate solutions: dissolve 5 g of the carbohydrate (D-mannitol or L-rhamnose or D-xylose) in 100 mL of distilled water. Sterilize by filtration. For each carbohydrate add aseptically 1 mL of the carbohydrate solution to 9 mL of the medium base.

   Appearance of prepared medium: purple.

3 Methods

Samples should be examined as soon as possible after receipt, preferably within 24 h. If they are highly perishable products (such as shellfish), testing should commence within 24 h of sampling.

In the case of impossibility of initiate the testing at time mentioned, the samples may be frozen at below −15 °C, preferably −18 °C, if the recovery of *L. monocytogenes* is not significantly impaired with the sample matrix concerned. Frozen samples should not be thawed until analysis.

3.1 Detection of L. monocytogenes

1. Weigh 25 g of analytical portions of solid food or 25 mL liquid foods into a sterile plastic bag. Add 225 mL of pre-enrichment medium broth (half Fraser base, BLEB or UVM I). Homogenize the mixture in a Blender or Stomacher for 1–3 min (*see* **Note 1**).

2. Incubate for 24 h at 30 °C.

3. After incubation transfer 0.1 mL of the pre-enrichment broth culture to the 10 mL of enrichment broth medium (Fraser).

4. Incubate for 24 h at 37 °C.

5. Streak a loop of pre-enrichment broth culture onto two selective solid media (*see* **Note 2**).

6. Incubate at 37 °C for 24–48 h (*see* **Note 3**).

7. Streak a loop of enrichment broth culture onto two selective solid media.

8. Incubate at 37 °C for 24–48 h.

9. Examine the dishes for the presence of typical colonies of *Listeria* spp. (*see* Table 1) and proceed to confirmation.

3.2 Enumeration of L. monocytogenes

1. Initial suspension (10^{-1} dilution)—Weigh 10 g of analytical portions of solid food or 10 mL liquid foods into a sterile plastic bag. Add 90 mL or g of diluent medium broth (Buffered peptone water or half Fraser base without the addition of selective agents) (*see* **Note 4**).

2. Homogenize the mixture in a Blender or Stomacher for 1–3 min.

3. Incubate for 1 h at 20 °C.

4. Prepare tenfold dilutions.

5. Transfer 0.1 mL of the liquid test sample or 0.1 mL of the initial suspension and dilutions onto dried ALOA plate (*see* **Note 5**).

6. Spread the inoculum over the surface of the agar plate with the aid of a sterile spreader (*see* **Note 6**).

7. Let the plates on the bench for 15 min for the inoculum to be absorbed into the agar.

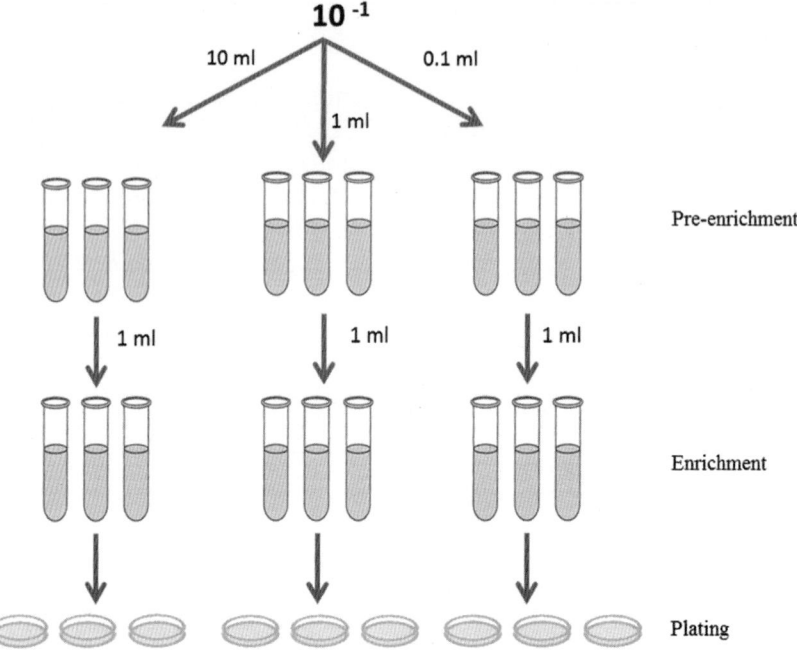

Fig. 2 Schematic representation of MPN method with three tubes dilutions

8. Invert dishes and incubate at 37 °C for 48 h.

9. Count all characteristic colonies presumed to be *L. monocytogenes* and proceed to confirmation (*see* **Note 7**).

3.3 Most Probable Number (MPN) of L. monocytogenes

1. Initial suspension (10^{-1} dilution). Weigh 10 g of analytical portions of solid food into a sterile plastic bag. Add 90 mL or g of diluent media broth (half Fraser base, BLEB or UVM I) (*see* Fig. 2).

2. Homogenize the mixture in a Blender or Stomacher for 1–3 min.

3. Transfer 10 mL of the liquid analytical portion or 10 mL of the initial suspension to three tubes containing 10 mL of double strength pre-enrichment (half Fraser base, BLEB or UVM I) (*see* **Note 8**).

4. Transfer 1 and 0.1 mL of the liquid analytical portion or 1 and 0.1 mL of the initial suspension to three tubes containing 10 mL of single strength pre-enrichment (half Fraser base, BLEB, or UVM I).

5. Incubate for 24 h at 30 °C.

6. Transfer 1 mL from each tube to 10 mL of enrichment media broth (Fraser broth).

7. Incubate for 24 h at 37 °C.

8. Streak a loop of the enrichment broth culture onto chromogenic selective solid medium.

Table 2
Biochemical tests to differentiate *Listeria* species

Species	Phospholipase C	Hemolysis	Production of acid from			CAMP test	
			D-Mannitol	L-Rhamnose	D-Xylose	*S. aureus*	*R. equi*
L. monocytogenes	+	+	–	+	–	+	–
L. innocua	–	–	–	V	–	–	–
L. ivanovii	+	++	–	–	+	–	+
L. seeligeri	–	(+)	–	–	+	(+)	–
L. welshimeri	–	–	–	V	+	–	–
L. grayi subsp. Grayi	–	–	–	+	–	–	–
L. grayi subsp. Murrayi	–	–	–	V	–	–	–

V: variable; (+): weak reaction; ++: strong positive reaction; +: >90 % positive reactions; –: negative reaction

9. Incubate at 37 °C for 24–48 h.

10. Examine the dishes for the presence of typical colonies of *Listeria* spp. (*see* Table 2) and proceed to confirmation.

3.4 Confirmation of Isolates

1. Select five colonies for confirmation that are representative of suspect colony types and isolate onto TSAYE (*see* **Note 9**).

2. Incubate TSAYE plates at 37 °C for 18–24 h.

3. For biochemical confirmation use only pure cultures. Perform the following classical tests: Gram stain, catalase, hemolysis, and carbohydrate fermentation (*see* Table 2).

4. Test typical colonies for catalase and Gram stain.

5. Inoculate carbohydrate broth (mannitol, rhamnose, and xylose).

6. Incubate at 37 °C 24–48 h (*see* **Note 10**).

7. Perform CAMP test as follows: streak a β-hemolytic *Staphylococcus aureus* and a *Rhodococcus equi* culture in parallel and diametrically opposite each other on a 5 % sheep blood agar plate (*see* **Note 11**).

8. Streak test cultures parallel to one another, but at right angles to and between the *S. aureus* and *R. equi* streaks (but not touching them).

9. Incubate at 37 °C for 24–48 h (*see* **Note 12**).

10. Read tests and interpret the results (*see* Table 2).

11. Report as present/absent in the case of *L. monocytogenes* detection; give a number of *L. monocytogenes* as cfu/g or mL; or as most probable number/g or mL in the case of the MPN method (*see* **Note 13**).

Alternatively, confirmation or identification of *Listeria* species can be performed using commercial kits: API *Listeria* (bioMerieux, Marcy-l'Etoile, France), MICRO-ID™ kit (bioMerieux, Hazelwood, MO; 1, 24), Phenotype MicroArray for *Listeria* (BiOLOG, Hayward, CA), or by Polymerase Chain Reaction (PCR; *see* Chapter 3).

4   Notes

1. If a different amount of sample is used, add a quantity of diluent equal to $9 \times m$ g or $9 \times V$ mL of pre-enrichment medium.

2. Choose media that are complementary, i.e., one chromogenic and one aesculin-containing medium.

3. In the case of use of chromogenic media, follow the manufacturer's instructions.

4. If a different amount of sample is used add a quantity of diluent equal to $9 \times m$ g or $9 \times V$ mL of pre-enrichment medium. Liquid samples could be inoculated directly onto selective agar.

5. If the sample has low numbers of *Listeria*, distribute 1 mL of the liquid test sample or the initial suspension on the surface of the agar medium in a 140 mm Petri dish or over the surface of three small Petri dishes. Other equivalent media can be used instead of ALOA. In this case follow the recommendations of manufacturer. Agar plates should be dried in an oven or in a laminar-flow cabinet between 25 and 50 °C until the droplets have disappeared from the surface of the medium.

6. It is possible to use the same spreader for the same sample if spreading is started from the higher dilution.

7. Count plates containing less than 150 characteristic or non-characteristic colonies.

8. Five tubes for each dilution can also be used; in case of liquid products prepare the first serial dilution in single strength pre-enrichment medium.

9. For confirmation of the typical colonies it is prescribed to streak isolated colonies from the selective plating medium onto TSAYE agar before performing the biochemical confirmation. However, this step is not necessary if well-isolated colonies (of a pure culture) are available on the selective plating medium. If this is the case, perform the biochemical confirmation directly on a typical (suspect), well-isolated colony of each selective plating medium.

10. Fermentation of carbohydrates usually occurs in 24 h. However, there are *Listeria* species that require more time of incubation so it is advisable to incubate up to 5 days.

11. Instead of the CAMP test commercially available lysin discs could be used.

12. The hemolytic activity of *L. monocytogenes* and to a lesser extent *L. seeligeri* is enhanced in the zone influenced by the *S. aureus* streak. The other species remain non-hemolytic. *L. ivanovii* hemolysis is enhanced in the vicinity of *R. equi*.

13. Use MPN tables to determine MPN value [12].

References

1. Hitchins AD, Jinneman K (2011) Detection and enumeration of Listeria monocytogenes in foods. In: Bacteriological analytical manual, Chapter 10. U.S. Food and Drug Administration. http://www.fda.gov/food/scienceresearch/laboratorymethods/bacteriologicalanalytical.manualbam/ucm071400.htm. Accessed 17 Mar 2013

2. USDA (2012) Isolation and identification of *Listeria monocytogenes* from red meat, poultry and egg products, and environmental samples. In: Microbiology laboratory guidebook, Method Number 8.08. United States Department of Agriculture Food Safety and Inspection Service, Office of Public Health Science. http://www.fsis.usda.gov/PDF/MLG-8.pdf. Accessed 17 Mar 2013

3. ISO (2004) ISO 11290-1. Microbiology of food and animal feeding stuffs—horizontal method for the detection and enumeration of *Listeria monocytogenes*—part 1: detection method amendment 1: modification of the isolation media and the haemolysis test and inclusion of precision data. International Organization for Standardization, Geneva

4. ISO (2004) ISO 11290-2. Microbiology of food and animal feeding stuffs—horizontal method for the detection and enumeration of *Listeria monocytogenes*—part 2: enumeration method amendment 1: modification of the enumeration medium. International Organization for Standardization, Geneva

5. EC (2005) Commission Regulation N° 2073/2005of 15 November 2005 on microbiological criteria for foodstuffs. Official Journal of the European Union L338:1–26. http://eurlex.europa.eu/LexUriServ/LexUriServ.do?uri=OJ:L:2005:338:0001:0026:EN:PDF. Accessed 17 Mar 2013

6. Miller FA, Brandão TRS, Teixeira P et al (2006) Recovery of heat-injured *Listeria innocua*. Int J Food Microbiol 112:261–265

7. Montville TJ, Matthews KR (2008) Factors that influence microbes in foods. In: Montville TJ, Matthews KR (eds) Food microbiology: an introduction, 2nd edn. ASM Press, Washington, DC, pp 17–19

8. ISO (1996) ISO 11290-1. Microbiology of food and animal feeding stuffs—horizontal method for the detection and enumeration of *Listeria monocytogenes*—part 1: detection method. International Organization for Standardization, Geneva

9. ISO (1998) ISO 11290-2. Microbiology of food and animal feeding stuffs—horizontal method for the detection and enumeration of *Listeria monocytogenes*—part 2: enumeration method. International Organization for Standardization, Geneva

10. Johansson T (1998) Enhanced detection and enumeration of *Listeria monocytogenes* from foodstuffs and food-processing environments. Int J Food Microbiol 40:77–85

11. Reissbrodt R (2004) New chromogenic plating media for detection and enumeration of pathogenic *Listeria* spp.—an overview. Int J Food Microbiol 95:1–9

12. USDA (2008) Most probable number procedure and tables. In: Laboratory guidebook, Appendix 2.03. United States Department of Agriculture Food Safety and Inspection Service, Office of Public Health Science. http://www.fsis.usda.gov/PDF/MLG_Appendix_2_03.pdf. Accessed 17 Mar 2013

13. Lovette J, Frances DW, Hunt JM (1987) *Listeria monocytogenes* in raw milk: detection, incidence and pathogenicity. J Food Prot 50:188–192

14. Donnelly CW, Baigent GJ (1986) Method for flow cytometric detection of *Listeria monocytogenes* in milk. Appl Environ Microbiol 52:689–695

15. Fraser JA, Sperber WH (1988) Rapid detection of *Listeria* spp. in food and environmental

samples by esculin hydrolysis. J Food Prot 51: 762–765

16. van Netten P, Perales I, van de Moosalijk A, Curtis GDW, Mossel DAA (1989) Liquid and solid selective differential media for the detection and enumeration of *L. monocytogenes* and other *Listeria* spp. Int J Food Microbiol 8: 299–317

17. Curtis GDW, Mitchell RG, King AF, Emma J (1989) A selective differential medium for the isolation of *Listeria monocytogenes*. Lett Appl Microbiol 8:95–98

Chapter 3

Confirmation of Isolates of Listeria by Conventional and Real-Time PCR

David Rodríguez-Lázaro and Marta Hernández

Abstract

Polymerase chain reaction (PCR) is an invaluable diagnostic technique in microbiology for rapid and specific detection and confirmation of microbial isolates from food and the environment. PCR is a simple, sensitive, specific, and reproducible assay and can be performed in conventional or in real-time formats. Here, we describe the application of real-time and conventional PCR-based methods for confirmation of presumptive *Listeria* isolates.

Key words Conventional PCR, Real-time PCR, Identification, *Listeria* spp, *Listeria monocytogenes*

1 Introduction

Polymerase chain reaction (PCR) is a simple, versatile, sensitive, specific, and reproducible assay [1]. It consists of an exponential amplification of a DNA fragment, and its principle is based on the mechanism of DNA replication in vivo: dsDNA is denatured to ssDNA, duplicated, and this process is repeated along the reaction. A development of the PCR, the real-time (q)PCR, represents a significant advance in many molecular procedures involving nucleic acids analysis. qPCR allows monitoring of the synthesis of new amplicon molecules during the amplification (i.e., in real time) by using fluorescence, and not only at the end of the reaction, as occurs in conventional PCR [2]. Major advantages of qPCR are the closed-tube format (that avoids risks of carry-over contamination), fast and easy analysis, an extremely wide dynamic range of quantification (more than eight orders of magnitude) and the significantly higher reliability and sensitivity of the results compared to conventional PCR [3]. Those advantages should foster its implementation in food laboratories and PCR has been predicted to be established as a routine reference [4].

Here, we described the complete analytical process for confirmation of presumptive *Listeria* isolates; identification of the

Kieran Jordan et al. (eds.), *Listeria monocytogenes: Methods and Protocols*, Methods in Molecular Biology, vol. 1157, DOI 10.1007/978-1-4939-0703-8_3, © Springer Science+Business Media New York 2014

isolates as *Listeria* spp. (genus level), and/or identification of the isolates at *Listeria* species level (only the three more relevant): *Listeria monocytogenes* (human and animal pathogen), *Listeria ivanovii* (animal pathogen), and *Listeria innocua* (apathogenic species sharing similar environments to *L. monocytogenes*). The process starts with the selection of the isolates to be confirmed and the obtaining of their DNA (two different protocols are described in Subheading 3.1). Finally, the protocols for PCR (both conventional and real-time formats) are described for confirmation of those different taxa.

2   Materials

Prepare all solutions using ultrapure water and molecular grade reagents. Prepare and store all reagents at room temperature (unless indicated otherwise).

2.1   Listeria DNA Extraction

1. Lysis buffer: the lysis buffer consists of a 1× Tris–EDTA buffer solution (TE; 10 mM Tris–HCl, 1 mM disodium EDTA, pH 8).

2. Lysozyme solution: an ultrapure water solution containing 150 mg/ml of Lysozyme.

3. Proteinase K solution: an ultrapure water solution containing 5 mg/ml of Proteinase K.

4. Chelex buffer: an ultrapure water solution containing 6 % of Chelex® 100 resin (Bio-Rad, Hercules, CA, USA).

5. 1.5 ml centrifuge tubes (e.g., Eppendorf).

6. Thermo-block with capacity to achieve 95 °C.

2.2   PCR and Real-Time PCR

1. Real-time PCR Master Mix (e.g., FastStart Universal Probe Master, Roche or TaqMan® Real-Time PCR Master Mix, Life Technologies).

2. PCR Master Mix (e.g., FastStart PCR Master, Roche).

3. Optical PCR plates and caps for real-time PCR according to the PCR device format (24, 48, 96, or 384 wells) (e.g., MicroAmp® Optical 96-Well Reaction Plate and Optical Adhesive Film, Applied Biosystems or LightCycler® 480 Multiwell Plate, Roche).

4. Optical PCR plates and caps for PCR (e.g., 4titude).

5. Oligonucleotides for detection of *Listeria* spp. by PCR. The oligonucleotides amplify a 77-bp region of the 23S rRNA gene of *Listeria* spp. [5] (Table 1).

6. Oligonucleotides for detection of *L. monocytogenes* by PCR. The oligonucleotides amplify a 64-bp region of the *hly* gene of *L. monocytogenes* [6] (Table 1).

Table 1
List of oligonucleotides for PCR confirmation

Taxa	Oligo-nucleotide	Sequence (5′–3′)	Length	Concentration	Reference
Listeria spp.	*L23SQF*	AGG ATA GGG AAT CGC ACG AA	77 bp	300 nM	[5]
	L23SQR	TTC GCG AGA AGC GGA TTT		300 nM	
	Lin23SQFR	TTC GCA AGA AGC GGA TTT G		300 nM	
	L23QP	FAM- TCT CAC ACT CAC TGC TTG GAC GC–BHQ		100 nM	
L. mono-cytogenes	*hlyQF*	CAT GGC ACC ACC AGC ATC T	64 bp	50 nM	[6]
	hlyQR	ATC CGC GTG TTT CTT TTC GA		50 nM	
	hlyQP	FAM—CGC CTG CAA GTC CTA AGA CGC CA—BHQ		100 nM	
L. ivanovii	*LivQF*	CGGTCATGCACGT CCACAT	62 bp	300 nM	[6]
	LivQR	CCACTGTGGTGACTT GGTATGC		300 nM	
	LivQP	FAM–ATGGCATAACAA AGTC–MGB		200 nM	
L. innocua	*lipHQF*	AAC CGG GCC GCT TAT GA	61 bp	50 nM	[7]
	lipHQR	CGA ACG CAA TTG GTC ACG		50 nM	
	lipHQP	FAM—TTC GAA TTG CTA GCG GCA CAC CAG T—BHQ		100 nM	

7. Oligonucleotides for detection of *L. ivanovii* by PCR. The oligonucleotides amplify a 61-bp region of the *smcL* gene of *L. ivanovii* [7] (Table 1).

8. Oligonucleotides for detection of *L. innocua* by PCR. The oligonucleotides amplify a 62-bp region of the *lin02483* gene of *L. innocua* [6] (Table 1).

9. Real-time PCR platform (e.g., LightCycler® 480 Instrument II, Roche or Applied Biosystems® 7500 Real-Time PCR System, Life Technologies).

10. PCR thermocycler (e.g., GeneAmp® PCR System 9700 Dual 96-Well, Life Technologies).

11. Electrophoresis cuvettes and power supplier.

3   Methods

Carry out all procedures at room temperature unless otherwise specified.

3.1   Listeria DNA Extraction: Rapid Lysis

1. Transfer one presumptive *Listeria* colony with a loop from a Petri dish into a 1.5-ml centrifuge tube containing 50 µl of rapid lysis buffer. This step can be repeated in separate 1.5-ml centrifuge tubes for confirmation of different colonies from a Petri dish—particularly if different morphology is observed in colonies isolated in *Listeria*-specific plates.

2. Mix using vortex.

3. Add 3 µl of Lysozyme solution.

4. Incubate at 37 °C for 45 min.

5. Add 2 µl of Proteinase K solution.

6. Incubate 1 h at 55 °C.

7. Stop the enzymatic reaction by incubation for15 min at 95 °C.

8. Centrifuge at $10,000 \times g$ for 5 min at 4 °C.

9. Transfer the supernatant carefully (up to 45 µl) to a fresh 1.5-ml centrifuge tube.

10. Store at 4 °C if used immediately or before 24 h; or store at –20 °C for longer periods.

3.2   Listeria DNA Extraction Using Chelex 100 Resin

1. Transfer one presumptive *Listeria* colony with a loop from a Petri dish into a 1.5-ml centrifuge tube containing 50 µl of Chelex buffer. This step can be repeated in separate 1.5-ml centrifuge tubes for confirmation of different colonies from a Petri dish—particularly if different morphology is observed in colonies isolated in *Listeria*-specific plates.

2. Mix thoroughly and incubate at 56 °C for 20 min in a thermoblock.

3. Mix thoroughly and incubate at 95 °C for 8 min.

4. Mix thoroughly by vortexing, and chill the mixture on ice.

5. Centrifuge at 4 °C for 5 min at $10,000 \times g$.

6. Transfer the supernatant gently (up to 40 µl) transferred to a fresh 1.5-ml centrifuge tube.

7. Store at 4 °C if used immediately or before 24 h; or store at –20 °C for longer periods.

3.3   Detection of Listeria by Real-Time PCR

1. Prepare the real-time PCR MIX containing 1× PCR commercial MasterMix, the adequate concentration of the specific primers and probes, and the adequate volume of water (*see* Table 2 as an example of a master mix for confirmation of *L. monocytogenes*)

Table 2
Conditions for preparation of the real-time PCR MasterMix
for confirmation of *Listeria monocytogenes*

Reagent	Working concentration	Final concentration	Volume (µl)
Mix	2×	1×	10
Primer hlyQF	1 µM	50 nM	1
Primer hlyQR	1 µM	50 nM	1
Probe hlyQP	1 µM	100 nM	2
Ultrapure water			1
Total volume of MIX			15
DNA sample			5
Final volume			20

It is calculated for one reaction; for more reactions, increase the amounts accordingly. For other species substitute the appropriate primers

(*see* **Note 1**). The appropriate concentrations of oligonucleotides to add to each reaction are shown in Table 1.

2. Aliquot 15 µl of real-time PCR MasterMix into each well of a real-time PCR plate.

3. Add 5 µl of the DNA extract into each well of a real-time PCR plate (*see* **Note 2**). For confirmation of each presumptive *Listeria* colony use at least 2 real-time PCR replicates, and 2 blank and 2 control positive real-time PCR replicates per each real-time PCR run (ultrapure water and DNA extracted from confirmed *Listeria* isolates, respectively).

4. Run the PCR on an conventional PCR thermocycler using the following program: 2 min at 50 °C, 10 min at 95 °C and 40 cycles of 15 s at 95 °C and 1 min at 60 °C for confirmation of *Listeria* spp., *L. innocua* or *L. ivanovii*, or use 50 °C, 10 min at 95 °C and 40 cycles of 15 s at 95 °C and 1 min at 63 °C for confirmation of *L. monocytogenes*.

5. Analyze the real-time PCR results using the software provided in the real-time PCR platform. Results can be considered as positive, i.e., confirmation of the isolate, when a positive amplification is observed, i.e., when the C_q values (*see* **Note 3**) are smaller than 40. Negative values or lack of amplification is considered for real-time PCRs with C_q values equal or higher than 40.

3.4 Detection of Listeria by Conventional PCR

1. Prepare a conventional mastermix containing 1× PCR MasterMix, the adequate concentration of the specific primers, and the adequate volume of water (*see* Table 3 as example for

Table 3
Conditions for preparation of the PCR MasterMix for confirmation of *Listeria monocytogenes*

Reagent	Working concentration	Final concentration	Volume (μl)
Mix	2×	1×	10
Primer hlyQF	1 μM	50 nM	1
Primer hlyQR	1 μM	50 nM	1
Ultrapure water			3
Total volume of MIX			15
DNA sample			5
Final volume			20

It is calculated for one reaction; for more reactions, increase the amounts accordingly. For other species substitute the appropriate primers

master mix for confirmation of *L. monocytogenes*) (*see* **Note 4**). The appropriate concentrations of oligonucleotides to each reaction are as follows: 300 nM *L23SQF/R* and *Lin23SQR* primers for confirmation of *Listeria* spp., 50 nM each *hlyQF/R* primer for confirmation of *L. monocytogenes*, 300 each nM LivQF/R primer for confirmation of *L. ivanovii*, and 50 nM each *lipHQF/R* primer for confirmation of *L. innocua*.

2. Aliquot 15 μl of PCR MasterMix into each well of a conventional PCR plate.

3. Add 5 μl of the DNA extraction solution into each well of a PCR plate (*see* **Note 2**). For confirmation of each presumptive *Listeria* colony use at least 2 PCR replicates, and 2 blank and 2 control positive PCR replicates per PCR run (ultrapure water and DNA extracted from confirmed *Listeria* isolates, respectively).

4. Run the PCR on an conventional PCR thermocycler using the following program: 2 min at 50 °C, 10 min at 95 °C and 40 cycles of 15 s at 95 °C and 1 min at 60 °C for confirmation of *Listeria* spp., *L. innocua* or *L. ivanovii*, or use 50 °C, 10 min at 95 °C and 40 cycles of 15 s at 95 °C and 1 min at 63 °C for confirmation of *L. monocytogenes*.

5. Mix 10 μl of the PCR product with 2 μl of commercial dye from each well and load the mix onto a 3 % agarose gel and apply a voltage of 1.5 V/cm until the dyes have run around two-third of the gel, i.e., approximately for 30 min.

6. Stain the gel with ethidium bromide (0.5 μg/ml) for 30 min after electrophoresis.

7. A positive PCR for confirmation of *Listeria* spp. must show a band of 77 bp, a positive PCR for confirmation of *L. monocytogenes* a band of 64 bp, a positive PCR for confirmation of *L. innocua* a band of 61 bp, and a positive PCR for confirmation of *L. ivanovii* a band of 62 bp (Table 1).

4   Notes

1. The preparation of the PCR MasterMix should be done in a room physically separated from that used for DNA extraction. It is advisable that the addition to the DNA solution to the mastermix is done in another separate room or a dedicated PCR cabinet to avoid any carry-over contamination. It is also advisable to use uracyl-N-glycosidase (UNG) to avoid that kind of contamination.

2. The standard volume of DNA extract for the PCR is 5 µl, but this volume can be increased if needed reducing the volume of water added, or using smaller volumes of oligonucleotides with higher concentrations. For example, instead of using 5 µl of DNA sample per reaction (*see* Tables 2 and 3), 9.6 µl could be used for real-time PCR detection if water is not added and only 0.1 µl of each nucleotide (at 10 µM) is added.

3. C_q value is the quantification cycle. This values is named differently in each real-time PCR devise, e.g., C_T (threshold cycle) or C_p (cycle to positivity). It defines the PCR cycle in which the amplification reaches a predefined cycle, and it is directly related to the initial amount of template in the real-time PCR; i.e., higher initial amounts, smaller C_q values. As the standard real-time PCR cycling conditions are defined to use only 40 cycles, samples will be considered as positive when the C_q value is smaller than 40, and negative when there is not any amplification, i.e., the C_q values equal or higher to 40.

4. If conventional PCR is used, a similar protocol to 3.3 must be followed, but use a specific mastermix for conventional PCR and only the forward and reverse primers but not the oligonucleotide fluorogenic probe.

References

1. Rodríguez-Lázaro D, Hernández M (2013) Real-time PCR in food science: introduction. Curr Issues Mol Biol 15:25–38

2. Rodríguez-Lázaro D, Lombard B, Smith H, Rzezutka A, D'Agostino M, Helmuth R, Schroeter A, Malorny B, Miko A, Guerra B, Davison J, Kobilinsky A, Hernández M, Bertheau Y, Cook N (2007) Trends in analytical methodology in food safety and quality: monitoring microorganisms and genetically modified organisms. Trends Food Sci Technol 18:306–319

3. Rodríguez-Lázaro D, Cook N, Hernandez M (2013) Real-time in food science: PCR diagnostics. Curr Issues Mol Biol 15:39–44

4. Hoorfar J, Cook N (2003) Critical aspects in standardization of PCR. In: Sachse K, Frey J

(eds) Methods in molecular biology: PCR detection of microbial pathogens. Humana Press, Totowa, pp 51–64

5. Rodríguez-Lázaro D, Hernández M, Pla M (2004) Simultaneous quantitative detection of *Listeria* spp. and *Listeria monocytogenes* using a duplex real time PCR-based assay. FEMS Microbiol Lett 233:257–267

6. Rodríguez-Lázaro D, Hernández M, Scortti M, Esteve T, Vázquez-Boland JA, Pla M (2004) Quantitative detection of *Listeria monocytogenes* and *Listeria innocua* by real-time PCR: assessment of *hly*, *iap* and lin02483 targets and AmpliFluor technology. Appl Environ Microbiol 70: 1366–1377

7. Rodríguez-Lázaro D, López-Enríquez L, Hernandez M (2010) *smcL* as a novel diagnostic marker for quantitative detection of *Listeria ivanovii* in biological samples. J Appl Microbiol 109:863–872

Part II

Characterization and Typing

Chapter 4

Serotype Assignment by Sero-Agglutination, ELISA, and PCR

Lisa Gorski

Abstract

For assessing isolates of *Listeria monocytogenes,* serotype designation is the foremost subtyping method used. Traditionally, serotyping has been done with agglutination reactions. In the last decade, alternative serotyping methods were described using enzyme-linked immunosorbent assay (ELISA) and polymerase chain reaction (PCR). Herein are described the three methods, and the advantages and disadvantages of each.

   Key words Listeria, Serotype, Agglutination, ELISA, PCR, Antigens, Antisera

1   Introduction

Several methods are used for subtyping *L. monocytogenes* isolates including DNA fingerprinting with specific or random primers, ribotyping, multi-locus sequence typing, multiple-locus variable number tandem repeat analysis, and pulsed-field gel electrophoresis. These methods and more are reviewed in Jadhav et al. [1]. All of these methods, however, are still compared to serotype analysis. Serotype analysis of isolates is necessary for outbreak studies and source tracking of isolates. The original method for serotype designation developed by Seeliger and Höhne [2] is still used and is based on the reactions by the cell-surface somatic (O) and flagellar (H) antigens to a set of high-quality antisera. Table 1 indicates the 13 serotypes, and the combinations of O- and H-antigens necessary for their designations. It is assumed that by the time a strain is ready for serotype analysis the isolate in question has been positively identified as *L. monocytogenes* by the various methods described earlier in this book.

About 95 % of the *L. monocytogenes* strains isolated from humans are of serotypes 4b, 1/2a, 1/2b, and 1/2c [3, 4]. There are genetically distinct lineages within the species with lineage I containing isolates of serotype 1/2b, 4b, 4e, and 4d; lineage II

Kieran Jordan et al. (eds.), *Listeria monocytogenes: Methods and Protocols*, Methods in Molecular Biology, vol. 1157, DOI 10.1007/978-1-4939-0703-8_4, © Springer Science+Business Media New York 2014

Table 1
Antigen components of *L. monocytogenes* serotypes

Serotype	O-Antigens[a]	H-Antigens
1/2a	I, II	A, B
1/2b	I, II	A, B, C
1/2c	I, II	B, D
3a	II, IV	A, B
3b	II, IV	A, B, C
3c	II, IV	B, D
4a	(V), VII, IX	A, B, C
4ab	V, VI, VII, IX	A, B, C
4b	V, VI	A, B, C
4c	V, VII	A, B, C
4d	(V), VI, VIII	A, B, C
4e	V, VI, (VIII), (IX)	A, B, C
7	(III), XII, XIII	A, B, C

[a]Antigens in parentheses may not be present in all isolates

containing isolates of serotype 1/2a, 1/2c, 3a, 3c, and some atypical 4b; and lineage III containing some strains of serotype 4a and 4c as well as some atypical strains of serotype 4b [5–7]. Recently, some lineage III strains were recharacterized into a lineage IV [8]. While most of the isolated strains are of serotypes 4b, 1/2a, 1/2b, and 1/2c, strains of the other serotypes have been isolated from food, environmental, and agricultural samples. Some atypical strains of *L. innocua* appear as serotype 4b [9].

Of the methods for determining serotype, the agglutination reaction is still used most commonly in health laboratories. This reaction depends on visual determination of agglutination by a set of antisera. It takes time to prepare the cultures, and the set of antisera is expensive. Also the reactions can be difficult to ascertain, and often a trained eye and/or a microscope is necessary to call weak reactions. A multicenter, agglutination-method validation study noted discrepancies in results between labs [10, 11]. Another method for determining serotype uses the same antisera as the agglutination method in an enzyme-linked immunosorbent assay (ELISA) with the data analyzed in a semiquantitative manner [12]. The ELISA assay is easier to read than agglutination, but it still requires that the cells be prepared in the same manner and still requires the same antisera set. However, the ELISA method uses about 100 times less antiserum than the agglutination method, so an expensive kit can last much longer. It should be noted that non-motile strains are occasionally isolated, and if they are non-motile because of a lack of flagella, then their flagellar H-antigens will not be typable with either the agglutination or ELISA methods (this is relevant only for serotypes 1/2 and 3 strains, Table 1). Finally, several PCR methods for serotype determination have been described [13–16]. A multiplex PCR protocol for serogroup designation was developed [13, 15], validated in multiple labs [17], and is used by many researchers. The primers for the multiplex PCR correspond roughly to lineage-specific genes. The results obtained by the PCR methods are easier to read than agglutination reactions, and they are faster and cheaper to use than both the agglutination and ELISA methods, so they are available to more laboratories. Non-motile isolates are more likely to give a product in this method of serotyping. However, none of the PCR protocols can distinguish between all 13 serotypes as the antisera-based methods can. The methodologies for the three protocols will be presented here. In the end, one may choose to use one or a combination of methods for determining serotype based on the needs and resources of the laboratory.

2   Materials

Prepare all solutions with ultrapure water. Prepare and store all reagents at room temperature, unless otherwise indicated. Follow all waste disposal and personal safety protocols when handling chemicals. Wear gloves when handling antisera and when setting up PCR reactions.

2.1   Agglutination Method

1. Rich liquid medium, agar plates, and soft agar plates for growing bacteria (*see* **Note 1**).

2. Antisera kit (Denka-Seiken, part number 294616) available from several distributors. Some distributors sell individual antisera from this manufacturer as well (*see* **Note 2**).

3. Test tubes (6 × 50 mm) for the tube method or clean microscope slides for the slide agglutination method.

4. 0.2 % (w/v) NaCl. Weigh 2 g of NaCl and dissolve in 1 L water. Autoclave and store at room temperature (RT).

5. Spectrophotometer

6. Water bath set at 48–50 °C.

7. 4 % Formaldehyde in 0.2 % NaCl. Weigh 2 g of NaCl and dissolve in 892 mL water. Add 108 mL of 37 % formaldehyde. Store at RT.

8. Centrifuge.

9. Autoclave.

2.2 ELISA Method

1. Brain heart infusion broth (BHI) and BHI soft agar (BHI broth + 0.4 % agar) (*see* **Note 1**).

2. Antisera kit (Denka-Seiken, part number 294616) available from several distributors. Some distributors sell individual antisera from this manufacturer as well (*see* **Note 2**).

3. Secondary antibody: alkaline-phosphatase conjugated anti-rabbit IgG.

4. 0.2 % (w/v) NaCl: Weigh 2 g of NaCl and dissolve in 1 L water. Autoclave and store at RT.

5. 4 % Formaldehyde in 0.2 % NaCl: Weigh 2 g of NaCl and dissolve in 892 mL water. Add 108 mL of 37 % formaldehyde. Store at RT.

6. 1 M Tris–HCl, pH 7.5: Add about 500 mL of water to a beaker or graduated cylinder. Weigh 121 g of Tris base and add to the water. Mix and adjust pH with HCl. Make up to 1 L with water. Store at room temperature (RT).

7. 10× Tris-Buffer-Salt (TBS): For 10× TBS, add about 500 mL of water to a beaker, add 100 mL of Tris–HCl, pH 7.5 and 87.7 g of NaCl to the beaker. Mix and add water to make a 1 L solution. Store at RT.

8. ELISA Wash buffer: 0.01 M Tris–HCl (pH 7.5), 0.15 M NaCl, 0.1 % Tween 20. To 900 mL of water add 100 mL of 10× TBS and 1 mL of Tween 20. Mix and store at RT.

9. Blocking solution: 0.01 M Tris–HCl (pH 7.5), 0.15 M NaCl, 0.5 % (w/v) casein, 0.031 M sodium azide. To 900 mL water add 100 mL 10× TBS, 5 g of casein, and 2 g of sodium azide. It takes 30–60 min for the casein to go into solution. The solution can be heated slightly to expedite this, but be sure to cool to RT before using. Store at RT (*see* **Note 3**).

10. Antiserum Dilution Buffer: 0.01 M Tris–HCl (pH 7.5), 0.15 M NaCl, 2.7 mM KCl, 10 % (w/v) Bovine Serum Albumin, 0.015 M sodium azide. To 900 mL of water add 100 mL of 10× TBS, 10 g of bovine serum albumin, 0.2 g of KCl, and 1 g of sodium azide (*see* **Note 3**).

11. Alkaline Phosphatase Substrate Buffer: 1 M Diethanolamine, 0.5 mM $MgCl_2$. To 500 mL of water add 105.2 mL of diethanolamine and 0.1 g of $MgCl_2 \cdot 6H_2O$. Adjust pH to 9.8 with HCl and bring the solution up to 1 L with water.

12. *Para*-nitrophenyl phosphate (PNP) substrate solution for ELISA: Prepare right before use by making a 1 mg/mL solution of PNP in Alkaline Phosphatase Substrate Buffer (above).

Prepare enough so that 0.1 mL can be used for each well being tested in the ELISA plate. The amount needed depends on the number of wells being read. For example, if one 96-well plate is being read then approximately 10 mL of PNP solution is needed. To make 10 mL of 1 mg/mL PNP solution, weigh out 10 mg of PNP and add it to 10 mL of Alkaline Phosphatase Substrate Buffer. Do not make a big stock of PNP solution to store because it will degrade and turn yellow on its own within 24 h.

13. Centrifuge.

14. Autoclave.

15. Spectrophotometer.

16. 96-Well flat bottom plates coated for ELISA assays or coated multiwall strip modules.

17. Oven set for 40 °C.

18. Multichannel pipettor.

19. A large squeeze bottle of distilled water for rinsing ELISA plates.

20. Spectrophotometer equipped to read 96-well plates.

21. Spreadsheet program for analysis and visualization of data and making graphs in order to call the antisera reactions.

2.3 PCR Method

1. Kit or method for preparation of genomic DNA from *L. monocytogenes* (*see* **Note 4**).

2. Tris–EDTA (TE) buffer pH 8.0: 10 mM Tris–HCl, pH 8.0, 1 mM EDTA. This can be purchased or made in the lab. To 988 mL of water add 10 mL of 1 M Tris–HCl, pH 8.0 and 2 mL of 0.5 M EDTA, pH 8.0. Autoclave at 121 °C for 20 min. Cool and store at RT.

3. Ultra-violet spectrophotometer for measuring DNA concentrations.

4. Nuclease-free water

5. Primers of specific sequence (Table 2).

6. Kit for PCR (*see* **Note 4**).

7. 10 mM solution of deoxynucleotides (dNTPs) for PCR: These can be purchased as a 10 mM solution of the four deoxynucleotides mixed together. Be sure to purchase a mixture that states it can be used for PCR. Alternatively, they are supplied with the kit.

8. Tris–Acetic Acid–EDTA buffer (TAE) at 1× strength or Tris–Boric Acid–EDTA buffer (TBE) at 0.5× strength for agarose gel electrophoresis: Both of these buffers can be purchased as concentrated solutions and then diluted for use. These can be supplied with the kit.

Table 2
Primers for multiplex PCR for serotype determination[a]

Primer name	Sequence (5′ → 3′)	Differentiation	Product size (bp)
lmo0737(F)	AGGGCTTCAAGGACTTACCC	1/2a, 1/2c, 3a, 3c	691
lmo0737(R)	ACGATTTCTGCTTGCCATTC		
lmo1118(F)	AGGGGTCTTAAATCCTGGAA	1/2c, 3c	906
lmo1118(R)	CGGCTTGTTCGGCATACTTA		
ORF2819(F)	AGCAAAATGCCAAAACTCGT	1/2b, 3b, 4b, 4d, 4e	471
ORF2819(R)	CATCACTAAAGCCTCCCATTG		
ORF2110(F)	AGTGGACAATTGATTGGTGAA	4b, 4d, 4e	597
ORF2110(R)	CATCCATCCCTTACTTTGGAC		
prs(F)	GCTGAAGAGATTGCGAAAGAAG	*Listeria* species	370
prs(R)	CAAAGAAACCTTGGATTTGCGG		

[a]From ref. [13]

9. 2.0 % (w/v) agarose: Weigh out 1 g of ultra-pure agarose and add to 50 mL of either 1× TAE or 0.5× TBE. Heat to boiling to dissolve (on a hot plate or a microwave), cool to 50 °C in a water bath. Add to the gel box.

10. DNA loading buffer: This can be purchased or made in the lab. One common formulation for a 10× loading buffer is 10× Tris–Acetic Acid–EDTA (or 5× Tris–Boric Acid–EDTA), pH 8, 2 % SDS, 0.2 % xylene cyanol, 0.25 % bromophenol blue, and 8 % Ficoll. To make this, add 50 mL of 10× TAE or 5× TBE (stock solutions of electrophoresis running buffer) to a beaker and add 2 g of SDS, 0.2 g of xylene cyanol, 0.25 g of bromophenol blue, and 8 g of Ficoll. Mix until in solution and then make the solution 100 mL with 10× TAE or 5× TBE.

11. 100 base pair DNA ladder for size standard: This can be purchased from multiple suppliers, and allows for discrimination of DNA from 100 to 1,000 bp in length.

12. Agarose electrophoresis system complete with gel box, casting tray, and power supply.

13. Dye for staining agarose gels containing DNA, such as ethidium bromide or a nontoxic substitute: An ethidium bromide staining bath can be made by purchasing a 10 mg/mL stock solution, and adding 20 μL of the stock solution into 400 mL of electrophoresis buffer, and placing the gel into the bath for 30 min. If using a nontoxic substitute, follow manufacturer's instructions.

14. Agarose gel visualization system.

3 Methods

Carry out all procedures at room temperature, unless otherwise indicated.

3.1 Agglutination Method

1. Cells are grown and prepared differently for assessment of O- and H-antigens. For O-antigen determination, *L. monocytogenes* strains can be grown in BHI, TSB, or TPB broth at 35–37 °C shaking at 100–120 rpm overnight. Centrifuge the cultures by adding culture to a centrifuge tube or microfuge tube and spin for 20 min at $800 \times g$ in a bench-top centrifuge. Wash the pellet by removing the supernatant and adding an equal volume of 0.2 % NaCl. Resuspend the pellet and spin again. Discard the supernatant and resuspend in an equal volume of fresh 0.2 % NaCl. Autoclave the suspension at 121 °C for 30 min, and cool to room temperature prior to use. Alternatively, colony material can be removed from an agar plate, and resuspended in TPB or 0.2 % NaCl to make a heavy cell suspension ($A_{600} > 0.55$, or turbidity > MacFarland standard 3; *see* **Note 5**). This suspension can be heated at 121 °C for 30 min and cooled to room temperature prior to use.

2. On a clean microscope slide(s) place 1 drop (25 μL) each of Denka-Seiken antisera O-I/II, O-V/VI, and 0.2 % NaCl. Place 25 μL of cell suspension into each drop with separate pipette tips. Mix each with the pipette tip or a disposable loop, and let rest at RT. After a few minutes tilt the slide back and forth and observe any agglutination pattern (Fig. 1).

3. If there is agglutination with O-I/II antiserum but not with O-V/VI, then repeat the agglutination test as in **step 2**, but use antisera O-I and O-IV as well as the 0.2 % NaCl control. Regardless of the results, these isolates will need to be tested

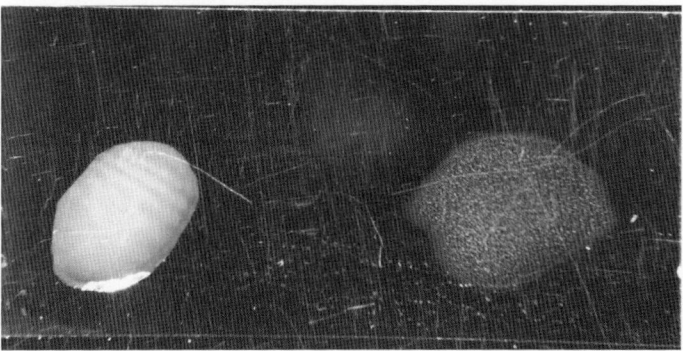

Fig. 1 Slide agglutination reactions. The spot on the *left* is the cell suspension mixed with 0.2 % NaCl, and it is a homogeneous suspension. The spot on the *right* is a positive agglutination reaction of cells mixed with antiserum

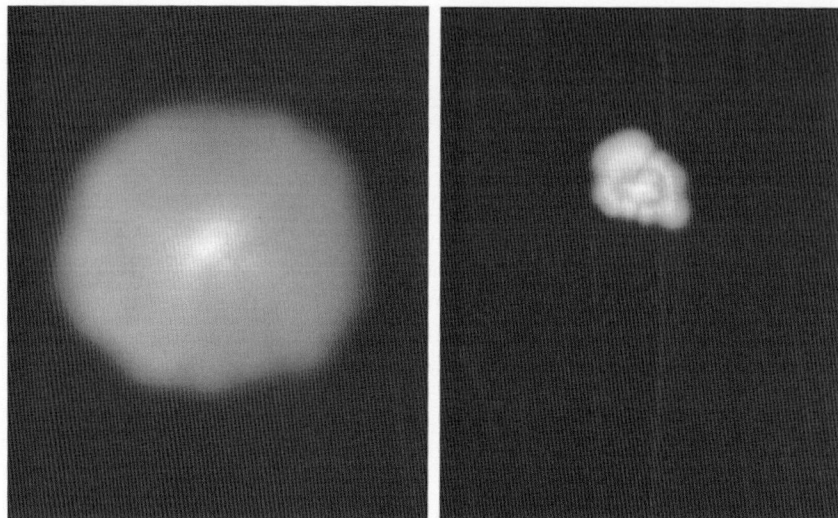

Fig. 2 Colonies on BHI + 0.4 % agar plates. The colony on the *left* is from a plate grown at 30 °C where *L. monocytogenes* is motile, and the colony on the *right* is from a plate grown at 37 °C where *L. monocytogenes* is not motile. The plates were incubated *upright* for 20 h

for reactions to flagellar H-antisera. If there is agglutination with O-V/VI and not O-I/II, then repeat agglutination test but use antisera O-VI, O-VII, O-VIII, and O-IX. Reactions with the flagellar H-antisera will not be needed for these strains (*see* **Note 6**).

4. Serotype 4 strains will be positive with O-V/VI and can be assigned serotype without determining flagellar-H antisera. Observe agglutination reactions with O-antisera VI, VII, VIII, and IX and compare with Table 1 to call the serotype. Serotype 1/2 and 3 strains will be positive with O-I/II antiserum and will need to be further differentiated with the H-antisera.

5. For flagellar H-antigen determination, the culture must be enriched for motile cells. This is done by passing the culture 1–3 times on semisolid media. Inoculate into Edge Motility Biological Medium (EB) or BHI soft agar by touching a colony with a sterile toothpick or inoculating loop and then dipping it into the center of a soft agar plate, and incubating the plate upright for 18–24 h at 25–30 °C. The colonies will be large as the cells will swim out (Fig. 2). With an inoculating loop, collect material from the colony edge and inoculate a fresh plate and incubate as above. Repeat up to 2 more times. Inoculate BHI broth with growth from the edge of a motile colony and incubate the culture at 25–30 °C without shaking for 18–24 h. Add an equal volume of 4 % formaldehyde in 0.2 % NaCl to the resulting culture. Mix gently (do not vortex). Incubate for 30 min at RT. Centrifuge 1 mL of the formaldehyde-treated

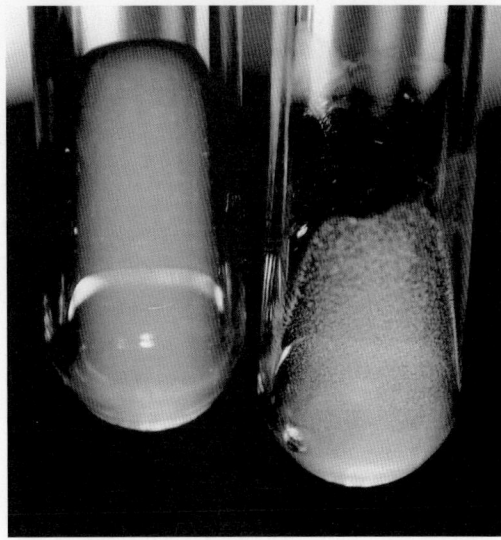

Fig. 3 Tube agglutination reactions. The control of cell suspension with 0.2 % NaCl is on the *left*. The positive agglutination reaction looks granular and is shown on the *right*

cell suspension in a microfuge at 3,000 rpm for 10 min. Discard supernatant and resuspend the pellet gently in 500 μL of 0.2 % NaCl (*see* **Note** 7).

6. H-antigen agglutination is done by the tube agglutination method. Place two drops (~50 μL) of each H-antisera into its own 6×50 mm tube. Include a control tube that contains 0.2 % NaCl. Place 50–100 μL of cell suspension into each tube, and place the tubes in a water bath set at 48–52 °C for up to 1 h. Observe the tubes for agglutination by tapping the tubes gently to resuspend the sediment. Do not shake the tubes too much as the agglutinant can break up. Hold up to a light background to observe agglutination (Fig. 3).

7. Record agglutination reactions and compare to antigenic structures in Table 1.

3.2 ELISA Method

It is important with each new kit of antisera to determine which dilutions are adequate for accurate reading of an ELISA assay. Generally we use the O-antisera at a dilution of 1:1,000 and the H-antisera at a dilution of 1:500; however, the titer of antisera changes from batch to batch. To determine the correct levels of antisera that will give a readable reaction after about 30 min of incubation, set up ELISAs as outlined below with strains of known serotype. Use dilutions of antisera such as 1:250, 1:500, 1:1,000, and 1:2,000, and select the dilution that works best. Also, you may need to run control experiments with various dilutions of secondary antibody if the manufacturer does not provide the concentration of the HP-conjugated antibody that is ideal for ELISA assays.

1. Cells are prepared differently for O- and H-antigen determinations. For O-antigen determination grow *L. monocytogenes* strains in BHI broth overnight at 37 °C in a shaking incubator (100–150 rpm). Measure the A_{600} of the cultures. This information will be needed later. Centrifuge 1 mL of the culture in a microfuge at 2300 × g for 15 min. Discard the supernatant and resuspend the pellet in 1 mL of 0.2 % NaCl. Autoclave the cell suspensions at 121 °C for 30 min and let cool to RT (*see* **Note 8**). Centrifuge the autoclaved suspensions for 15 min at 2300 × g in a bench-top centrifuge, discard the supernatant, and resuspend the pellet in 1 mL of 0.2 % NaCl. Dilute this cell suspension in 0.2 % NaCl to achieve so that the A_{600} is between 0.1 and 0.3.

2. Not all strains will need to be tested for H-antisera reactions. For example, serotype 4 strains all share the same H-antigens (Table 1) (*see* **Note 9**). For those strains where H-antigen reactions are necessary, enrich the *L. monocytogenes* strains in BHI soft agar as indicated for the agglutination method by three passages of the culture on BHI soft agar and overnight incubation upright at 25–30 °C. After the third passage, inoculate a portion of the colony edge into BHI broth and incubate overnight at 25–30 °C without shaking. Measure the A_{600} of the cultures. Add an equal volume of 4 % formaldehyde in 0.2 % NaCl, and let the suspension sit at room temperature for 30 min. Centrifuge the suspension at 2300 × g for 15 min in a bench-top centrifuge, and discard the supernatant. Gently resuspend the pellet in 0.2 % NaCl (do not vortex) to an A_{600} between 0.1 and 0.3.

3. Wear gloves whenever handling ELISA plates and antisera. Fill ELISA wells with 70 µL of cell suspension. Use a coated ELISA plate or coated ELISA strips to set up for the assay (*see* **Note 10**). There are eight O-antisera, so set up a vertical row of 8 wells with an aliquot each of the cell suspension. See the schematic in Fig. 4 for setting up the plate. One vertical row will cover the O-antisera for 1 strain. There are four H-antisera, so one vertical row of eight will cover two replicates of one strain, or two strains if duplicates are not being done. It is a good idea to set up extra rows of cell suspensions for controls such that there is a row that receives no primary antisera and another that receives no secondary antisera. If these rows show reactions then the assay was compromised and should be started over.

4. Place ELISA plates into a 40 °C oven, and let the cell suspensions dry overnight. If not performing the ELISA on the next day, place the plates into resealable plastic bags and leave at

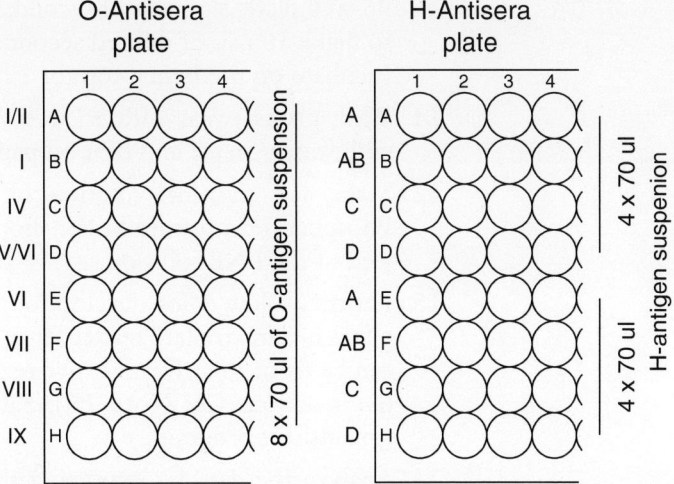

Fig. 4 Sections of 96-well ELISA plates with suggested layout of cell suspensions and antisera

room temperature until use. They should not sit for more than between a few days and up to a week.

5. Rinse dried plates twice with water using a large squeeze bottle. Blot on paper towels (*see* **Note 11**).

6. Fill each well with 200 μL each of casein blocking solution (*see* **Note 12**). Let the plates incubate on the bench top for at least 1 h. Do not let them dry.

7. Shake out the blocking solution and rinse twice with water. Blot on paper towels.

8. Make dilutions of the primary antisera in Antiserum Dilution Buffer (*see* **Note 13**). Group the antisera according to horizontal rows as suggested in Fig. 4. For a full 96-well plate each of the O-antisera will be used in 12 wells, and each H-antiserum will be used in 24 wells. Examples of how to dilute the antisera follow. If a 1:1,000 dilution is used, for each 96-well plate add 1.3 μL of each antiserum to 1.3 mL of Antiserum Dilution Buffer. If a 1:500 dilution is used, for each 96-well plate add 2.6 μL of antiserum to 2.6 mL of Antiserum Dilution Buffer. Add 100 μL of each diluted primary antisera to their assigned wells. Incubate plates on the bench top for 1–2 h.

9. Wash plates twice with ELISA wash buffer and then twice with water (*see* **Note 14**). Shake and blot on paper towels.

10. Add 100 μL/well of secondary antibody diluted in Antiserum Dilution Buffer according to control titration experiments or manufacturer's instructions packaged with the alkaline-phosphatase conjugated antibody (*see* **Note 15**). For each

96-well plate, 96 wells of secondary antiserum will be needed, so make 10 mL of diluted secondary antiserum for each plate. Incubate on the bench top for 1 h.

11. Wash plates twice with ELISA Wash Buffer and then twice with water. Shake and blot on paper towels.

12. Make a 1 mg/mL solution of PNP substrate in Alkaline Phosphatase Substrate Buffer shortly before use. Add 100 μL/well of the PNP solution.

13. Let the yellow color develop for approximately 30 min. Read plates in a microplate reader set at 405 nm. The range of values can be from ≤0.1 for negative reactions to 0.5 to ≥1 for positive reactions (*see* **Note 15**). Save the data for analysis in a spreadsheet program.

14. Analyze the data by importing the spectrophotometric readings into a spreadsheet program (e.g., Microsoft Excel, Quattro Pro, Lotus, or other online or cloud based spreadsheet programs). The O- and H-antigen readings are analyzed separately from each other. If duplicate wells were set up, the A_{405} values are averaged. Negative controls should have very low values. Calculate the percentage of maximum values for each strain for the O- and H-antisera separately by dividing the A_{405} value of a well by the largest A_{405} value for that strain in its column. Example data analysis is shown in Fig. 5. Graph these values for each strain for the O- and H-antisera with the Percent Maximum A_{405} on the *y*-axis, and the name of each antiserum on the *x*-axis (see examples in Figs. 6 and 7). One value will be 100 %, and no values should be higher than 100 %. Analyze the O-antisera data according to the flow chart in Fig. 8. For serotype four strains only the O-antisera reactions are necessary for calling serotype. Compare the bars on the graphs for antisera O-VI, O-VII, O-VIII, and O-IX. Count a reaction as positive if the Percent Maximum A_{405} is ≥60 % (Fig. 6 and Table 1). If necessary set up strains for H-antigen testing and analyze Percent Maximum A_{405} for H-A, H-AB, H-C, and H-D (Fig. 7 and Table 1) (*see* **Notes 6** and **9**).

3.3 PCR Method

Set up PCR reactions on ice, and run agarose gels at room temperature.

1. When primers arrive from the manufacturer dissolve in nuclease-free water to a concentration of 100 μM. The amount of water added will depend on the amount of the primer delivered. This information is usually given in an insert enclosed with the primers, and often it will say the amount of liquid to add to obtain a 100 μM solution. This will be a master stock solution from which dilutions will be made for working stocks of primer solutions. Once diluted, store primers at –20 °C.

Raw Data (A_{405})-- O Antisera

	Strain 1	Strain 1	Strain 2	Strain 2	Strain 3	Strain 3
I/II	1.719	1.607	1.307	1.249	0.265	0.29
I	0.731	0.804	0.873	0.805	0.517	0.603
IV	0.197	0.22	0.279	0.299	0.241	0.299
V/VI	0.234	0.224	0.23	0.218	2.201	2.775
VI	0.584	0.511	0.542	0.558	2.081	2.851
VII	0.251	0.276	0.272	0.287	0.409	0.495
VIII	0.287	0.311	0.292	0.268	0.548	0.619
IX	0.234	0.217	0.198	0.228	0.224	0.26

AVERAGES-- O Antisera

	Strain 1	Strain 2	Strain 3
I/II	1.663	1.278	0.2775
I	0.7675	0.839	0.56
IV	0.2085	0.289	0.27
V/VI	0.229	0.224	2.488
VI	0.5475	0.55	2.466
VII	0.2635	0.2795	0.452
VIII	0.299	0.28	0.5835
IX	0.2255	0.213	0.242

% MAXIMUM ABSORBANCE -- O Antisera

	Strain 1	Strain 2	Strain 3
I/II	100.00	100.00	11.15
I	46.15	65.65	22.51
IV	12.54	22.61	10.85
V/VI	13.77	17.53	100.00
VI	32.92	43.04	99.12
VII	15.84	21.87	18.17
VIII	17.98	21.91	23.45
IX	13.56	16.67	9.73

Fig. 5 Example of data analysis for somatic O-antisera. Raw data (A_{405}) for duplicates of three strains is shown in the top table. In the middle table the A_{405} values were averaged. In the bottom table the value of each well in the middle table was divided by the maximum value in its column, and multiplied by 100 to obtain the Percent Maximum Value. The data in the bottom table is used to make the graphs. Flagellar H-antisera data will be analyzed similarly

2. Make working stocks of primer mixes from the 100 µM master stocks. The working stocks for the primers *lmo*0737 F and R, *lmo*1118 F and R, ORF2819 F and R, and ORF2110 R and F are 10 µM, and the working stocks for *prs* F and R are 1 µM. To make 10 µM stocks of *lmo*0737 primers, for example, add 50 µL of 100 µM *lmo*0737(F) and 50 µL of 100 µM *lmo*0737(R)

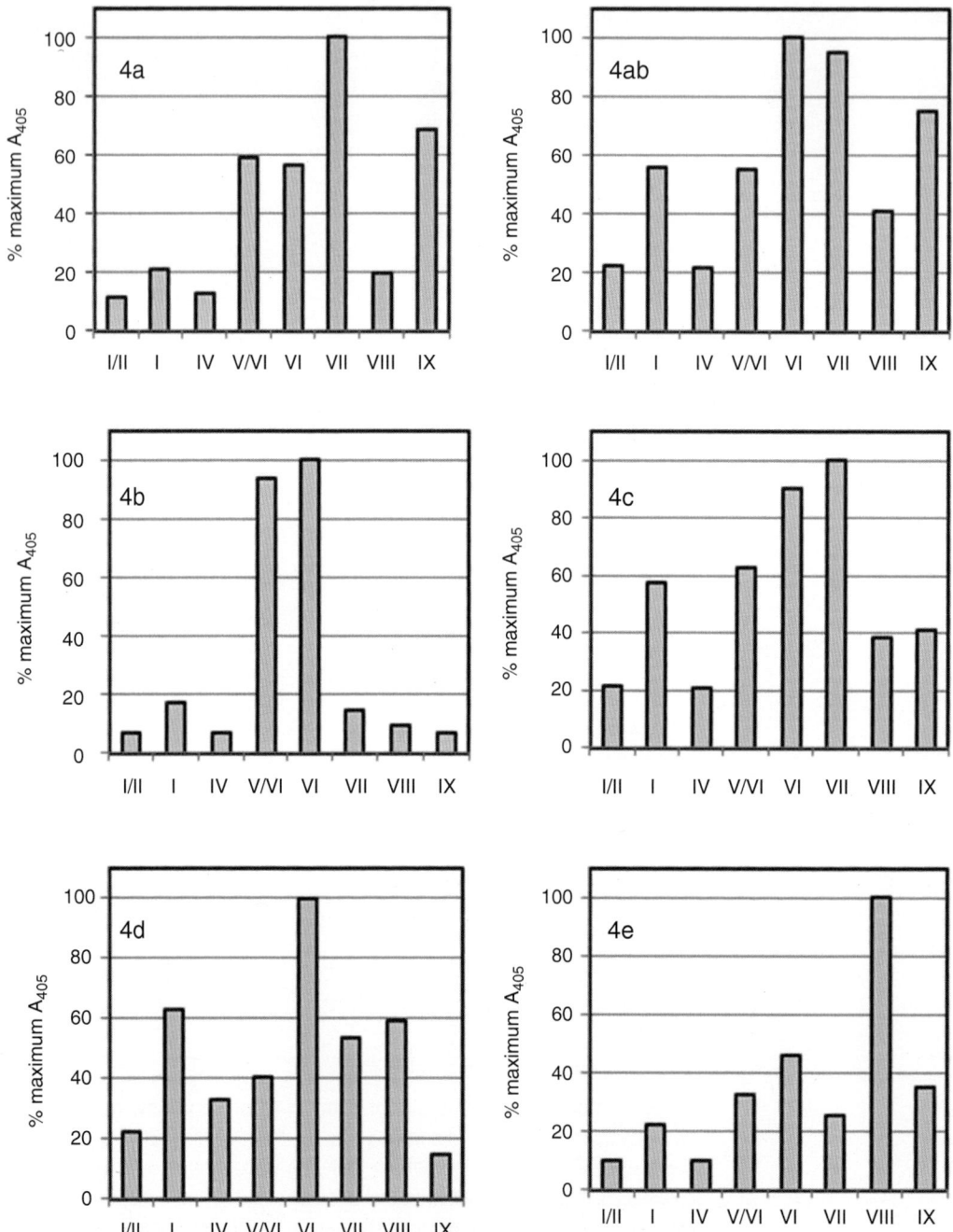

Fig. 6 Examples of ELISA reactions for serotype 4 strains. The serotype is indicated on the graph. Compare reactions to antigenic designations in Table 1. Note that there is strain-to-strain variation, even among isolates of the same serotype, and these graphs serve as examples not absolute values

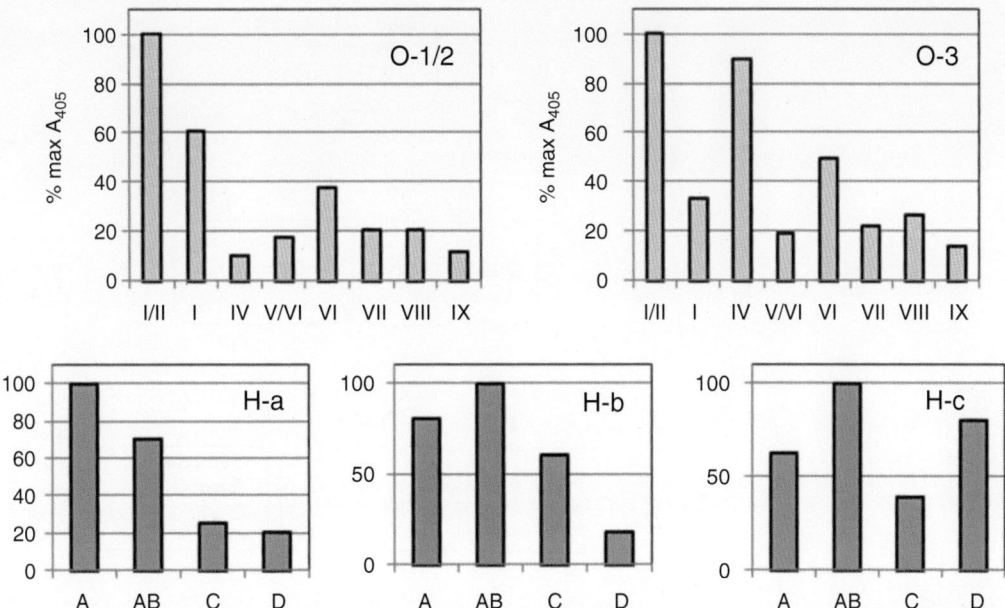

Fig. 7 Examples of ELISA reactions for O-antigen serotypes 1/2 and 3, and possible H-antisera reactions. Once the somatic O-antigen is called as 1/2 or 3, the flagellar-H-antigen needs be determined. The top two graphs are examples of reactions against O-antisera for serotypes 1/2 and 3, and the bottom three graphs are examples of H-antisera reactions that would be labeled with the antigenic designation of a, b, or c. The value of the *y*-axis is Percent Maximum A_{405}. Compare reactions to antigenic designations in Table 1. There can be strain-to-strain variation, even among isolates of the same serotype, and these graphs serve as examples not absolute values

to 400 µL of nuclease-free water and mix. This will be the *lmo*0737 primer mix working stock. Make working stocks in a similar fashion for the *lmo*1118, ORF2819, and ORF2110 primer pairs. For the *prs* working stock primer mix add 5 µL of 100 µM *prs*(F) and 5 µL of 100 µM *prs*(R) to 490 µL of nuclease-free water and mix. Store all working stocks at –20 °C when not in use, and keep on ice when preparing PCR tubes.

3. Prepare genomic DNA from *L. monocytogenes* strains according to methods described earlier in this book or by using a commercial kit. Once the DNA is purified, prepare a 1 µg/µL solution in nuclease-free water or in TE buffer. The recipe for this will depend on the final DNA concentration of the preparation. For example, if the final concentration of DNA is measured to be 100 µg/µL, then a 1 µg/µL solution is made by 1 µL of the 100 µg/µL solution to 99 µL of TE buffer or water.

4. Prepare 25 µL PCR reactions for each DNA preparation by the following recipe: 2.5 µL *lmo*0737 primer mix, 2.5 µL ORF2819 primer mix, 2.5 µL ORF2110 primer mix, 3.75 µL *lmo*1118 primer mix, 5.0 µL *prs* primer mix, 5.0 µL of 5× PCR buffer

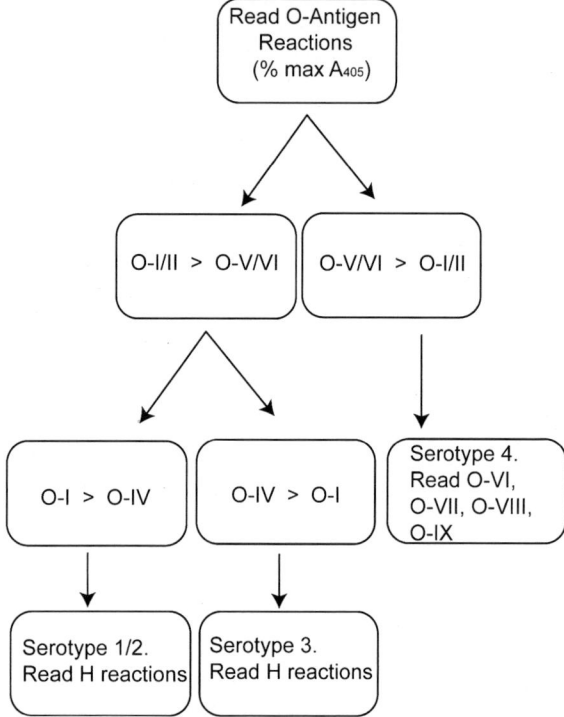

Fig. 8 Flowchart for determination of somatic O-antigen based on graphs made using the ELISA method. Note: Serotype 7 strains that react only occasionally with O-III antisera are extremely rare

with MgCl₂ (supplied in PCR kits), 2.0 μL of 10 mM dNTP solution, 0.25 μL of *Taq* polymerase, 0.55 μL nuclease-free water, and 1.0 μL of 1 μg/μL DNA solution. These volumes will give a final concentration in each PCR reaction of 1 μM of each of the primers for *lmo*0737, ORF2819, and ORF2110, 1.5 μM *lmo*1118 primers, 0.2 μM *prs* primers, 0.8 μM dNTPs, 1 U of Taq polymerase, and 1× buffer in 25 μL reactions. You may need to adjust volumes of buffer, enzyme, and water depending on the concentrations of buffer and *Taq* polymerase supplied with your PCR enzyme (*see* **Note 16**).

5. Seal the PCR tubes, strips, or plates. Mix the contents of the tubes and ensure the solutions are in the bottoms of the tubes and not in the caps by doing a quick spin (3–5 s) in a microcentrifuge.

6. Run PCR in a thermocycler according to the following protocol: an initial denaturation at 94C for 3 min; 35 cycles of 94 °C for 24 s, 53 °C for 75 s, and 72 °C for 75 s; and a final incubation at 72 °C for 7 min before an infinite hold at 4 °C until the reactions are removed from the machine.

7. Prepare an agarose gel with 2.0 % agarose (in either TAE or TBE buffer) in a gel box with a comb inserted that will be able

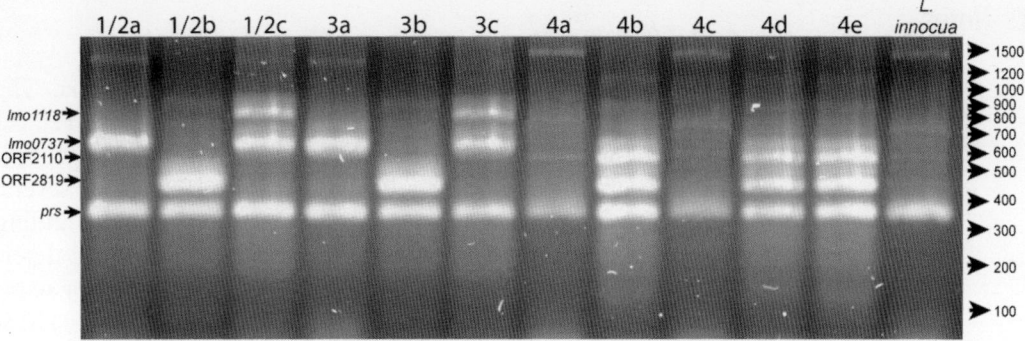

Fig. 9 Agarose gel electrophoresis of multiplex PCR reactions for strains of different serotypes

to handle all the PCR reactions being run. Fill the reservoir and cover the gel with either TAE or TBE buffer (the same as used to make the gel).

8. Mix 9 μL of PCR reaction with loading buffer. The amount will depend on the concentration of DNA loading dye used (For 10× loading dye, add 1 μL; for 5× loading dye, add 2 μL). Load into 2.0 % agarose gel. Allow an empty well on either end to load size standards mixed with loading buffer. Connect the gel box to a power supply and run at 100 V. The smallest band will be 370 bp, so run the gel long enough to separate the bands, but not so long that the smallest band will run off the gel. This will depend on the size of gel you run (*see* **Note 17**).

9. Stain with ethidium bromide or a comparable nucleic acid stain by immersing the gel into a bath of stain for approximately 30 min.

10. Visualize the bands with an agarose gel visualization system.

11. Compare the bands in each lane with those shown in Fig. 9 and the information in Table 2 to determine serogroup. All *Listeria* species will give a product with the *prs* primers, so all the lanes should have a 370 bp product. If this band is absent in any of the lanes, regardless of the presence of other bands, then the PCR did not work or the strain under study is not *Listeria*.

12. This multiplex PCR method will not differentiate between 1/2a and 3a strains, 1/2b and 3b strains, and 1/2c and 3c strains. However, serotype 1/2 is much more common than serotype 3 strains. Also, the banding patterns for serotypes 4b, 4d, and 4e are identical, but serotype 4b is much more common than the others. Also, the rare serotypes 4a and 4c only show the *prs* control *Listeria* band (*see* **Note 18**). Strains of the very rare serotypes 4ab and 7 were not tested with this multiplex PCR. Some atypical serotype 4b strains may react with the Lineage II-specific primers (*lmo0737*) and give an additional 691 bp product [6].

4 Notes

1. BHI medium is used commonly for *L. monocytogenes*. The antisera manufacturer's instructions call for using BHI for somatic O-antigen determination. BHI agar plates (1.5 % w/v agar) can be made using BHI broth (available from several manufacturers) according to label instructions and adding 15 g/L agar before autoclaving. For flagellar H-antigen determination the antisera manufacturer's instructions call for semi-solid agar medium. In my lab we use BHI soft agar plates [BHI broth with 0.4 % (w/v) agar], which are prepared by adding 4 g/L of agar to BHI broth before autoclaving. DO NOT INVERT soft agar plates. We have also substituted Trypticase Soy Broth (TSB, also available from several manufacturers) for BHI and use the same agar concentrations as above, using label instructions for making TSB broth. The US Food and Drug Administration *Bacterial Analytical Manual* (FDA-BAM) [18] suggests using Tryptose phosphate broth (TPB) for growing cells for somatic O-antigen determination. TPB contains 20 g/L tryptose, 2 g/L dextrose, 5 g/L NaCl, 2.5 g/L Na_2HPO_4, and sterilizing by filtration through a 0.2 μm filter membrane. In place of BHI soft agar, the FDA-BAM [18] suggests using EB motility medium (3 g/L beef extract, 10 g/L peptone, 5 g/L NaCl, and 4 g/L agar mixed together in 1 L water, heated to boiling to dissolve the agar, aliquotting 8 mL portions into 16×150 mm screw-cap tubes, and autoclaving for 15 min at 121 °C).

2. This is the only commercially available kit that contains all of the typing antisera. The sera are prepared in rabbits. The kit contains eight O-antisera (O-I/II, O-I, I-IV, O-V/VI, O-VI, O-VII, O-VIII, and O-IX) and four H-antisera (H-A, H-AB, H-C, and H-D). Other manufacturers sell polyvalent anti-Listeria antiserum, anti-O-antiserum I, and anti-O-antiserum IV, but these sera will not give complete serotype information.

3. Sodium azide is a poison. Wear personal protective equipment such as gloves, masks, and lab coat. Rinse spatulas before discarding. Azide is used so that the solution can be stored at RT without threat of contamination.

4. There are many published protocols for making genomic DNA from *L. monocytogenes*. Additionally there are commercial kits that can be adapted for use with *L. monocytogenes*. *Taq* polymerase and kits for PCR are available from many vendors.

5. In our experience cell suspensions at an A_{600} of approximately 1.0 work best for seeing agglutination.

6. If agglutination occurs with both O-I/II and O-V/VI, then the cultures are contaminated with more than one type of

L. monocytogenes, and the strain should be streaked out for purification by single colony subculture. Agglutination should not occur with saline, and if it does then the assay should be discarded and started over with fresh cultures. Serotype 7 strains are extremely rare and cannot be differentiated accurately with this collection of antisera.

7. There are only 1–4 flagella on the surface of the cells, which is why better results are obtained when the strains are enriched for motile cells by passage in semisolid media. It is essential that cultures be grown at 25–30 °C because flagella production in *L. monocytogenes* is temperature controlled, and they are not made at 37 °C. Flagella can be removed from the surface of cells easily, so the cultures are not shaken and the suspensions are not vortexed.

8. We have obtained good results in my lab by incubating aliquots of cultures in microfuge tubes in a dry bath set for at least 100 °C for 30 min. However, using autoclaved cultures is better.

9. In my lab we set up O-antigen typing for all strains to be tested, and only do H-antigen testing for those strains that are not serotype 4. All serotype 4 strains have the same H-antigens, so there is no need for H-antigen testing for those strains.

10. My lab uses Immunosorp strips which are two rows of eight wells, and they fit into plate frames which can then be inserted into 96-well plate readers.

11. Do all the washes over the sink with the plate in one hand and the squeeze bottle of water in the other. The cells are dead at this point, so there is no worry about contamination. Remember to wear gloves whenever handling antisera.

12. Blocking solution with bovine serum albumin can also be used if you prefer.

13. While we do a titration with each new antisera kit, generally we use dilutions of O-Antisera of 1:1,000 and H-Antisera of 1:500.

14. Generally we will wash twice with 200 μL in each well of ELISA Wash Buffer added with a multichannel pipettor. Some people use a second wash bottle with ELISA Wash Buffer along with the wash bottle of water.

15. The amount of time necessary for this level of color to develop depends on the amount of secondary antibody used. The dilution used will depend on the supplier of the secondary antibody. We have used secondary antibody at dilutions from 1:1,000 up to 1:30,000, depending on the source. Some manufacturers include instructions with their AP-conjugated anti-rabbit IgG of dilutions ideal for various

applications. If the manufacturer does not suggest dilutions then they will have to be determined experimentally. An ideal is to aim for positive yellow PNP reactions that fall in the range of A_{405} of 0.5–1.5 after about 30 min of incubation. Times for adequate color development that are shorter than 30 min have reactions too fast to capture accurate readings in the spectrophotometer (especially if you are reading a large number of plates).

16. In my lab we usually make a master mix of the PCR solution containing everything except the template DNA. Make enough of this master mix for each PCR reaction you are setting up plus about 5 more tubes so that you do not run out when aliquotting. Add 24 µL to each PCR reaction tube. Then aliquot the template DNA into each tube. You may even need to adjust the level of template by adding more or less concentrated DNA if you get too little or too much product. This will depend on several variables including the method to make DNA, the type and brand of *Taq* polymerase, and the thermocycler used. However, the recipe given in this step is a good starting point from which to optimize the reaction in your lab.

17. We usually run gels that are roughly 10 cm long for these reactions. Our loading buffer has two dyes, one of which runs at 500 bp. We usually run the gel until this dye front is about 2 cm from the bottom of the gel, and it takes about 30 min.

18. In my lab we usually use the ELISA method to call the somatic O-antigen serotype and then do multiplex PCR for those strains that are of serotype 1/2 or 3.

Acknowledgement

This work was supported by US Department of Agriculture, Agricultural Research Service CRIS project 5325-42000-046-00D. Thanks to J. Palumbo for helpful discussions.

References

1. Jadhav S, Bhave M, Palombo EA (2012) Methods used for the detection and subtyping of *Listeria monocytogenes*. J Microbiol Meth 88:327–341

2. Seeliger HPR, Hohne K (1979) Serotyping of *Listeria monocytogenes* and related species. Methods Microbiol 13:31–49

3. Kathariou S (2002) *Listeria monocytogenes* virulence and pathogenicity, a food safety perspective. J Food Prot 65:1811–1829

4. Tappero JW, Schuchat A, Deaver KA et al (1995) Reduction in the incidence of human listeriosis in the United States. JAMA 273:1118–1122

5. Wiedmann M, Bruce JL, Keating C et al (1997) Ribotypes and virulence gene polymorphisms suggest three distinct *Listeria monocytogenes* lineages with differences in their pathogenic potential. Infect Immun 65:2707–2716

6. Lee S, Ward TJ, Graves LM et al (2012) Atypical *Listeria monocytogenes* serotype 4b strains harboring a lineage II-specific gene cassette. Appl Environ Microbiol 78:660–667

7. Liu D, Lawrence ML, Gorski L et al (2006) *Listeria monocytogenes* serotype 4b strains belonging to lineages I and III possess distinct molecular features. J Clin Microbiol 44(1):214–217

8. Ward TJ, Ducey TF, Usgaard T et al (2008) Multilocus genotyping assays for single nucleotide polymorphism-based subtyping of *Listeria monocytogenes* isolates. Appl Environ Microbiol 74:7629–7642

9. Lan Z, Fiedler F, Kathariou S (2000) A sheep in wolf's clothing: *Listeria innocua* strains with teichoic acid-associated surface antigens and genes characteristic of *Listeria monocytogenes* serogroup 4. J Bacteriol 182:6161–6168

10. Bille J, Rocourt J (1996) WHO international multicenter Listeria monocytogenes subtyping study-rationale and set-up of the study. Int J Food Microbiol 32:251–262

11. Schönberg A, Bannerman E, Courtieu AL et al (1996) Serotyping of 80 strains from the WHO multicentre international typing study of *Listeria monocytogenes*. Int J Food Microbiol 32:279–287

12. Palumbo JD, Borucki MK, Mandrell RE et al (2003) Serotyping of *Listeria monocytogenes* by enzyme-linked immunosorbent assay and identification of mixed-serotype cultures by colony immunoblotting. J Clin Microbiol 41:564–571

13. Doumith M, Buchrieser C, Glaser P et al (2004) Differentiation of the major *Listeria monocytogenes* serovars by multiplex PCR. J Clin Microbiol 42:3819–3822

14. Borucki MK, Call DR (2003) *Listeria monocytogenes* serotype identification by PCR. J Clin Microbiol 41:5537–5540

15. D'Agostino M, Wagner M, Vazquez-Boland JA et al (2004) A validated PCR-based method to detect *Listeria monocytogenes* using raw milk as a food model—towards an international standard. J Food Prot 67:1646–1655

16. Kérouanton A, Marault M, Petit L et al (2010) Evaluation of a multiplex PCR assay as an alternative method for *Listeria monocytogenes* serotyping. J Microbiol Meth 80:134–137

17. Doumith M, Jacquet C, Gerner-Smidt P et al (2005) Multicenter validation of a multiplex PCR assay for differentiating the major *Listeria monocytogenes* serovars 1/2a, 1/2b, 1/2c, and 4b: toward an international standard. J Food Prot 68:2648–2650

18. Bennet RW, Weaver RE (2001) BAM: serodiagnosis of *Listeria monocytogenes*. http://www.fda.gov/Food/ScienceResearch/Laboratory Methods/BacteriologicalAnalytical ManualBAM/ucm071418.htm. Accessed 13 Feb 2013

Chapter 5

Pulsed-Field Gel Electrophoresis (PFGE) Analysis of *Listeria monocytogenes*

Marion Dalmasso and Kieran Jordan

Abstract

PFGE is a valuable tool for assessing *Listeria monocytogenes* strain interrelatedness. It is based on the study of total bacterial DNA restriction patterns. Cells are embedded in agarose plugs before being lysed. The released DNA is then digested into large fragments by restriction enzymes. As DNA fragments are too large to be separated by traditional electrophoresis in an agarose gel, changes in the direction of the electrical current are periodically applied in order to allow the proper migration of large DNA fragments. Strains are characterized by the obtained DNA fragment patterns or pulsotypes which vary depending on the number and size of bands.

Key words PFGE, *Listeria monocytogenes*, Electrophoresis, Restriction, Pulsotype

1 Introduction

PFGE was first developed by Schwartz and Cantor at Columbia University in 1984 [1]. It has made a tremendous impact in the field of molecular biology by making possible the separation of large DNA fragments. In conventional agarose gel electrophoresis, DNA molecules bigger than 40–50 kb in size fail to migrate efficiently and appear in the gel as a single large diffuse band due to their size-independent co-migration, known as reptation [2, 3]. By periodically applying changes in the direction of the electrical field in which large DNA molecules are suspended, PFGE allows the separation of DNA molecules over 1,000 kb.

The principle of PFGE is as follows: bacteria are embedded in agarose plugs in which cells are lysed in order to release DNA into the agarose plug. DNA is then cleaved into large fragments by restriction enzymes. The restricted DNA fragments are then separated in a horizontal agarose gel using an electrical current with periodic changes of direction termed pulsed field. This results in DNA fragment patterns or pulsotypes which differ from one strain to the other depending on the number and size of DNA fragments

Kieran Jordan et al. (eds.), *Listeria monocytogenes: Methods and Protocols*, Methods in Molecular Biology, vol. 1157, DOI 10.1007/978-1-4939-0703-8_5, © Springer Science+Business Media New York 2014

that compose them. Pulsotypes are used to identify and to compare bacterial strains such as *Listeria monocytogenes* after analysis of the gel image using specific software.

After more than 20 years of use, PFGE is currently accepted as the "gold standard" for assessing epidemiological relationships for most clinically relevant bacteria [4]. It is also a valuable tool in tracing the *L. monocytogenes* strain similarities and putative transfer routes in food and food processing facilities [5, 6].

The PulseNet International network (www.pulsenetinternational. org, page last updated on 25/08/2011) proposes several standardized PFGE protocols for the study of food borne pathogenic bacteria including *L. monocytogenes*. This allows the creation of databases for the comparison of strains worldwide [7, 8]. The PFGE methodology described in this chapter is directly based on the PulseNet protocol for *L. monocytogenes* with slight modifications [9].

2   Materials

Prepare all solutions using ultrapure water (prepared by purifying deionized water to attain a sensitivity of 18 MΩ cm at 25 °C).

2.1   Preparation of Plugs

Except for the lysozyme solution, all solutions for preparation of plugs must be autoclaved at 121 °C for 15 min and can be stored at room temperature after sterilization.

1. Brain Heart Infusion (BHI) agar plates (catalogue number 70138-500G, Sigma-Aldrich, St. Louis, MO, USA) (*see* **Note 1**).

2. 50-Well disposable plug molds (catalogue number 170-3713, Bio-Rad, Hercules, CA, USA).

3. SeaKem® Gold agarose (catalogue reference 50150, Lonza, Basel, Switzerland).

4. Proteinase K solution, 20 mg/mL (catalogue number 7326348, Bio-Rad).

5. Trizma–HCl (Tris–HCl) solution: 1 M Tris–HCl, pH 8.0. Weigh 121.1 g Tris and transfer to a 1 L glass beaker. Add 800 mL 18 MΩ water and dissolve Tris by agitation with a magnetic stirrer. Adjust pH with HCl (*see* **Note 2**). Make up to 1 L with 18 MΩ water.

6. Ethylenediaminetetraacetic acid (EDTA) solution: 0.5 M EDTA, pH 8.0. Weigh 73.06 g EDTA and transfer to a beaker. Add 400 mL 18 MΩ water and dissolve the EDTA by agitation with a magnetic stirring and heating the solution to 40 °C. Once all EDTA is dissolved, allow the solution to cool to room temperature before adjusting pH as described in previous step. Make up to 500 mL with 18 MΩ water.

7. TE buffer: 10 mM Tris–HCl, 1 mM EDTA, pH 8.0. Add 10 mL of 1 M Tris–HCl solution, 2 mL of EDTA solution and make up to 1 L with 18 MΩ water.

8. *N*-Lauroylsarcosine sodium salt (sarcosyl): 10 % solution in 18 MΩ water.

9. Cell lysis buffer: 50 mM Tris–HCl, 50 mM EDTA, 1 % sarcosyl, pH 8.0. Add 50 mL Tris–HCl solution, 100 mL EDTA solution, 100 mL 10 % sarcosyl solution and make up to 1 L with 18 MΩ water.

10. Sodium dodecyl sulfate (SDS): 20 % solution in 18 MΩ water.

11. Lysozyme: 2 % solution in TE buffer. Swirl to mix, aliquot 500 µL amounts into microcentrifuge tubes and freeze at −20 °C for future use. Do not autoclave.

2.2 Restriction Digestion of Plugs

1. *Asc*I-restriction master mix (quantities for the digestion of one plug slice): Add 1.5 µL *Asc*I (FastDigest™, catalogue number FD1894, Thermo Fischer Scientific, Waltham, MA, USA), 20 µL FastDigest™ buffer, and 178.5 µL 18 MΩ water (*see* **Notes 3–5**).

2. *Apa*I-restriction master mix (quantities for the digestion of one plug slice): Add 1 µl *Apa*I (FastDigest™, catalogue number FD1414, Thermo Fischer Scientific), 20 µL FastDigest™ buffer, 2 µL BSA, and 177 µL 18 MΩ water (*see* **Notes 3–5**).

3. *Xba*I-restriction master mix (quantities for the digestion of one plug slice): Add 1.5 µL *Xba*I (FastDigest™, catalogue number FD1894, Thermo Fischer Scientific), 20 µL FastDigest™ buffer, and 178.5 µL 18 MΩ water (*see* **Notes 3–5**).

2.3 Casting the Gel

1. 15-Well comb (catalogue number 170-4324, Bio-Rad).

2. Standard casting stand (catalogue number 170-3689, Bio-Rad).

3. Thiourea solution: 5 mM thiourea in 18 MΩ water.

4. 0.5× Tris–Borate–EDTA (TBE) buffer: Add 150 mL of 10× TBE buffer (catalogue number 161-770, Bio-Rad), 300 µL of thiourea solution, and 2,850 mL 18 MΩ water.

5. SeaKem® Gold agarose (Lonza).

2.4 Eelectrophoresis and Staining

1. Electrophoresis equipment: CHEF-DR® III Variable Angle System (catalogue number 170-3700, Bio-Rad), cooling module (catalogue numbers 170-3654, 170-3688 or 170-3655 depending on country input voltage range, Bio-Rad) and variable-speed pump (catalogue number 170-3644, Bio-Rad).

2. Ethidium bromide solution, 10 mg/mL (catalogue number 161-0433, Bio-Rad).

3 Methods

3.1 Preparation of Plugs from Agar Cultures

D-1 day.

1. Streak an isolated colony from test cultures onto BHI agar plates using an inoculation loop and incubate at 37 °C for 14–18 h.

D day.

2. Turn on a shaking water bath or incubator at 54 °C, a stationary water bath at 55–60 °C, a stationary water bath at 50 °C and a spectrophotometer (wavelength: 610 nm).

3. Prepare 1 % agarose for plugs by weighing 0.5 g SeaKem® Gold agarose into a 250 mL screw-cap flask and by adding 49.5 mL TE buffer. Swirl gently to disperse the agarose. Loosen the cap and microwave for 1 min, mix gently and repeat for 10-s intervals until the agarose is completely melted (see Note 6). Place the flask into the 55–60 °C stationary water bath for 5 min before adding 2.5 mL SDS solution. Mix gently and replace the agarose with SDS in the water bath until use.

4. Label 1.5 mL microcentrifuge tubes with culture numbers.

5. Transfer 1 mL of TE buffer to labelled microcentrifuge tubes.

6. Use a sterile cotton swab that has been premoistened with sterile TE buffer to remove some of the growth from the BHI agar plates. Suspend the cells in TE buffer contained in the microcentrifuge tubes by spinning the swab gently to evenly disperse the cells (see Note 7).

7. Adjust the concentration of cells to an optical density of 1.0 ± 0.2 at 610 nm by diluting with TE buffer or by adding additional growth from agar plates if necessary using a sterile cotton swab.

8. Label sterile microcentrifuge tubes with culture numbers and transfer 400 μL of the cell suspension to each appropriate tube (see Note 8).

9. Add 20 μL lysozyme solution to each tube and mix gently by pipetting up and down several times (see Note 7).

10. Place the microcentrifuge tubes in the 55–60 °C stationary water bath for 10–20 min.

11. Label the wells of plug molds with the culture numbers (see Note 9).

12. Add 20 μL proteinase K solution to each tube and mix gently by pipetting up and down several times (see Note 7).

13. Add 400 μL of molten 1 % agarose to each microcentrifuge tube containing cell suspension, mix gently by pipetting up and down three times without creating bubbles and immediately dispense the agarose suspension mixture into assigned wells in

the plug molds (*see* **Note 10**). Do not allow bubbles to form (*see* **Note 11**).

14. Allow the plugs to solidify at room temperature for 10–15 min or at 4 °C for 5–10 min (*see* **Notes 12** and **13**).

3.2 Cell Lysis in Agarose Plugs

1. Label 15 mL polypropylene screw-cap centrifuge tubes with culture numbers.

2. For each tube, prepare proteinase K/cell lysis buffer as follows: 25 µL proteinase K solution and 5 mL of cell lysis buffer are needed per tube (*see* **Note 14**).

3. With a scalpel, trim excess agarose from the top of the plugs in the wells of the molds.

4. Carefully open the molds and, using a 6-mm wide spatula, transfer each plug to the appropriately labelled tube containing the proteinase K/cell lysis buffer (*see* **Note 15**).

5. Place the tubes in a rack and incubate in the 54 °C shaking water bath or incubator for 2 h with constant and vigorous shaking at 160–200 rpm (*see* **Notes 16** and **17**).

6. Place some sterile 18 MΩ water and TE buffer in the 50 °C stationary water bath in preparation for the next step.

3.3 Washing of Agarose Plugs After Cell Lysis

1. Remove the tubes containing plugs from the water bath or incubator, and carefully discard lysis buffer. In the meantime cool down the shaking water bath to 50 °C.

2. Add approximately 5 mL sterile 18 MΩ (preheated to 50 °C) to each tube, and place in the 50 °C shaking water bath for 10–15 min with shaking at 160 rpm (*see* **Note 18**).

3. Discard the water from the tubes and repeat **step 2**.

4. Discard the water, add approximately 5 mL TE buffer (preheated to 50 °C) and allow the plugs to shake in the 50 °C water bath for 10–15 min.

5. Discard the TE buffer from the tubes and repeat **step 4** two more times (three washes in TE buffer in total).

6. Discard the TE buffer from the tubes, suspend the washed plugs in fresh TE buffer and store at 4 °C (*see* **Note 19**).

3.4 Restriction Digestion of DNA in Agarose Plugs

1. Turn on the water baths or heating blocks at 30 °C and at 37 °C.

2. Label 1.5 mL microcentrifuge tubes with the culture numbers and three tubes for *Salmonella* ser Braenderup H9812 standards (*see* ref. [10]) in the case of a 15-well gel (*see* **Note 20**). Add or remove one additional *S.* ser Braenderup H9812 standard depending on the size of comb used.

3. For each strain, cautiously remove a plug from TE buffer with a spatula and place it on a clean Petri dish.

4. Cut a 2-mm wide plug slice with a sterile disposable scalpel (*see* **Note 21**).

5. Using a spatula, transfer the plug slice to the appropriate microcentrifuge tube.

6. Replace the rest of the plug in the original tube containing TE buffer and store at 4 °C.

7. Add 20 µL of the relevant restriction enzyme buffer to each tube, make up to 200 µL with 18 MΩ water, make sure that the entire plug slice is immersed by gently tapping the tube bottom, and incubate at room temperature for 5–10 min (*see* **Note 22**).

8. Discard the buffer, and add 200 µL of the relevant enzyme restriction master mix into each microcentrifuge tube. Close the tube and tap the bottom of each tube gently to completely submerge the plug slice.

9. Incubate the plug slices at 37 °C (*Xba*I and *Asc*I) for 2–3 h or at 30 °C (*Apa*I) overnight in water baths or heating blocks.

3.5 Casting Agarose Gel

The next steps are to be started approximately 1 h before the enzyme restriction reaction is finished.

1. Turn on a water bath at 54 °C.

2. Make 1 % SeaKem® Gold agarose in 0.5× TBE buffer by weighing 1.5 g SeaKem® Gold agarose into a 500 mL screw-cap flask, adding 150 mL 0.5× TBE buffer and swirling gently to disperse the agarose. Microwave for 3 min after loosening the cap of the flask, mix gently and repeat for 15 s intervals until the agarose is completely melted. Place the flask in the 54 °C water bath until use.

3. Put the appropriate black gel frame in the electrophoresis chamber (*see* **Note 23**).

4. Fill the electrophoresis chamber with approximately 2.5 L 0.5× TBE buffer.

5. Start cooling the 0.5× TBE buffer by turning on the power supply, cooling module (set on 14 °C) and pump (set at 70 to achieve a flow rate of 1 L/min) approximately 30–60 min before electrophoresis is to be run.

6. Assemble the appropriate size gel form and the black casting platform, making sure it is perfectly levelled before pouring gel.

7. Position the comb holder so that the front part (side with small metal screws) and teeth face the bottom of the gel and the bottom edge of the comb teeth is 2 mm above the surface of the gel platform.

8. Carefully pour the agarose into gel form fitted with the comb avoiding bubble formation and allow the gel to solidify undisturbed for 30–60 min at room temperature (*see* **Note 24**). Make sure to keep 2–3 mL of molten agarose in the flask in the 54 °C water bath for sealing wells after loading plug slices into the gel wells (Subheading 3.6, **step 4**).

3.6 Loading Restricted Plugs into the Wells

1. Remove the comb after the gel has solidified.

2. Remove the microcentrifuge tubes containing the restricted plug slices from the water baths or heating blocks.

3. Remove restricted plug slices from microcentrifuge tubes with a spatula and load into appropriate wells by gently pushing the plug slices to the bottom of wells. Make sure no bubbles are formed between plug slice and well bottom (*see* **Note 25**).

4. Fill in wells with molten agarose that has been kept in the 54 °C water bath and allow to solidify for 2–3 min at room temperature.

5. Carefully disassemble the gel form and slide the black casting platform holding the gel out the side.

6. Remove loose agarose from the back and edges of the black casting platform holding the gel with a tissue as loose agarose can clog the electrophoresis system.

7. Place the black casting platform holding the gel inside the black gel frame in the electrophoresis chamber and close the cover of the chamber.

3.7 Electrophoresis Conditions

1. Program the following electrophoresis parameters into the CHEF-DR® III Variable Angle System:
   - Initial switch time: 4.0 s.
   - Final switch time: 40.0 s.
   - Voltage: 6 V.
   - Included Angle: 120°.
   - Run time: 21 h.

2. Start the electrophoresis run.

3.8 Staining and Documentation of PFGE Agarose Gel

1. When the electrophoresis run is over, turn off equipment and remove the gel from the electrophoresis chamber.

2. Add 50 µL ethidium bromide solution to 500 mL 18 MΩ water in a staining container. Mix well.

3. Immerse the gel in the staining solution.

4. Agitate the gel gently at room temperature for at least 30 min.

5. Destain the gel in approximately 500 mL 18 MΩ water for 30–60 min.

6. Capture the image of the gel following directions given with the imaging equipment and export the image to .TIFF format (*see* **Notes 26** and **27**).

7. Analyze the image using BioNumerics software (Applied Maths, Belgium) (*see* **Note 28**).

4 Notes

1. Use of trade names and commercial sources in this section is to make easier the identification of products only. This implies that equivalent products might be provided by other suppliers.

2. Concentrated HCl (12 N) can be used at first to narrow the gap between the starting pH and the required pH. Then, it would be better to use HCl solutions (e.g., 6 N or 1 N) with lower ionic strengths to avoid a sudden drop in the pH below the required pH value.

3. Keep the vial of restriction enzyme on ice at all times as restriction enzymes are labile and extremely vulnerable to temperature fluctuations.

4. The addition of BSA to the restriction master mix may vary according to the supplier of the enzyme.

5. The enzyme master mix is to be prepared just before the plug digestion as it cannot be stored more than a few minutes on ice.

6. Molten agarose should be free from agarose particles but should not be boiled excessively.

7. Do not vortex in order to avoid the formation of aerosols.

8. Place cell suspension aliquots on ice if the next step is not immediately performed. In this case, place cell suspensions in a heating block at 37 °C for 1–2 min before starting the next step.

9. Label generally 5 wells per culture.

10. Maintain the temperature of molten agarose by keeping the flask in a beaker of warm water (55–60 °C) or in the water bath (55–60 °C).

11. To avoid the formation of bubbles, press gently the extremity of the pipette tip against one side of the well so that the agarose will slip slowly along the side of the well.

12. The generation of cell suspension and the subsequent casting of the plugs should be performed as rapidly as possible in order to minimize premature cell lysis. If large numbers of samples are being prepared, it is recommended that they be processed in batches of about 10 samples at a time. Once one batch of samples is at the cell lysis incubation step, then start preparing cell suspensions for the next group of samples. All batches can

be washed together, since a minimum of 2 h is needed for lysis and additional lysis time will not affect the initial batches.

13. Unused agarose can be kept at room temperature and reused three times before discarding.

14. The total volume of proteinase K/cell lysis buffer mix needed can be prepared, and 5 mL mix can be added to each tube.

15. Ensure the plugs are completely submerged in the proteinase K/cell lysis buffer and not on the side of the tube.

16. Ensure the level of water in the water bath is above the level of proteinase K/cell lysis buffer in the tube.

17. The plugs may be lysed overnight.

18. The plugs from each tube can also be transferred with a spatula to a compartment of a sterile square 25-compartment Petri dish. The washing steps can then be performed directly in the square Petri dish which can also be used to store the plugs once washed.

19. At this stage, plugs can be kept under refrigeration for 2–3 months. They can also be transferred to smaller tubes or square 25-compatment Petri dishes for storage.

20. The preparation of *Salmonella* ser Braenderup H9812 plugs is the same as for *L. monocytogenes* plugs except for **step 9** in Subheading 3.1 where no lysozyme is added to the cell suspension.

21. For *S.* ser Braenderup H9812 standard, the plug slices should be 4-mm wide.

22. *S.* ser Braenderup H9812 standard is restricted with *Xba*I and *L. monocytogenes* are restricted with *Asc*I and *Apa*I in two different microcentrifuge tubes.

23. Make sure that the electrophoresis chamber is perfectly levelled.

24. Uneven solidification and surface dehydration of the gel will distort the appearance of bands. Therefore, do not cast the gel more than 90 min before loading. Keep the gel covered as much as possible to prevent surface desiccation and dust deposition.

25. Load *S.* ser. Braenderup H9812 standard in wells 1, 8, 15 (15-well gel) and load samples in remaining wells (Fig. 1).

26. If too much background is observed destain the gel in 18 MΩ water for an additional 30–60 min.

27. The gel image should fill the entire window of the imaging equipment screen (without cutting off wells or lower bands). Ensure that the image is in focus and that there is little to no saturation (over-exposure) in the bands.

28. Comparison of strains should combine both *Asc*I and *Apa*I pulsotypes to accurately discriminate strains.

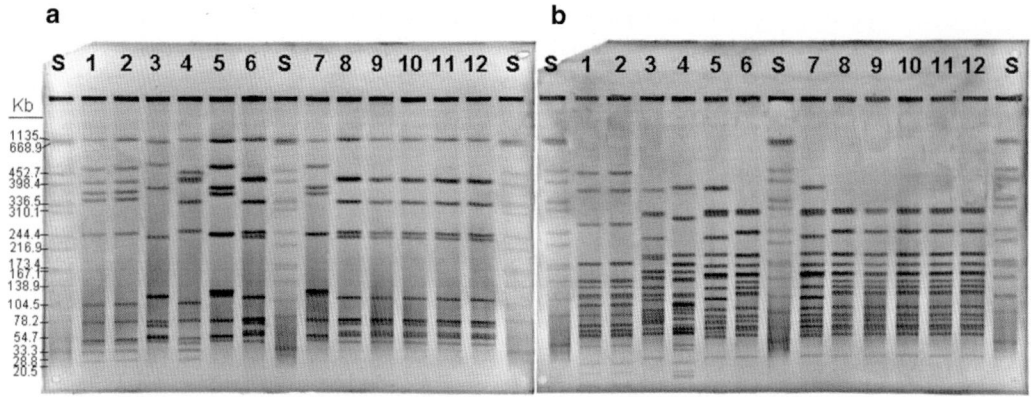

Fig. 1 Example of *L. monocytogenes* pulsotypes (No. 1–12) restricted with *Asc*I (**a**) or *Apa*I (**b**), and run under the PFGE protocol described in this chapter. S: *Salmonella* ser Braenderup H9812 standard restricted with *Xba*I

References

1. Schwartz DC, Cantor CR (1984) Separation of yeast chromosome-sized DNAs by pulsed field gradient gel electrophoresis. Cell 37:67–75

2. Singh A, Goering RV, Simjee S, Foley SL, Zervos MJ (2006) Application of molecular techniques to the study of hospital infection. Clin Microbiol Rev 19:512–530

3. Slater GW (2009) DNA gel electrophoresis: the reptation model(s). Electrophoresis 30:S181–S187

4. Goering RV (2010) Pulsed field gel electrophoresis: a review of application and interpretation in the molecular epidemiology of infectious disease. Infect Genet Evol 10:866–875

5. Fox E, Hunt K, O'Brien M, Jordan K (2011) *Listeria monocytogenes* in Irish farmhouse cheese processing environments. Int J Food Microbiol 145:S39–S45

6. Fox EM, Leonard N, Jordan K (2011) Molecular diversity of *Listeria monocytogenes* isolated from Irish dairy farms. Foodborne Pathog Dis 8:635–641

7. Fox EM, deLappe N, Garvey P, McKeown P, Cormican M, Leonard N, Jordan K (2012) PFGE analysis of *Listeria monocytogenes* isolates of clinical, animal, food and environmental origin from Ireland. J Med Microbiol 61:540–547

8. Martin P, Jacquet C, Goulet V, Vaillant V, De Valk H (2006) Pulsed-field gel electrophoresis of *Listeria monocytogenes* strains: the PulseNet Europe feasibility study. Foodborne Pathog Dis 3:303–308

9. PulseNet USA (2009) One-day (24–28 h) standardized laboratory protocol for molecular subtyping of *Listeria monocytogenes* by Pulsed Field Gel Electrophoresis (PFGE). http://www.pulsenetinternational.org/assets/PulseNet/uploads/pfge/5.3_2009_PNetStandProtLMonocytogenes.pdf. Accessed 12 Oct 2013

10. Hunter SB, Vauterin P, Lambert-Fair MA, Van Duyne MS, Kubota K, Graves L, Wrigley D, Barrett T, Ribot E (2005) Establishment of a universal size standard strain for use with the PulseNet standardized pulsed-field gel electrophoresis protocols: converting the national databases to the new size standard. J Clin Microbiol 43:1045–1050

Chapter 6

Multilocus Sequence Typing (MLST) of *Listeria monocytogenes*

Beatrix Stessl, Irene Rückerl, and Martin Wagner

Abstract

Nucleotide sequence-based methods focusing on the single nucleotide polymorphisms (SNPs) of *Listeria monocytogenes* housekeeping genes facilitate the rapid and interlaboratory comparison on open accessible databases, such as the multilocus sequence typing (MLST) databases that are available. MLST has advantages over other methods as it can reconstruct ancestral and evolutionary linkage between *L. monocytogenes* isolates. MLST detects all genetic variations within the amplified housekeeping gene that accumulate slowly. This chapter describes how to undertake MLST.

Key words *Listeria monocytogenes*, Housekeeping genes, Multilocus sequence typing, MLST

1 Introduction

Nucleotide sequence-based methods focusing on the single nucleotide polymorphisms (SNPs) of *Listeria monocytogenes* housekeeping genes facilitate the rapid and interlaboratory comparison on open accessible databases as the multilocus sequence typing (MLST) database of the Institute Pasteur (data are publicly available at: http://www.pasteur.fr/recherche/genopole/PF8/mlst/Lmono.html; accessed on: 09.10.2013). In *L. monocytogenes* MLST, sequences of seven housekeeping genes (length 399–537 bp), which are spread across dispersed genomic locations, are amplified by PCR and sequenced by universal primers. The unique alleles (haplotypes) are allocated to allelic profiles and subsequently the sequence type (ST) is determined [1–3]. Currently there are 2652 *L. monocytogenes* isolates and 641 *L. monocytogenes* MLST profiles available at the Institute Pasteur database. The latter is curated by T. Cantinelli and V. Chenal-Francisque (French National Reference Center for Listeria and WHO Collaborative Center for listeriosis, Institute Pasteur), S. Brisse and J.-M. Thiberge (Genotyping of Pathogens and Public Health, Institute Pasteur), and J. Haase (Environmental Research Institute, Cork, Ireland).

Kieran Jordan et al. (eds.), *Listeria monocytogenes: Methods and Protocols*, Methods in Molecular Biology, vol. 1157, DOI 10.1007/978-1-4939-0703-8_6, © Springer Science+Business Media New York 2014

A further open access web-based tool was published by Vangay et al. [4] (http://www.foodmicrobetracker.com; accessed on: 10.10.2013). The platform provides beside DNA sequence-based data, also pulsed field-gel electrophoresis (PFGE) and ribotyping patterns, or biochemical profiles for the comparison of a broad range of food-associated microbes.

The first MLST protocol was published by Maiden et al. [5] to overcome the lower discriminatory power and poor reproducibility of, e.g., ribotyping, Polymerase chain reaction-restriction fragment length polymorphism (PCR-RFLP), or PFGE typing. One of the first authors who reported better results for the differentiation of 175 *L. monocytogenes* strains after the application of MLST in comparison to PFGE was Revashivili et al. [6].

In the same year an alternative sequence typing approach including virulence (*prfA*, *inlB*, and *inlC*) and virulence associated genes was published by Zhang et al. [7]. The comparison of multi-virulence-locus sequence typing (MVLST) with MLST analysis showed a higher discriminatory power for *L. monocytogenes* serotype 1/2a and 4b strains. MVLST was rarely applied in literature due to small strain sets for the method evaluation, high costs, and the lack of international enforcement [3, 8–10]. Cantinelli et al. [11] compared 125 *L. monocytogenes* isolates with PFGE, MLST, and MVLST and reported that PFGE had the highest discriminatory power. MLST and MVLST resulted in similar phylogenetic clusters and the combination of datasets did not improve the depth of focus.

Generally, the major advantage of MLST is to reconstruct ancestral and evolutionary linkage between *L. monocytogenes* isolates. MLST detects all genetic variations within the amplified housekeeping gene that accumulate slowly, whereas PFGE only illustrates the variations in the cleavage sites for a particular restriction enzyme [3, 12, 13]. Therefore, PFGE is more applicable to investigate large-scale genomic rearrangements in genomic islands, insertion sequences, and mobile elements resulting in the gain or loss of restriction sites [14–16].

The initial *L. monocytogenes* MLST scheme was published by Salcedo et al. [15]. The actual MLST protocol refers to Ragon et al. [2], who shortened the analyzed regions of six loci (ABC transporter (*abcZ*), d-amino acid aminotransferase (*dat*), catalase (*cat*), succinyl diaminopimelate dessucinylase (*dapE*), beta-glucosidase (*bglA*), and histidine kinase (*lhkA*)) and extended the 5′ side of the lactate dehydrogenase (*ldh*) gene, to improve isolate discrimination. The discriminatory index (D. I.) >0.8 is for most of MLST housekeeping genes very high, except for *lhkA* (D. I. 0.5). Additionally, rare events of recombination were noticed for *abcZ* and *dapE* [1].

Ragon et al. [2] published the phylogenentic structure of 360 *L. monocytogenes* representatives by MLST typing. A long-term genetic stability of multilocus genotypes over time was evident among clinical-associated isolates. Interestingly, the majority of STs (85) could be assigned to seven major clones/clonal complexes (CC). Furthermore a relation between serotype/genetic lineage and CC could be realized: *L. monocytogenes* clone 1, 2, and 4 was attributed to serotype 4b (including 4d and 4e isolates), and clone 3 and 5 was linked to serotype 1/2b (including 3b and 7 isolates), all representing genetic lineage I. CC 7 was related to 1/2a and CC9 to 1/2c serotypes, both included in genetic lineage II. Other frequently present clones were CC6, CC8, and CC101 [2]. These findings underline previous results concerning the phylogenetic evolution of serotype 4b from 1/2b, the ancestral serotype of lineage I. Serotype 1/2c originated from 1/2a [15, 17].

A further large-scale MLST comparison of *L. monocytogenes* food and environmental isolates followed [1]. The predominant CCs illustrated in a minimum spanning tree were CC1, CC2 (lineage I, serotype 4b/4e), and CC3 (lineage I, serotype 1/2b, 3b). Linage II was more heterogeneous and comprised CC8, CC29, CC121, and CC182 (serotype 1/2a and 3a) and CC9 (serotype 1/2c). Den Bakker et al. [18] compared 36 European, Canadian, and US listeriosis outbreak strains obtained from the Centers for Disease Control and Prevention (CDC), defined as epidemic clones (EC) which showed higher transmission rates. The in silico MLST sequences were provided by the Broad Institute (http://www.broadinstitute.org/annotation/genome/listeria_group/MultiHome.html). The majority of outbreak strains clustered into a well-supported clade within *L. monocytogenes* lineage I (ST 29, ST1, and ST93 (both related to CC1), ST17, ST24, and ST33).

Many authors described this predominance of Lineage I strains among outbreak-related strains [19, 20] due to a better adaption towards toxic metals (as arsenic) propagated by plasmids. Lineage II strains showed a better resistance to environmental bacteriocins, cadmium and benzalkonium (BC) resistance and virulence attenuation (premature stop codon mutations in *inlA* (ST 9, serotype 1/2c; ST 121, serotype 1/2a) and *prfA*).

In the previous years it has been recognized that outbreaks have been more often linked to *L. monocytogenes* serotype 1/2a. Knabel et al. [9] reported that the majority of Canadian human listeriosis outbreaks (1988–2010) have been caused by members of *L. monocytogenes* CC8 (ST 120). The *L. monocytogenes* Austrian Quargel outbreak clones 1 and 2 were also confirmed as serotype 1/2a [21]. Furthermore, an US listeriosis outbreak linked to Mexican style cheese in 2008/2009, and an outbreak in Belgium linked to the consumption of hard cheese, were both caused serotype 1/2a strains [22, 23]. Lomonaco et al. [10] showed a global widely dissemination of *L. monocytogenes* epidemic clones involved in the 2011 multistate cantaloupe-associated outbreak and also prevalent in US chicken processing plants (ST7, serotype 1/2a; ST5, serotype 1/2b).

The first study targeting the global clonal diversity of *L. monocytogenes* included a large dataset of 300 isolates from five continents was described by Chenal-Fransisque et al. [24]. The authors demonstrated the worldwide dominance of the following clonal complexes arranged in descending order: CC2 → CC1 → CC3 → CC9 → CC7 → CC59 → CC121 → CC288 → CC199 → CC8 → CC101 → CC155 → CC5 → CC6.

Recently an increasing prevalence of *L. monocytogenes* genotypes, e.g., ST 6 among adults with listerial meningitis with an unfavorable outcome has been recognized [25]. Therefore, most experts in the research field of genotyping postulate a harmonization in terminology (definition of epidemic clones) and techniques to improve outbreak investigations and listeriosis surveillance [11, 26]. An attractive alternative, the multilocus variable number of tandem repeat (VNTR) analysis (MLVA) subtyping scheme was improved by experts of Institute Pasteur [26]. The method utilizes the naturally occurring variation in the number of tandem repeated DNA sequences found in many different loci in the genome of *L. monocytogenes*. Subsequently, PCR targeting the VNTR loci follows, by accurate sizing of the PCR products on an automated DNA sequencer (http://www.mlva.net/; http://www.pulsenetinternational.org/protocols/mlva/; both accessed on 10.10.2013). The discriminatory power is lower compared to PFGE or MLST, but the high reproducibility and high sample throughput is an attractive methodological tool to support the gold standard analysis PFGE.

The application of helpful tools such as next-generation sequencing technologies in large outbreaks was shown by Gilmour et al. [27] and Laksanalamai et al. [28]. However, whole-genome sequencing (WGS) is still too laborious and time consuming to obtain useful data in routine outbreak clarification [3].

2 Materials

2.1 Cultivation and Storage of L. monocytogenes Isolates

1. Cryocultivation: Brain Heart Infusion (BHI; Merck KgA, Darmstadt, Germany); 15 % glycerol (Merck KgA).

2. Alternative: Microbank™ Bacterial & Fungal Preservation System (Pro-Lab Diagnostics, Round Rock, TX, USA).

2.2 DNA Extraction

1. Nucleospin Tissue Kit, Machery-Nagl, GmbH & Co, KG, Düren, Germany.

2. Alternative: DNeasy Blood & Tissue Kit, Quiagen, Hilden Germany).

3. Prelysis buffer (20 mM Tris(hydroxymethyl)-aminomethan (Tris)/hydrochloric acid (HCl); 2 mM Ethylenediaminetetraacetic acid (EDTA); 1 % Triton X-100; pH 8; all components: Sigma Aldrich, St. Louis, MO, USA).

4. Lysozyme from chicken egg white (Sigma Aldrich).

2.3 PCR Amplification	1. PCR primers can be ordered by all companies providing high-quality primers-oligos-nucleotides: e.g., http://www.lifetechnologies.com; http://www.eurofinsgenomics.eu/.
	2. Components of the PCR mastermix:

- Deionized, diethylpyrocarbonate (DEPC)–water (Thermo-Fisher Scientific, Inc., Fermentas, Waltham, MA, USA).

- dNTP Mix and Platinum® Taq DNA Polymerase (Life Technologies Corporation, Carlsbad, CA, USA).

3. PCR thermocycler as T100 Thermal Cyler (Bio-Rad Laboratories, Hercules, CA, USA).

2.4 DNA Purification

1. DNA purification kit, e.g., GenElute™ PCR Clean-Up (Bio-Rad Laboratories).

2. Alternative: select in cases of external MLST sequencing the manufacturers offer on DNA purification (e.g., http://www.lgcgenomics.com/; http://www.macrogen.com/; http://www.microsynth.ch/).

2.5 Quality Control of Specific PCR Amplicons by Agarose Gel Electrophoresis

1. Tris–Borate–EDTA (TBE) buffer 10× (Carl Roth, Karlsruhe, Germany).

2. Agarose for electrophoresis (Sigma Aldrich).

3. Sybr-Safe DNA gel stain (Life Technologies Corporation, Carlsbad, CA, USA) molecular weight standard (Brome Phenol Blue Gene Ruler 100 bp Ladder; Thermo Fisher Scientific, Inc., Fermentas).

4. Loading Dye Solution (33 % glycerine, 0.07 % bromophenol blue; Thermo Fisher Scientific, Inc., Fermentas).

5. Gel casting platform and gel combs (Bio-Rad Laboratories).

6. Electrophoresis unit (Power supply Power Pac 1000, Bio-Rad Laboratories).

7. UV light imaging system (e.g., Gel Doc, Bio-Rad Laboratories).

---

3 Methods

3.1 Cultivation and Storage of L. monocytogenes Isolates

L. monocytogenes isolates should be stored in cryocultures.

Examples:

1. Cryocultivation in Brain Heart Infusion containing 15 % glycerol stored in a –80 °C freezer.

2. Microbank™ Bacterial & Fungal Preservation System for storage at –80 °C. It is composed of a unique cryovial system incorporating treated beads and a special cryopreservative solution.

3. Activate *L. monocytogenes* isolates in Brain Heart Infusion broth at 37 °C for 18–24 h.

4. Include 1 mL of *L. monocytogenes* in the DNA extraction (2.2).

3.2 DNA Extraction

1. DNA-template: isolate chromosomal DNA using silica-based kits applying lysozyme in a prelysis step.

2. Resuspend the pelleted cells in a prelysis buffer supplemented with 20 mg/mL lysozyme and incubate the samples for 30–60 min at 37 °C.

3. Add 25 µL Proteinase K, incubate at 56 °C until complete lysis is obtained.

4. Proceed with the DNA extraction protocol according to manufacturer's instructions.

3.3 PCR Amplification

1. The *L. monocytogenes* MLST housekeeping alleles according to Institute Pasteur and Ragon et al. [23] are shown in Table 1.

2. The PCR mastermix should be prepared separate for each target allele as shown in Table 2.

   Details for PCR conditions in a conventional thermocycler are listed in Table 3.

3. Keep all components for the PCR reaction on ice.

3.4 DNA Purification

1. Use a DNA purification kit for rapid purification of PCR amplification products (100 bp to 10 kb) from other components in the reactions, such as excess primers, nucleotides, DNA polymerase, oil, and salts.

2. Alternative: select in cases of external MLST sequencing the manufacturers offer on DNA purification (*see* **Notes 2** and **4**).

3.5 Quality Control of Specific PCR Amplicons by Agarose Gel Electrophoresis

1. Prepare a 1.5 % agarose gel: in 100 mL 1× TBE Buffer using a microwave, add 3.5 µL Sybr-Safe DNA gel stain.

2. The gel should be poured into a gel casting platform and gel combs have to be inserted.

3. After 30 min of gel solidification combs can be removed and the gel can be transferred in an electrophoresis chamber, which is filled with 1× TBE buffer.

4. Add 5 µL of molecular weight standard into the first and the last well of every row.

5. Mix 5 µL of each sample and positive and negative reaction controls gently with a drop of sample loading buffer and transfer the volume into the wells.

6. Electrophoresis run conditions: 120 V for 30 min.

7. Finally, visualize the gel with a UV light imaging system.

8. Store files in tiff format.

Table 1
L. monocytogenes MLST target housekeeping alleles and primer sequences [23]

Target allele	Primer code	Sequence (5'–3')	Amplicon length (bp)
abcZ (ABC transporter)	*abcZoF* / *abcZoR*	GTTTTCCCAGTCACGACGTTGTATCGCTGCTGCCACTTTTATCCA / TTGTGAGCGGATAACAATTTCTCAAGGTCGCCGTTTAGAG	537
bglA (beta glucosidase)	*bglAoF* / *bglAoR*	GTTTTCCCAGTCACGACGTTGTAGCCGACTTTTTATGGGGTGGAG / TTGTGAGCGGATAACAATTTCCGATTAAATACGGTGCGGACATA	399
cat (catalase)	*catoF* / *catoR*	GTTTTCCCAGTCACGACGTTGTAATTGGCGCATTTTGATAGAGA / TTGTGAGCGGATAACAATTTCAGATTGACGATTCCTGCTTTTG	486
dapE (succinyl diaminopimelate desuccinylase)	*dapEoF* / *dapEoR*	GTTTTCCCAGTCACGACGTTGTACGACTAATGGGCATGAAGAACAAG / TTGTGAGCGGATAACAATTTCATCGAACTATGGGCATTTTACC	462
dat (D-amino acid aminotransferase)	*datoF* / *datoR*	GTTTTCCCAGTCACGACGTTGTAGAAAGAGAAGATGCCACAGTTGA / TTGTGAGCGGATAACAATTTCTGCGTCCATAATACACCATCTTT	471
ldh (L-lactate dehydrogenase)	*ldhoF* / *ldhoR*	GTTTTCCCAGTCACGACGTTGTAGTATGATTGACATAGATAAAGA / TTGTGAGCGGATAACAATTTCTATAAATGTCGTTCATACCAT	453
lhkA (histidine kinase)	*lhkAoF* / *lhkAoR*	GTTTTCCCAGTCACGACGTTGTAAGAATGCCAACGACGAAACC / TTGTGAGCGGATAACAATTTCTGGGAAACATCAGCAATAAAC	480

Table 2
Composition of the PCR mastermix

Components	Final concentration		Stock concentration		µl/reaction
DEPC water					34.3
10× PCR buffer	1×				5
MgCl₂	2.5	mM	50	mM	2.5
PrimerF[a]	200	nM	5,000	nM	2
PrimerR[a]	200	nM	5,000	nM	2
dNTP's	200	µM	5,000	µM	2
Taqman polymerase	1	U	5	U/µl	0.2
Mastermix					48
Template					2
Reaction volume					50

[a]Seven PCR reactions have to be prepared separately including each allelic primer pair (forward and reverse primers for *abcZ, bglA, cat, dapE, dat, ldh, lhkA*)

Table 3
PCR conditions for each target housekeeping gene

PCR step	Temperature	Time
Initial denaturation	94 °C	4 min
Denaturation	94 °C	30 s
Annealing[a]	52 °C/*bgl*A: 45 °C	30 s
Elongation	72 °C	2 min
Final elongation	72 °C	10 min
	4 °C	Hold

[a]35 Cycles acc. to Ragon et al. [2]

3.6 Sequencing and Cluster Analysis

1. The primers for the seven *L. monocytogenes* housekeeping genes (Table 1) have universal sequencing tails to sequence all genes with the same forward and reverse sequencing primers (*see* **Notes 1–4**).

2. Sequences can be trimmed and assembled with free software tools as http://multalin.toulouse.inra.fr/multalin/ or imported and assembled with Bionumerics software (actual version seven; Applied Maths, Sint-Martens-Latem, Belgium).

3. Alleles and sequence type (ST) can be assigned by submitting the DNA sequences to the *Listeria* MLST database at the Pasteur Institute France (www.pasteur.fr/mlst) (*see* **Note 5**).

4. For each *L. monocytogenes* isolate the allelic combination at seven housekeeping loci results in an allelic profile and ST.

5. For the comparison of several STs a Minimum Spanning Tree from allelic profile data is helpful (a plugin for Minimum Spanning Tree drawing is provided by Institute Pasteur or the Bionumerics software).

6. Neighbor-joining tree analysis can be performed using MEGA 4.0 (*see* ref. [29]) or the SplitsTree Software (*see* ref. [30]; http://www.splitstree.org/).

7. A clonal complex (CC) is defined based on eBURST algorithm with member STs differing by only one of the 7 MLST genes (*see* ref. [31]).

4   Notes

1. The sequencing of *L. monocytogenes* housekeeping genes is provided by several companies and time saving and cost-reducing (e.g., http://www.lgcgenomics.com/; http://www.macrogen.com/; http://www.microsynth.ch/).

2. An important step is to select in an extra order position the universal primers for sequencing of the purified PCR products (oF: GTT TTC CCA GTC ACG ACG TTG TA; oR: TTG TGA GCG GAT AAC AAT TTC).

3. Example: Macrogen, Inc., Seoul, Korea-Sequencing is conducted under BigDyeTM terminator cycling conditions and products purified using ethanol precipitation and run using automatic sequencer ABI3730XL (Life Technologies Corporation, Applied Biosystems).

4. Data analysis is finished after 1–2 weeks; ABI files, text-, pdf, and report files are provided for quality tests.

5. A great advantage of the *Listeria* MLST database is the integrated MLST BLAST search against all seven MLST loci with untrimmed sequences.

Acknowledgements

The authors wish to acknowledge the Institute Pasteur for providing the MLST database of *L. monocytogenes* in the Genotyping of Pathogens and Public Health Platform.

References

1. Parisi A, Latorre L, Normanno G, Miccolupo A, Fraccalvieri R, Lorusso V, Santagada G (2010) Amplified fragment length polymorphism and multi-locus sequence typing for high-resolution genotyping of *Listeria monocytogenes* from foods and the environment. Food Microbiol 27:101–108

2. Ragon M, Wirth T, Hollandt F, Lavenir R, Lecuit M, Le Monnier A, Brisse S (2008) A new perspective on *Listeria monocytogenes* evolution. PLOS Pathog 4:e1000146

3. Sabat A, Budimir A, Nashev D, Sa-Leao R, van Dijl J, Laurent F, Grundmann H, Friedrich A (2013) Overview of molecular typing methods for outbreak detection and epidemiological surveillance. Euro Surveillance 18:1–15

4. Vangay P, Fugett EB, Sun Q, Wiedmann M (2013) Food microbe tracker: a web-based tool for storage and comparison of food-associated microbes. J Food Protect 76:283–294

5. Maiden MC, Bygraves JA, Feil E, Morelli G, Russell JE, Urwin R, Zhang Q, Zhou J, Zurth K, Caugant DA (1998) Multilocus sequence typing: a portable approach to the identification of clones within populations of pathogenic microorganisms. Proc Natl Acad Sci 95:3140–3145

6. Revazishvili T, Kotetishvili M, Stine OC, Kreger AS, Morris JG, Sulakvelidze A (2004) Comparative analysis of multilocus sequence typing and pulsed-field gel electrophoresis for characterizing *Listeria monocytogenes* strains isolated from environmental and clinical sources. J Clin Microbiol 42:276–285

7. Zhang W, Jayarao BM, Knabel SJ (2004) Multi-virulence-locus sequence typing of *Listeria monocytogenes*. Appl Environ Microbiol 70:913–920

8. Chen Y, Zhang W, Knabel SJ (2007) Multi-virulence-locus sequence typing identifies single nucleotide polymorphisms which differentiate epidemic clones and outbreak strains of Listeria monocytogenes. J Clin Microbiol 45:835–846

9. Knabel SJ, Reimer A, Verghese B, Lok M, Ziegler J, Farber J, Pagotto F, Graham M, Nadon CA, Gilmour MW (2012) Sequence typing confirms that a predominant *Listeria monocytogenes* clone caused human listeriosis cases and outbreaks in Canada from 1988 to 2010. J Clin Microbiol 50:1748–1751

10. Lomonaco S, Verghese B, Gerner-Smidt P, Tarr C, Gladney L, Joseph L, Katz L, Turnsek M, Frace M, Chen Y (2013) Novel epidemic clones of *Listeria monocytogenes*, United States, 2011. Emerg Infect Dis 19:147

11. Cantinelli T, Chenal-Francisque V, Diancourt L, Frezal L, Leclercq A, Wirth T, Lecuit M, Brisse S (2013) Epidemic clones of *Listeria monocytogenes* are widespread and ancient clonal groups. J Clin Microbiol 51:3770–3779

12. Liu D (2006) Identification, subtyping and virulence determination of *Listeria monocytogenes*, an important foodborne pathogen. J Med Microbiol 55:645–659

13. Jiang L, Chen J, Xu J, Zhang X, Wang S, Zhao H, Vongxay K, Fang W (2008) Virulence characterization and genotypic analyses of *Listeria monocytogenes* isolates from food and processing environments in eastern China. Int J Food Microbiol 121:53–59

14. Cooper JE, Feil EJ (2004) Multilocus sequence typing—what is resolved? Trends Microbiol 12:373–377

15. Salcedo C, Arreaza L, Alcala B, De La Fuente L, Vazquez J (2003) Development of a multilocus sequence typing method for analysis of *Listeria monocytogenes* clones. J Clin Microbiol 41:757–762

16. Spratt BG (1999) Multilocus sequence typing: molecular typing of bacterial pathogens in an era of rapid DNA sequencing and the internet. Curr Opin Microbiol 2:312–316

17. Doumith M, Cazalet C, Simoes N, Frangeul L, Jacquet C, Kunst F, Martin P, Cossart P, Glaser P, Buchrieser C (2004) New aspects regarding evolution and virulence of *Listeria monocytogenes* revealed by comparative genomics and DNA arrays. Infect Immun 72:1072–1083

18. den Bakker HC, Fortes ED, Wiedmann M (2010) Multilocus sequence typing of outbreak-associated *Listeria monocytogenes* isolates to identify epidemic clones. Foodborne Pathog Dis 7:257–265

19. Kathariou S (2002) *Listeria monocytogenes* virulence and pathogenicity, a food safety perspective. J Food Protect 65:1811–1829

20. Orsi RH, Bakker HCD, Wiedmann M (2011) Listeria monocytogenes lineages: genomics, evolution, ecology, and phenotypic characteristics. Int J Med Microbiol 301:79–96

21. Fretz R, Pichler J, Sagel U, Much P, Ruppitsch W, Pietzka AT, Stoger A, Huhulescu S, Heuberger S, Appl G, Werber D, Stark K, Prager R, Flieger A, Karpiskova R, Pfaff G, Allerberger F (2010) Update: multinational listeriosis outbreak due to "Quargel," a sour milk curd cheese, caused by two different *L. monocytogenes* serotype 1/2a strains, 2009–2010. Euro Surveill 15:19543

22. Jackson KA, Biggerstaff M, Tobin-D'Angelo M, Sweat D, Klos R, Nosari J, Garrison O,

Boothe E, Saathoff-Huber L, Hainstock L, Fagan RP (2011) Multistate outbreak of *Listeria monocytogenes* associated with Mexican-style cheese made from pasteurized milk among pregnant, Hispanic women. J Food Protect 74:949–953

23. Yde M, Naranjo M, Mattheus W, Stragier P, Pochet B, Beulens K, De Schrijver K, Van den Branden D, Laisnez V, Flipse W, Leclercq A, Lecuit M, Dierick K, Bertrand S (2012) Usefulness of the European Epidemic Intelligence Information System in the management of an outbreak of listeriosis, Belgium, 2011. Euro Surveill 17:20279

24. Chenal-Francisque V, Lopez J, Cantinelli T, Caro V, Tran C, Leclercq A, Lecuit M, Brisse S (2011) Worldwide distribution of major clones of *Listeria monocytogenes*. Emerg Infect Dis 17:1110

25. Koopmans MM, Brouwer MC, Bijlsma MW, Bovenkerk S, Keijzers W, van der Ende A, van de Beek D (2013) *Listeria monocytogenes* sequence type 6 and increased rate of unfavorable outcome in meningitis: epidemiologic cohort study. Clin Infect Dis 57:247–253

26. Chenal-Francisque V, Diancourt L, Cantinelli T, Passet V, Tran-Hykes C, Bracq-Dieye H, Leclercq A, Pourcel C, Lecuit M, Brisse S (2013) Optimized multilocus variable-number tandem-repeat analysis assay and its complementarity with pulsed-field gel electrophoresis and multilocus sequence typing for *Listeria monocytogenes* clone identification and surveillance. J Clin Microbiol 51:1868–1880

27. Gilmour M, Graham M, Van Domselaar G, Tyler S, Kent H, Trout-Yakel K, Larios O, Allen V, Lee B, Nadon C (2010) High-throughput genome sequencing of two *Listeria monocytogenes* clinical isolates during a large foodborne outbreak. BMC Genomics 11:120

28. Laksanalamai P, Joseph LA, Silk BJ, Burall LS, L Tarr C, Gerner-Smidt P, Datta AR (2012) Genomic characterization of *Listeria monocytogenes* strains involved in a multistate listeriosis outbreak associated with cantaloupe in US. PloS One 7:e42448

29. Tamura K, Dudley J, Nei M, Kumar S (2007) MEGA4: molecular evolutionary genetics analysis (MEGA) software version 4.0. Mol Biol Evol 24:1596–1599

30. Huson DH, Bryant D (2006) Application of phylogenetic networks in evolutionary studies. Mol Biol Evol 23:254–267

31. Feil EJ, Li BC, Aanensen DM, Hanage WP, Spratt BG (2004) eBURST: inferring patterns of evolutionary descent among clusters of related bacterial genotypes from multilocus sequence typing data. J Bacteriol 186: 1518–1530

Chapter 7

Ribotyping and Automated Ribotyping of *Listeria monocytogenes*

Mazin Matloob and Mansel Griffiths

Abstract

Ribotyping is a molecular technique that allows identification and typing of bacteria to the strain level. It is based on restriction endonuclease cleavage of total genomic DNA followed by electrophoretic separation, Southern blot transfer, and hybridization of transferred DNA fragments with a radiolabeled ribosomal operon probe. Following autoradiography, only those bands containing a portion of the ribosomal operon are visualized. The number of fragments generated by ribotyping is a reflection of the multiplicity of rRNA operons present in a bacterial species.

Automated Ribotyping—AR (RiboPrinter) is a commercially available instrument with a high level of reproducibility and standardization. The RiboPrinter automates all of the steps in the process from cell lysis to data capture and database comparisons. Further, reagent cassettes including the enzymes, enzyme conjugated-hybridization probe, electrophoretic gel, and membrane have been developed to deliver consistent performance. Data capture is accomplished via a CCD camera and the gel patterns obtained stored in a digitized format, making it easier to compare results among laboratories and to exchange data electronically.

Key words Ribotyping, RiboPrinter, rRNA, Operon, Probe, Hybridization

1 Introduction

Ribotyping is a form of Restriction Fragment Length Polymorphism (RFLP) analysis that relies on differences in the location and number of ribosomal RNA (rRNA) gene sequences present in the bacterial genome for genotyping [1, 2]. Ribotyping targets conserved chromosomal genetic elements and thus allows for reliable grouping of isolates [3, 4]. It is a molecular technique that allows identification and typing of bacteria to the strain level by the analysis of band pattern differences obtained through hybridization of labelled rRNA or ribosomal DNA (rDNA) with DNA fragments produced by cleavage of total DNA with an endonuclease [5, 6]. Most importantly for the purpose of ribotyping, the DNA sequences encoding rRNA should be highly conserved even between different

Kieran Jordan et al. (eds.), *Listeria monocytogenes: Methods and Protocols*, Methods in Molecular Biology, vol. 1157, DOI 10.1007/978-1-4939-0703-8_7, © Springer Science+Business Media New York 2014

bacterial species, but the flanking and spacer sequences may vary [6]. Variations in the flanking sequences give rise to variations in ribopatterns generated by ribotyping [1].

A completely automated system for ribotyping (the RiboPrinter Microbial Characterization System) has been developed by Qualicon, Inc. (Wilmington, DE), providing a degree of standardization that makes it possible to combine the results carried out on different instruments. The RiboPrinter system includes identification libraries with over 6,900 RiboPrinter patterns, representing 219 bacterial genera and more than 1,440 species and serotypes that are critical to the food and pharmaceutical industries.

Automated ribotyping (AR) is often considered the DNA subtyping method of choice for many large-scale studies as well as for industrial applications, as PFGE is more time-consuming and labor-intensive, and requires more personnel with a higher level of technical expertise [7]. This technique is also highly reproducible, making the profiles obtained amenable to computer-based analysis and storage. It is then possible to compare the ribotypes found by other users of the RiboPrinter [8, 9].

Generally, AR is rapid, with the analysis being completed in less than 8 h, so that the RiboPrinter can process up to 32 samples per day [10, 11]. Interestingly, the ribopatterns arise generally from sequence differences within rDNA. The changes in flanking sequences, which are composed of housekeeping genes that evolve through point mutations, alter the ribotype of an organism [1].

For *Listeria monocytogenes,* automated ribotyping has been used for the discrimination and for the characterization of virulence gene polymorphism lineage [12]. *Eco*RI ribotyping has been confirmed as a reliable and rapid method for *L. monocytogenes* typing, and ribotypes appear to be good markers for *L. monocytogenes* clones that vary in virulence for particular host species [8, 13]. Table 1 shows the RiboPrinter Patterns and the DuPont—ID for *L. monocytogenes* bacteria from different sources using *EcoRI* enzyme.

2   Materials

Brain Heart Infusion (BHI) agar (Catalogue number 241830-500G, BD-Difco, Sparks, MD, USA). Except sample buffer, all plasticware must be autoclaved at 121 °C for 15 min and can be stored at room temperature after sterilization.

2.1   The RiboPrinter Microbial Characterization System

The system includes the following segments:

1. *Mixer/Collector:* routine laboratory methods were used to culture the bacterial sample. Bacterial colonies were captured

Table 1
Shows the RiboPrinter Patterns and the DuPont—ID for *L. monocytogenes* bacteria from different sources using *EcoRI* enzyme

EcoRI pattern #	Source/site	RiboPrinter pattern	DuPont -ID
1	Cheese		DUP-1038
2	Ice cream		DUP-1045
3	Washing water/ dairy plant		DUP-18598
4	Equipment swab		DUP-1039
5	Farm		DUP-1042
6	Floor/dairy plant		DUP-19170

through special collector and mixed with buffer before transferring to the sample carrier.

2. *Heat Treatment Station:* in order to prevent the instrument from possible biohazard contamination, sample is treated with high temperature of 80 °C to inactivate any viable cells.

3. *Microbial Characterization Unit:* this unit consists of the following internal components:

   DNA Preparation: bacteria are lysed through series of chemical reactions to prepare and have DNA available for further processing. The DNA is digested with a specific restriction enzyme.

   Separation and Transfer: the application of electrophoresis is utilized to separate the DNA fragments on agarose gel. The DNA fragments are then immobilized and transferred to a nylon membrane.

   Membrane Processing: by exposing the nylon membrane to a series of enzymatic and chemical reactions, the DNA fragments chemiluminescent.

4. *Workstation PC:* to store and retrieve information, and backup data.

5. *Laser Printer:* to generate reports and prepare hard copies of the RiboPrinter patterns.

2.2 Disposables

1. Colony Picks, used to collect samples from incubated plates.

2. Sample Carrier, used for processing.

3. Lysing Agent A and B, used for sample preparation.

4. DNA Prep Carrier and Enzyme, used to lyse bacteria and digest DNA to prepare fragments.

5. Gel Cassette, used to separate and transfer the DNA fragments.

6. Membrane, used to hold the DNA fragments after separation.

7. Membrane Processing (MP) Base and Inserts, contained substances (conjugate, substrate, probe, and liquid membrane processing reagents) to create chemiluminescence in the processed DNA fragments.

8. Purified water, used for processing and cleaning purposes.

---

3   Methods

3.1 Preparation of Sample and Instrument

1. Pure isolated colonies should be used to streak BHI agar plate for the preparation of lawn and/or single colonies.

2. Incubate plates for 18–30 h in a humidified incubator (*see* **Note 1**).

3. Conduct Gram stain test to conform the Gram reactivity of your sample (*see* **Note 2**).

4. Transfer 5 mL of sample buffer to a sterile disposable 15 mL tube (*see* **Note 3**).

5. Autoclave the capped microcentrifuge tubes (0.65 mL) at 121 °C for 15 min.

6. Place one tube in each hole in the lower row of the sample rack.

7. Add 40 µL of sample buffer from the 15 mL tube for the Gram-positive samples, and 200 µL for the Gram-negative samples. Close the tubes lids.

8. Use an autoclaved colony pick to harvest one of the single colonies or bacteria from the lawn (*see* **Notes 4** and **5**).

9. Mix the sample by mixer for 5 s in a safe area to prevent possible aerosolization (*see* **Note 6**).

10. Place the pick into one of the filled microcentrifuge tubes (two picks for Gram-positive and one pick for Gram-negative samples). The sample liquid should appear turbid.

11. Repeat the steps for harvesting and mixing samples in the original tube (for Gram-positive samples only).

12. Place the filled microcentrifuge tube on the top row of the sample preparation rack, this shows that the tube is filled (*see* **Note 7**).

13. Discard the used picks in a biohazard waste bag.

14. Use 100 µL pipetter to pipette 30 µL of sample from the microcentrifuge tube into the well of the sample carrier. Close the lid cover for the well and repeat the procedures for other samples.

15. Locate the sample carrier into the heat treatment station, and ensure that the sample is identified and recorded properly (*see* **Note 8**). The display on the heat treatment station will show "Insert," and after insertion, the display shows "Press Button."

16. Press the button (*see* **Note 9**). The display changes to "Heat" and counts down from 13 min after reaching the operating temperature (*see* **Note 10**).

17. The display message will change to "Cool," and count down from 9 min, indicating the remaining time in the cooling cycle.

18. The heat treatment step is completed when the indicating message shows "Ready" and counts down from 90 min. The display will flash and beep three times, and will beep every 10 min until the sample is removed or 90 min is over.

19. Use 10 µL pipette to transfer 5 µL of lysing agent A and B to each well in the sample carrier (*see* **Notes 11–13**).

20. Ensure that the instrument is turned on and load the sample carrier into the characterization unit. Record the sample information (*see* **Notes 14** and **15**).

21. Place the sealed carrier into the labelled slot on the right side of the characterization unit (*see* **Note 16**).

22. Remove the DNA Prep enzyme from the freezer and tap it cautiously against the laboratory bench several times (*see* **Note 17**).

23. Remove the cap from the vial and insert it into the carrier.

24. Place the DNA Prep carriers into the slot labelled "Reagent" located at the left side of the sample carrier slot. Push the DNA Prep carrier down until it fits securely.

25. Place probe reagents, conjugate and substrate into the MP Base station and after lifting the cover of the MP Reagent, locate the MP Base into the carousel then close the cover.

26. Take out the gel cassette from its original package and use one end of the rubber comb to pull the comb away from the cassette.

27. Fold the handle of the cassette to your direction, this movement will simplify installation and recovering of the cassette (*see* **Note 18**).

28. Check out the edge of the gel (*see* **Note 19**).

29. Insert the gel cassette into the slot marked "Gel Bay" (*see* **Note 20**).

30. Press the gel cassette forward until it fits in place.

31. Hold the membrane cautiously and insert it in the labelled front slot (*see* **Note 21**).

32. Close up all doors, after loading the Sample Carrier, DNA Prep Carrier, Gel Cassette, Membrane, and MP Base. The system will proceed with running the first batch (*see* **Note 22**).

3.2 Creating Standard Batches

1. Generate a standard batch by clicking on "Operation" button (*see* **Note 23**).

2. Click on "Create a standard batch."

3. Use the view menu to observe all items and possibly remove unnecessary items (*see* **Note 24**).

4. Click on "Use Default Libraries" to generate a minimum view then enter a DuPont ID and Custom ID (*see* **Notes 25–27**).

5. Complete the blanks for every field that you display. Otherwise, the system will not proceed to the next window.

6. Select the type (Sample, Empty, or Control-QC Number) for each well in the sample carrier (*see* **Note 28**).

7. Fill in the name you specify for your sample. This information will be considered as Sample Label in the Data Analysis software screens (*see* **Note 29**).

8. Repeat the same steps for the remaining items (*see* **Note 30**).

9. Click on "File" menu, then select "Save this Batch" (*see* **Note 31**).

10. Click on "Yes—Submit Batch to Instrument" (*see* **Note 32**).

11. Follow the screen instructions to load disposables and check the DNA Prep waste (*see* **Notes 33** and **34**), the displayed icons will flash and direct you to remove or load an item. The screen directs you to use the suitable chamber for the membrane and gel cassette (*see* **Note 35**).

12. Click "YES-Submit Batch to Instrument" button on the load disposable window (*see* **Note 36**).

3.3 Retrieve the Batch Report

1. Upon completion of the processing, the system automatically conducts a series of analysis and produces a Batch Information Report. No action is required from the operator.

2. The standard report includes the instrument number, accession number, event log, starting date, starting time, operator's name, batch status, the number of the sample, the sample label, the DuPont identification label, characterization information, and the RiboPrinter pattern for each sample (*see* **Note 37**).

3. You may choose to have a set of eight sample reports as a result of the automatic process, or the image diagnostic report which displays the batch image in addition to the analytical information of the batch.

4 Notes

1. The optimal incubation temperature for *E. coli*, *Salmonella*, and *Staphylococcus* is 37 °C and for *Bacillus* and *Listeria* is 30 °C.

2. The procedures for sample preparation vary for Gram-positive and Gram-negative bacteria.

3. Do not pipette the sample buffer directly from the bottle to the sample tubes to eliminate cross-contamination of the sample buffer.

4. It is recommended to harvest colonies from the lawn area of the plate, and normally with large colonies, a single colony will be sufficient.

5. Do not reuse the same pick twice. Adequate amount of sample is required for proper operation. Over sampling causes inaccurate results.

6. Working underclass-2 biosafety cabinet is highly recommended.

7. It is recommended to have a back-up BHI plate, streaked by the tested pick.

8. Try to transfer the sample carrier to the heat treatment station within 2 h. After 2 h delay, the sample carrier should be discarded.

9. The screen indicates "Warm up" and it takes 10 min for the station to warm up. The normal come up time is about 4 min.

10. This represents each minute of heat treatment.

11. It is recommended to remove the lysing agent A and B from the freezer 10 min prior to the usage. Lysing agents should be kept frozen (–20 °C) while not being used within 2 h.

12. Use a clean 200 µL pipette tip to disperse any bubbles in the sample carrier wells.

13. Do not add the lysing agents to the sample carrier before the heat treatment cycle is finished.

14. Do not insert a sample carrier into the instrument unless it has been processed in the heat treatment station.

15. You may place a new carrier any time after removing the sample carrier from the heat treatment station.

16. Ensure the liquid existence in the bottom of the wells without any bubbles, prior proceeding to the next step.

17. This will release any liquid which might have attached to the cap.

18. A click will be heard as the handle fits into place.

19. A new cassette should be used in case of excessive build up of gel on the front edge of the cassette or in case of possible shrinkage of gel away from the cassette.

20. The RiboPrinter system stops you from inserting the cassette into the wrong slot.

21. Ensure that the two large slots are on the top and the square hole on the side faces your left while you insert the membrane.

22. The system will warn you to correct any action upon detecting any disposable error during its check before processing. It requires 2 h before loading the next batch.

23. Three types of standard batches, *Eco*RI, *Pst*I, and *Pvu*II are supported by RiboPrinter system and each enzyme requires a separate protocol.

24. Type of the sample and ribogroup library information is required by the system for each sample.

25. Click on "View," and then click on "Sample Items" when entering information regarding the sample.

26. Click on "Presumptive ID," if you wish to enter a presumptive ID for the sample. Consecutively, fields for presumptive ID Genus/Species/Subspecies appear for each sample.

27. By clicking on the Presumptive ID Genus, a pop-up menu becomes visible, showing all options. Scroll through the list to choose the appropriate genus/species/subspecies.

28. The system default to Sample, the DuPont ID and Ribogroup are automatically entered as DUP and RIBO1 when you select Sample.

29. The Ribogroup library name can be changed and saved to the pop up list for future use.

30. When running less than eight samples in a batch, choose "Empty" from the "Sample Type" menu for each well not having any sample. Otherwise, the batch data may be lost when the system begins processing.

31. If you do not fill in information completely for all of the fields on the screen, a message pops up "Data Entry Error." Click "Acknowledge" to remove this message then you have option to either fill in the information or change the view.

32. The language changes depending on protocol.

33. The DNA Prep waste container must be removed out every time a new batch started.

34. Autoclave the container with the waste (121 °C for 60 min), discard the waste and reuse the container.

35. Do not load disposables unless you are directed by the system. The alarm will sound while the doors are open and once the doors are close the alarm stops.

36. You can create one of the three types of standard batches each time.

37. In case the system does not identify the sample with an item in the DuPont identification library, the report will print "None."

References

1. Bouchet V, Huot H, Goldstein R (2008) Molecular genetic basis of ribotyping. Clin Microbiol Rev 21:262–273

2. Foley S, Lynne A, Nayak R (2009) Molecular typing methodologies for microbial source tracking and epidemiological investigations of Gram-negative bacterial foodborne pathogens. Infect Genet Evol 9:430–440

3. Sauders B, Fortes E, Morse D, Dumas N, Kiehlbauch J, Schukken Y, Hibbs J, Wiedmann M (2003) Molecular subtyping to detect human listeriosis clusters. Emerg Infect Dis 9:672–680

4. Klaeboe H, Lunestad B, Borlaug K, Paulauskas A, Rosef O (2010) Persistence and diversity of Listeria monocytogenes isolates in Norwegian processing plants. Vet Med Zoot 50:42–47

5. Regnault B, Grimont F, Grimont P (1997) Universal ribotyping method using a chemically labelled oligonucleotide probe mixture. Res Microbiol 148:649–659

6. Bingen E, Denamur E, Elion J (1994) Use of ribotyping in epidemiological surveillance of nosocomial outbreaks. Clin Microbiol Rev 7:311–327

7. Wiedmann M (2002) Molecular subtyping methods for *Listeria monocytogenes*. J AOAC Int 85:524–531

8. Manfreda G, De Cesare A, Stella S, Cozzi M, Cantoni C (2005) Occurrence and ribotypes of *Listeria monocytogenes* in Gorgonzola cheeses. Int J Food Microbiol 102:287–293

9. Meloni D, Galluzzo B, Mureddu A, Piras F, Griffiths M, Mazzette R (2009) *Listeria monocytogenes* in RTE marketed in Italy: prevalence and automated EcoRI ribotyping of the isolates. Int J Food Microbiol 129: 166–173

10. Ito Y, Iinuma M, Baba H (2003) Evaluation on automated ribotyping system for characterization and identification of verocytotoxin-producing *Escherichia coli* isolated in Japan. Jpn J Infect Dis 56:200–204

11. Pavlic M, Griffiths M (2009) Principles, application, and limitations of automated ribotyping as a rapid method in food safety. Foodborne Pathog Dis 6:1047–1055

12. Gendel S (2004) Riboprint analysis of *Listeria monocytogenes* isolates obtained by FDA from 1999 to 2003. Food Microbiol 21:187–191

13. Blais B, Martinez-Perez A, Gauthier M, Allain R, Pagotto F, Tyler K (2008) Development of unique bacterial strains for use as positive controls in the food microbiology testing laboratory. J Food Protect 71: 2301–2306

Chapter 8

Fluorescent Amplified Fragment Length Polymorphism (fAFLP) Analysis of *Listeria monocytogenes*

Corinne Amar

Abstract

Fluorescent amplified fragment length polymorphism (fAFLP) is based on the selective PCR amplification of restriction fragments from a digest of total genomic DNA. Genomic DNA extracted from a purified bacterial isolate is completely digested with two endonucleases generating fragments which are ligated to specific double-stranded adaptors. The ligated fragments are then amplified by PCR using fluorescently labelled primers. Fluorescent amplified fragments are separated by size on an automated sequencer with a size standard. fAFLP is a rapid, highly reproducible technique which can be used to discriminate and subtype *Listeria monocytogenes* strains.

Key words AFLP, *Listeria monocytogenes*, Fragment analysis, Restriction

1 Introduction

Amplified fragment length polymorphism (AFLP) was developed by Zabeau and Vos in 1993 [1] and is based on the selective PCR amplification of restriction fragments from a digest of total genomic DNA. AFLP is a robust and a highly reproducible whole-genome DNA fingerprint technique. Since 1993, AFLP has successfully been applied for the rapid typing of a range of different bacterial species [2–4] including *Listeria monocytogenes* [5].

To perform AFLP, genomic DNA extracted from a purified bacterial isolate is completely digested with a restriction enzyme, generating fragments that are then ligated to double-stranded adaptors. These adaptors are designed to be complementary to the base sequence of the restriction site but in such way, that when ligated, they do not reconstitute the site of restriction. The ligated fragments are then amplified by PCR using primers that are complementary to the adaptor sequence, and gel electrophoresis is performed to separate the resulting fragments and thus creates a DNA fingerprint. When bacterial genomic DNA is treated in this way, many hundreds of different fragments can result.

Kieran Jordan et al. (eds.), *Listeria monocytogenes: Methods and Protocols*, Methods in Molecular Biology, vol. 1157, DOI 10.1007/978-1-4939-0703-8_8, © Springer Science+Business Media New York 2014

Therefore, to amplify only a subset of the fragments, primers are extended by usually one selected nucleotide running into the unknown part of the fragments sequence. A primer extension of one nucleotide reduces the number of amplified fragments by a factor of 4. AFLP using the enzyme *Eco*RI and the selective nucleotide G incorporated into the primer was developed for the typing of *L. monocytogenes* isolates by Guerra and colleagues [5].

The use of AFLP for the typing of bacteria is fast, reproducible and high throughput but is hampered by the time spent in the preparation and performing gel electrophoresis (4–4½ h) and the subjective interpretation and analysis of the banding patterns. Fluorescent detection instrumentation has facilitated the further refinement of AFLP and permitted the use of internal size standards enabling the accurate, reproducible sizing of amplified fragments. In fluorescent AFLP (fAFLP), the fluorescent amplified fragments are detected and sized by the laser reader of an automated sequencer and primers tagged with fluorophores, such as 5-carboxyfluorescein (5-FAM), NED™, JOE™, or Cy5™ are used. fAFLP methods using different two endonucleases combinations have been developed for high throughput, highly reproducible, and highly discriminatory typing assays for many bacterial species [6–8] including *L. monocytogenes* [9–13].

The fAFLP method described here has been used since 2007 by the UK National Reference Laboratory for *Listeria* (UK-NRL, Public Health England PHE) for the routine typing of *L. monocytogenes* isolated from human cases, foods and food production environments. This technique was developed using a modification of a protocol previously described for *Campylobacter*, by Desai et al. [8]. The two restriction enzymes, *Hha*I and *Hin*dIII, were selected for *L. monocytogenes* genome digestion using the in silico analysis AFLP Fragment Predictor Program ALFIE (Underwood A, http://www.hpa-bioinformatics.org.uk/cgi-bin/ALFIE/index.cgi, PHE London) to give approximately 50–85 fragments of between 60 and 600 bp in size. Digestion and ligation steps are simultaneously performed in the presence of RNase A and bovine serum albumin. FAM-labelled *Hin*dIII-A and a non-labelled *Hha*I-A selective primers are used for fragment amplification by PCR. The UK-NRL sends PCR products to the PHE Sequencing Department to be separated on an automated sequencer with a size standard. Chromatographs are emailed to the UK-NRL and visualized on appropriate software such as PEAK SCANNER™ v1.0 (Life Technologies™) and exported as Excel files into Bionumerics v6.1 (Applied Maths) where they are visualized as virtual electrophoresis gels and analyzed.

2   Materials

All enzymes are stored at –20 °C and kept in a cooling block while in use.

Digestion/ligation reactions and PCR occur in 0.2 mL thin wall PCR tubes or whenever possible in 96-well PCR plates.

Thermocyclers with blocks fitting 0.2 mL tubes or 96-well plates are used.

2.1   Digestion and Ligation

1. *Hin*dIII enzyme 10,000 units (catalogue number NEB R0104S, New England BioLabs).

2. *Hha*I enzyme 2,000 units (catalogue number NEB R0139S, New England BioLabs).

3. 10× NEBuffer2, supplied with 100× BSA (catalogue number B7002S, New England BioLabs).

4. T4 DNA ligase 20,000 units supplied with 10× Buffer (catalogue number NEB M0202S, New England BioLabs) (*see* **Note 1**).

5. RNAse A 50 mg (catalogue number R-4642, Sigma-Aldrich).

2.2   Adaptors Oligonucleotides

1. *Hin*dIII double-stranded adaptors (S1 and S2):

*Hin*dIII-S1	5′ CTCGTAGACTGCGTACC 3′
*Hin*dIII-S2	5′ AGCTGGTACGCAGTC 3′

2. *Hha*I double-stranded adaptors (S1 and S2):

*Hha*I-S1	5′ GACGATGAGTCCTGATCG 3′
*Hha*I-S2	5′ ATCAGGACTCATCG 3′

2.3   Amplification

1. Sterile, nuclease-free distilled water.

2. Mega-Mix-GOLD© 2× (reference 2MMG, Microzone Ltd) (*see* **Note 2**).

3. Selective primer oligonucleotides:

*Hin*dIII-A	5′ FAM-GACTGCGTACCAGCTTA 3′
*Hha*III-A	5′ GATGAGTCCTGATCGCA 3′

2.4   Preparation of PCR Products for Fragment Analysis (for Used with Applied Biosystems ABi, Life Technologies™ Capillary Electrophoresis Systems)

PCR products can be prepared and performed by the genomic service company of choice offering a fragments analysis service.

1. Sterile, nuclease-free distilled water for dilution of PCR products.

2. Hi-Di™ Formamide (reference 4311320, Life Technologies™).

3. GeneScan™- 600 LIZ® Size standard (reference 4408399, Life Technologies™).

2.5 Analysis of Fragments	1. Softwares for reading generated scans. PeakScanner™ or GeneMapper® Analysis Software (©Applied Biosystems).
	2. Software for importing and analyzing scans. BioNumerics v.6.1 AppliedMaths.

3   Methods (*See* Note 3)

3.1 Preparation of Double-Stranded Adaptors	This can be performed in batches with the double-stranded oligonucleotides stored in 25 μL aliquots at –20 °C until further use.
	1. Prepare the following mixtures for each of the adaptors as in Table 1, and distribute in 25 μL aliquots in 0.2 mL tubes.
	2. In a thermocycler, denature the oligonucleotides at 95 °C for 10 min then let them cool slowly at room temperature for 15 min and store at –20 °C until further use.
3.2 Digestion and Ligation	1. Prepare the following mixture (total volume of mixture per reaction is 55 μL) in the same order as shown in Table 2. Distribute 55 μL of mixture in 0.2 mL PCR tubes or 96-well PCR plates.
	2. Add 5 μL of each DNA extract to each of the 55 μL mixture (*see* **Note 4**).
	3. Place the tubes in a thermocycler. The cycling conditions are 37 °C for 4 h for the digestion, followed by 16 °C for 3 h for the ligation followed by 65 °C for 10 min to deactivate the enzymes and a pause at 4 °C. Proceed as soon as possible to the amplification step. Otherwise, freeze the ligated digests at –20 °C.
3.3 Amplification of Digests	1. Pre-heat the thermocycler at 95 °C to minimize the amplification of nonspecific products.

Table 1
Master mixture for the preparation of double-stranded adaptors

Ingredients	(Initial)	Volume (μL)	(Final)
*Hind*III *adaptors*			
Water	–	576	–
*Hind*III-S1	100 μM	12	2 μM
*Hind*III-S2	100 μM	12	2 μM
*Hha*I *adaptors*			
Water	–	360	–
*Hha*I-S1	100 μM	120	20 μM
*Hha*I-S2	100 μM	120	20 μM

Table 2
Master mixture for the simultaneous digestion and ligation reactions of *L. monocytogenes* genomic DNA

Ingredients	(Initial)	Volume (µL)	(Final)
Water		44.4	
NEBuffer2	10×	3	0.5×
BSA	100×	0.3	0.5×
T4 DNA Ligase Buffer	10×	5	0.8×
*Hin*dIII-S1/S2 adaptors	2 µM	0.5	0.02 µM
*Hha*I-S1/S2 adaptors	20 µM	0.5	0.2 µM
T4 DNA Ligase	400 U/µL	0.2	80 U
RNase A	30 mg/mL	0.5	15 µg
*Hin*dIII enzyme	20 U/µL	0.3	5 U
*Hha*I enzyme	20 U/µL	0.3	5 U

Table 3
Master mixture for PCR amplification of digested ligated fragments

Ingredients	(Initial)	Volume (µL)	(Final)
Water	–	2.5	–
MegaMix-GOLD©	2×	12.5	1×
*Hin*dIII-A	10 µM	2.5	1 µM
*Hha*I-A	10 µM	2.5	1 µM

2. Prepare the PCR mixture as shown in Table 3 and distribute 20 µL of PCR master mixture in 0.2 mL PCR tubes.

3. Add 5 µL of ligated fragments to 20 µL of PCR mixture.

4. Place the tubes in the pre-heated thermocycler. Stop the pre-heat program and immediately run the touch-down PCR cycling conditions which are as follow:

Polymerase activation	94 °C	5 min		1 Cycle
Denaturation	94 °C	20 s		
Primer annealing	**	30 s		30 Cycles
Primer extension	72 °C	2 min		
Final extension	60 °C	30 min		1 Cycle

**Touchdown PCR Primer annealing temperatures are

First cycle	66 °C	30 s
9 cycles	Temperature decreased by 1 °C at each cycle	
Then 20 cycles	56 °C	30 s

3.4 Preparation of Amplified Digest for Fragment Analysis

1. Dilute the PCR products (*see* **Note 5**).

2. Prepare (*see* **Note 6**) 10.5 μL of 10.0 μL of Hi-Di Formamide and 0.5 μL of GeneScan™- 600 LIZ® Size standard Add 1.0 μL of diluted amplified digests.

3.5 Analysis of fAFLP

Chromatographs are visualized on PEAK SCANNER™ v1.0 (Life Technologies™) or GeneMapper® Analysis Software and exported as Excel files into Bionumerics v6.1 (Applied Maths) where they are visualized as virtual electrophoresis gels and analyzed.

4 Notes

1. The author evaluated enzymes from a number of various suppliers and those supplied by New England BioLabs were selected.

2. The Mega-Mix-GOLD© was chosen by the author because it contains a "Hot-start" PCR enzyme.

3. Master mixtures for digestion/ligation reactions and PCR can be prepared in batches for the number of reactions plus two or more to compensate for loss of volume when pipetting. The mixtures can be distributed to thin wall 0.2 mL PCR tubes using, whenever possible, multichannel pipettes.

4. The author found that DNA does not need to be highly purified, but most importantly should not be extensively sheared. Therefore, the author recommends single colony DNA extraction using products such as MicroLYSIS® from Microzone Ltd.

5. It is always necessary to dilute the PCR product. The dilution factor is usually between 1/3 and 1/10. The optimal dilution has to be determined by trial and error.

6. The Genomic Company providing fragment analysis service might perform this step.

References

1. Vos P, Hogers R, Bleeker M, Reijans M, van de Lee T, Hornes M, Frijters A, Pot J, Peleman J, Kuiper M (1995) AFLP: a new technique for DNA fingerprinting. Nucleic Acids Res 23:4407–4414

2. Keim P, Kalif A, Schupp J, Hill K, Travis SE, Richmond K, Adair DM, Hugh-Jones M, Kuske CR, Jackson P (1997) Molecular evolution and diversity in *Bacillus anthracis* as

detected by amplified fragment length polymorphism markers. J Bacteriol 179:818–824

3. Lin JJ, Kuo J, Ma J (1996) A PCR-based DNA fingerprinting technique: AFLP for molecular typing of bacteria. Nucleic Acids Res 24:3649–3650

4. Janssen P, Coopman R, Huys G (1996) Evaluation of the DNA fingerprinting method AFLP as a new tool in bacterial taxonomy. Microbiology 142:1881–1893

5. Guerra MM, Bernardo F, McLauchlin J (2002) Amplified fragment length polymorphism (AFLP) analysis of *Listeria monocytogenes*. Syst Appl Microbiol 25:456–461

6. Desai M, Efstratiou A, George R, Stanley J (1999) High-resolution genotyping of *Streptococcus pyogenes* serotype M1 isolates by fluorescent amplified-fragment length polymorphism analysis. J Clinical Microbiol 37:1948–1952

7. Tamada Y, Nakaoka Y, Nishimori K (2001) Molecular typing and epidemiological study of *Salmonella enterica* serotype typhimurium isolates from cattle by fluorescent amplified-fragment length polymorphism. J Clinical Microbiol 39:1057–1066

8. Desai M, Logan JMJ (2001) Sequence-based fluorescent amplified fragment length polymorphism of *Campylobacter jejuni*, its relationship to serotyping, and its implications for epidemiological. J Clinical Microbiol 39:3823–3829

9. Roussel S, Félix B, Grant K, Tam Dao T, Brisabois A, Amar C (2013) Fluorescence amplified fragment length polymorphism compared to pulsed field Gel electrophoresis for *Listeria monocytogenes* subtyping. BMC Microbiol 13:14

10. Aarts HJ, Hakemulder LE, Van Hoef AM (1999) Genomic typing of *Listeria monocytogenes* strains by automated laser fluorescence analysis of amplified fragment length polymorphism fingerprint patterns. Int J Food Microbiol 49:95–102

11. Fonnesbech Vogel B, Fussing V, Ojeniyi B, Gram L, Ahrens P (2004) High-resolution genotyping of *Listeria monocytogenes* by fluorescent amplified fragment length polymorphism analysis compared to pulsed-field Gel electrophoresis, random amplified polymorphic DNA analysis, ribotyping, and PCR-restriction fragment length Po. J Food Prot 67:1656–1665

12. Keto-Timonen R, Tolvanen R, Lundén J, Korkeala H (2007) An 8-year surveillance of the diversity and persistence of *Listeria monocytogenes* in a chilled food processing plant analyzed by amplified fragment length polymorphism. J Food Prot 70:1866–1873

13. Lomonaco S, Nucera D, Parisi A, Normanno G, Bottero MT (2011) Comparison of Two AFLP methods and PFGE using strains of *Listeria monocytogenes* isolated from environmental and food samples obtained from piedmont, Italy. Int J Food Microbiol 149:177–182

Chapter 9

High-Throughput Characterization of *Listeria monocytogenes* Using the OmniLog Phenotypic Microarray

Edward M. Fox and Kieran Jordan

Abstract

High-throughput biochemical screening techniques are an important tool in phenotypic analysis of bacteria. New methods, simultaneously measuring many phenotype responses, increase the output of such investigations and allow a more complete overview of the bacterial phenotype, and how this may relate its genotype. This chapter describes the application of OmniLog Phenotype Microarray analysis, a high-throughput assay for the phenotypic characterization of bacterial strains across a variety of different traits, including nutrient utilization and antimicrobial sensitivity, to *Listeria* species.

Key words OmniLog, Phenotypic microarray, Biochemical characterization, Nutrient utilization, Antimicrobial sensitivity

1 Introduction

Phenotypic characterization of bacteria is an important tool with applications across many aspects of microbiology. It is an integral component of species differentiation and underpins the formulation of the growth media used for their isolation and propagation. It is used in tandem with genomic comparisons and manipulation of bacterial genomes as a key component to elucidate gene function. To these ends, high-throughput phenotypic characterization is integral to maximizing the information gained from such investigations. Traditional phenotypic investigations were often limited by the labor and time demanded, however, new high-throughput technologies have advanced these capabilities. Examples of applications of simultaneous phenotypic characterization includes species identification, e.g., the API® system from bioMérieux, Inc., which uses a combination of up to 20 biochemical tests for the identification and/or differentiation of Gram-positive and Gram-negative bacteria.

High-throughput genomic characterization, including the advent of "Next-Generation Sequencing," has paved the way for

Kieran Jordan et al. (eds.), *Listeria monocytogenes: Methods and Protocols*, Methods in Molecular Biology, vol. 1157, DOI 10.1007/978-1-4939-0703-8_9, © Springer Science+Business Media New York 2014

more accessible, in-depth investigation of bacterial genomes [1]. In order to maximize the output potential from these large data sets, advances in complimentary techniques such as phenotypic characterization are key. Biolog's OmniLog platform allows simultaneous characterization of up to 1,920 different growth conditions, covering both nutrient utilization and antimicrobial sensitivities, yielding a greater knowledge of various phenotypic traits of the bacterial strain in question [2].

The principle of the OmniLog phenotype screening is based on measuring cell metabolism using a redox dye. As the cells metabolize, reduction of the dye results in a colorimetric reaction, and this increase in color is imaged by a camera in the OmniLog instrument and quantified using the software. The system uses twenty 96-well plates (called PM 1-20). Plates constitute related substrates or inhibitors, for example PM1 contains different carbon sources, while PM9 contains various osmolytes, for screening different pH and growth conditions. In that way, a selected number of plates can be used for a specific purpose.

This technique can be applied to a variety of functions, such as identification of bacterial species, relating genotype with phenotype, or determining antibiotic sensitivities. OmniLog analysis has been applied to *Listeria* species for identification purposes [3], to investigate phenotype role in persistence [4], and to assess cytotoxicity of *Listeria* species [5].

2 Materials

1. *Listeria monocytogenes* culture, 81 % transmittance (*see* **Note 1**).

2. Spectrophotometer to equalize the OD of cultures.

3. OmniLog plate reader (*see* **Note 2**).

4. PM plates (*see* **Note 3**).

5. Solution A: Tricarballylic acid, pH 7.1. Add 14.088 g tricarballylic acid to 55 mL of water and then add 20 mL of 25 % NaOH. The pH should read 6.5 on a pH meter—adjust with NaOH if necessary. Make up to a final volume of 100 mL with distilled water.

6. Solution B: $MgCl_2 \cdot 6H_2O$. $CaCl_2 \cdot 2H_2O$, as described in Table 1.

7. Solution C: L-Arginine·HCl, L-glutamic acid, as described in Table 1.

8. Solution D: L-Cystine, pH 8.55'-UMP.2Na, pH 8.5, as described in Table 1.

9. Solution E: Yeast extract, as described in Table 1.

10. Solution F: Tween 80 (*see* **Note 4**). Use undiluted.

Table 1
Recipes for each of the stock solutions used in the PM additive solutions, and what volumes of each are combined to make the different PM additive solutions for the different PM plates

Solution	Constituent	Concentration		Formula weight	Grams/100 mL	PM plate				
		1×	40× or 120×			1, 2	3, 6, 7, 8	4	5	9–20
A	Tricarballylic acid, pH 7.1[a]	20 mM	800 mM	171.6	14.088	–	30 mL	30 mL	30 mL	–
B	MgCl$_2$·6H$_2$O	2 mM	240 mM	203.3	4.88	10 mL	10 mL	10 mL	10 mL	10 mL
	CaCl$_2$·2H$_2$O	1 mM	120 mM	147.0	1.76					10 mL
C	L-Arginine·HCl	25 µM	3 mM	210.7	0.063	10 mL	–	10 mL	–	–
	L-Glutamic acid	50 µM	6 mM	169.1	0.101					
D	L-Cystine, pH8.5	12.5 µM	0.5 mM	240.3	0.012	30 mL	30 mL	–	–	–
	5′-UMP.2Na, pH8.5	25 µM	1 mM	368.1	0.037					
E	Yeast extract	0.005 %	0.60 %	–	0.6	10 mL	10 mL	10 mL	–	10 mL
F	Tween 80[b]	0.005 %	0.60 %	–	0.6	10 mL	10 mL	10 mL	–	10 mL
G	D-Glucose	2.5 mM	300 mM	180.2	5.4	–	10 mL	10 mL	10 mL	10 mL
	Pyruvate-Na	5 mM	600 mM	110.0	6.6					
H	Sterile water	–				30 mL	–	20 mL	50 mL	60 mL
	12× PM additive solution—total volume					100 mL	100 mL	100 mL	100 mL	100 mL

[a]Prepare by adding 14.088 g to 55 mL of water and then add 20 mL of 25 % NaOH. The pH should read 6.5 on a pH meter—adjust with NaOH if necessary. Make up to a final volume of 100 mL with water
[b]Some species may perform better with Tween 40

11. Solution G: D-Glucose, as described in Table 1.

12. Solution H: Pyruvate·Na, as described in Table 1.

13. PM Stock solutions IF-0a GN/GP (1.2×), IF-10b GP/GP (1.2×), Dye mix F (100×), PM Additive (12×) (*see* **Note 5**).

3 Methods

3.1 Preparation of Stock Solutions

Table 1 lists the recipes for all the stock solutions used for the various different PM plates. All these solutions should be filter-sterilized and stored at 4 °C.

3.2 Preparation of Cell Suspensions

1. Grow the *Listeria* strain to be tested by streaking for single colonies on a TSAYE agar plate. Incubate for 18 ± 2 h at 37 °C.

2. Subculture a second time, by streaking a single colony on a TSAYE agar plate. Incubate for 18 ± 2 h at 37 °C.

3. From this plate, using a sterile swab, prepare a cell suspension with a turbidity of 81 % transmittance, in 20 mL of fluid IF-0a. Swirl to mix (do not vortex).

3.3 Inoculation and Incubation of PM Plates

For PM3-8 plates, well A-1 (the control well) can sometimes produce a false-positive color result. To eliminate this, add potassium ferricyanide to a final concentration of 0.3 mM in these wells, i.e., the 120× stock solution concentration would be 36 mM (*see* **Note 6**).

1. Table 2 lists the composition of each of the inoculating fluids used for the various different PM plates. Prepare all inoculating fluids needed for the PM plates to be assayed.

2. Inoculate 100 µL of the relevant inoculating fluid into each well of the PM plate.

3. Enter the run details onto the OmniLog software program.

4. Load the OmniLog instrument with each of the PM plates, ensuring each plate has been loaded in the correct drawer.

5. Incubate the plates for 48 h at 37 °C (*see* **Note 7**).

6. Data are analyzed using the OmniLog analysis software (*see* **Note 8**). Figure 1 shows a typical result output following comparison of growth of two *L. monocytogenes* strains.

4 Notes

1. To obtain a culture of 81 % transmittance, use a sterile swab to transfer cells from an agar plate into 20 mL of 1× IF-0a. Ensure the cell suspension is uniform by stirring (do not vortex—use gentle mixing). Adjust as necessary until the transmittance reads 81 % on a turbidimeter.

Table 2
Recipes for the PM inoculating fluids

PM stock solution	PM plate (mL)				
	1, 2	3, 6, 7, 8	4	5	9–20
IF-0a GN/GP (1.2×)	20	40	10	10	–
IF-10b GP/GP (1.2×)	–	–	–	–	110
Dye mix F (100×)	0.24	0.48	0.12	0.12	1.32
PM additive (12×)	2	4	1	1	11
Cells (81 % Turbidity)	1.76	3.52	0.88	0.88	9.68
Inoculating fluid—total volume	24	48	12	12	132

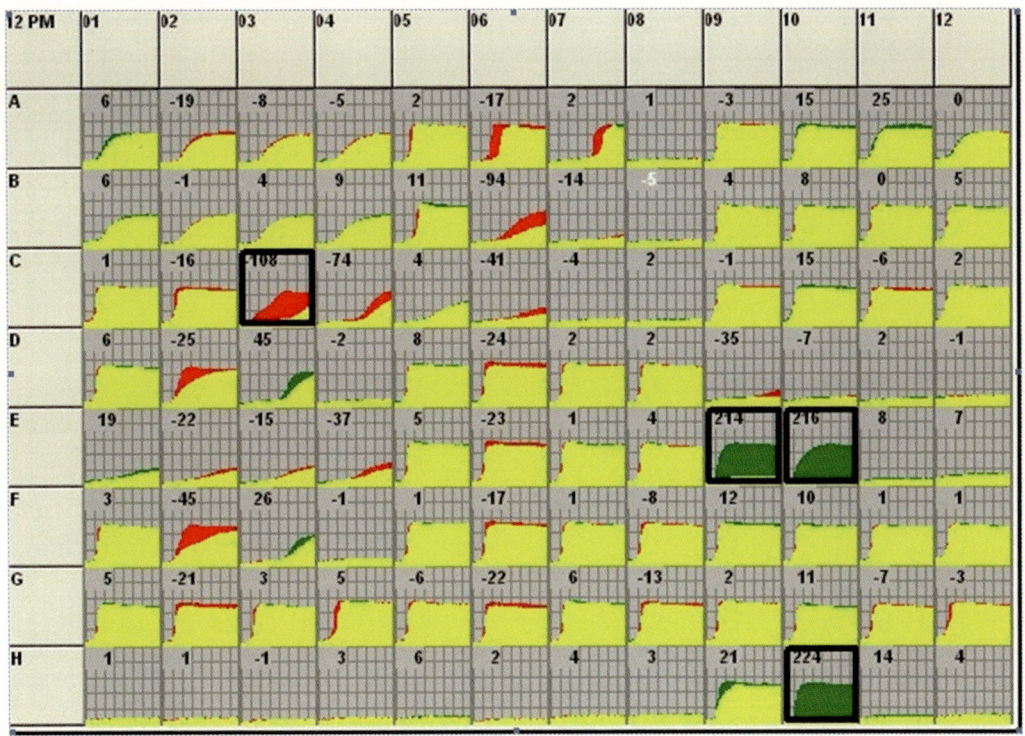

Fig. 1 This figure is a comparison of results obtained from two strains of *L. monocytogenes* screened on PM plate 12. Each well of the 96-well PM plate is represented (A01 through H12). Each of the wells displays growth/metabolism of the two strains: strain A (the reference strain) is represented by a *red growth curve*, and strain B (the test strain) is represented by a *green growth curve*. Where both growth curves overlap, a *yellow* color is displayed. The numerical value displayed in each well represents the difference in growth of the test strain, when compared with that of the reference strain, in OmniLog units (a unit of measurement used by the software to express growth/metabolism). Four wells have been highlighted with a *black border*: well C03, where the test strain showed decreased growth rate when compared with the reference (as indicated by the *red growth curve*); and wells E09, E10, and H10, where the test strain showed increased growth when compared with the reference strain (as indicated by the *green growth curve*)

2. The OmniLog plate reader is manufactured by Biolog (Hayward, CA, USA).

3. The PM plate system contains plates designed for assaying a variety of phenotypes, from nutrient source metabolism to antimicrobial screening. The suite of PM plates can be selected based on the requirements of the investigation.

4. Some species may perform better with Tween 40.

5. See the Biolog website for a listing of local distributors.

6. A concentration of 0.3 mM in the control wells was successful for the strains we have tested (as per ref. 2); however, this may need to be optimized for other *Listeria* strains if a false-positive result is resulting. Reducing the concentration of the glucose and pyruvate in Solution G by half may also help avoid a false-positive result.

7. This time may be tailored to the specific needs of the investigation (e.g., 24 h could be used).

8. We have used the OmniLog platform for screening other compounds using 96-well plate format assays. In general, concentrations of the compound to be tested (e.g., an antimicrobial or other nutrient source). We do this by weighing-out the appropriate amount of the compound to be screened, dissolving it in the relevant inoculating fluid, and adding it to well(s) of a sterile 96-well plate as described above. This plate is then run as normal on the OmniLog platform, and data interpreted as described above.

References

1. den Bakker HC, Cummings CA, Ferreira V, Vatta P, Orsi RH, Degoricija L, Barker M, Petrauskene O, Furtado MR, Wiedmann M (2011) Comparative genomics of the bacterial genus *Listeria*: genome evolution is characterized by limited gene acquisition and limited gene loss. BMC Genomics 11:688

2. Bochner BR (2009) Global phenotypic characterization of bacteria. FEMS Microbiol Rev 33:191–205

3. den Bakker HC, Manuel CS, Fortes ED, Wiedmann M, Nightingale KK (2013) Genome sequencing identifies *Listeria fleischmannii* subsp. Coloradensis subsp. nov., a novel *Listeria fleischmannii* subspecies isolated from a ranch in Colorado. Int J Syst Evol Microbiol 63:3257–3268. doi:10.1099/ijs.0.048587-0

4. Fox EM, Leonard N, Jordan K (2011) Physiological and transcriptional characterization of persistent and nonpersistent *Listeria monocytogenes* isolates. Appl Environ Microbiol 77:6559–6569

5. Lei XH, Bochner BR (2013) Using phenotype MicroArrays to determine culture conditions that induce or repress toxin production by *Clostridium difficile* and other microorganisms. PLoS One 8:e56545

Chapter 10

Analysis of *Listeria monocytogenes* Subproteomes

Michel Hébraud

Abstract

The proteomic approaches have considerably evolved over the past two decades. This opened the doors for larger scale and deeper explorations of cellular physiology. Like for other living organisms, using the tools of proteomics has undoubtedly improved knowledge about the foodborne pathogen *Listeria monocytogenes*. Among the different technologies and approaches permanently evolving in the field of proteomics, the 2-DE is an analytical separation method of choice to resolve thousands of proteins simultaneously in a single gel, allowing their quantification, the study of their posttranslational modifications and the understanding of their biological function. In this, 2-DE remains a perfectly complementary technique to the new high-throughput techniques such as shotgun proteomics approaches. Moreover, in order to gain in analysis depth and improve knowledge about the target of action and the function of proteins in relation to their subcellular location, it is necessary to explore more specifically the different subcellular proteomes. Thus, the subproteomic analyses became essential and dramatically increased these last years, particularly on proteins secreted into the extracellular milieu, named exoproteome, or on cell envelope proteins (cell wall and membrane proteins) which are involved in the interactions with the surrounding environment. Here, the extraction and separation of *L. monocytogenes* subproteomes are described based on cell fractionation and 2-DE techniques. This chapter gives a workflow to obtain the exoproteome, the intracellular proteome, the cell wall, and membrane proteomes of the Gram-positive bacterium *L. monocytogenes*. The different steps of 2-DE technology, composed of a first dimension based on the separation of proteins according to their charge, an equilibration step, then a second dimension based on the separation of proteins according to their mass, and finally the staining of proteins in the gel are detailed. Emerging technologies to extract the exoproteome or the cell surface proteome after enzymatic shaving and to analyze them by shotgun method are also discussed briefly.

Key words Subproteome extractions, Two-dimensional gel electrophoresis (2-DE), Isoelectric focusing (IEF), Sodium dodecyl sulfate-polyacrylamide gel electrophoresis (SDS-PAGE), Protein staining

1 Introduction

Proteomic approaches are composed of powerful and advanced technologies for large-scale studies of proteins in living organisms. The proteomic research domain provides global and accurate information on the abundance, posttranslational modifications, function and interactions of proteins and an integrated view of organism processes, and their regulatory networks.

Kieran Jordan et al. (eds.), *Listeria monocytogenes: Methods and Protocols*, Methods in Molecular Biology, vol. 1157,
DOI 10.1007/978-1-4939-0703-8_10, © Springer Science+Business Media New York 2014

In this field, the two-dimensional gel electrophoresis (2-DE) technique constitutes the cornerstone of proteomics. In the mid-1970s, the idea of using two properties of proteins, their charge and their molecular mass, to combine two in-gel successive separation techniques, isoelectric focusing (IEF) and SDS-PAGE, gave birth to the beginnings of proteomics. Although 2-DE has long been described as a "comtemplative" method, due to the difficulty in characterizing which protein was present in each spot, it was the most suitable technology to separate and view thousands of proteins simultaneously. To date, there have been many advances in 2-DE, both in the extraction and solubilization of proteins and in techniques and materials used (immobilized pH gradient, new apparatus for IEF (first dimension), and SDS-PAGE (second dimension)), which led to improved protein separation and resolution as well as reproducibility and implementation of the techniques. Parallel to these developments, other advances led to a considerable impetus in proteomic analyses, i.e., (1) the availability of ever growing amounts of genomic sequence data and (2) important advances in mass spectrometry technology for ionization and detection of large molecules such as peptides and proteins. Indeed, data obtained with mass spectrometry analysis, namely peptide mass fingerprinting or fragmentation, can be matched against databases of all known gene products and thus greatly facilitate protein identification. Thus, 2-DE has become the classical method of choice to separate, quantify, identify, and compare complex cellular protein extracts, which is particularly well suited for protein profiling studies in all domains of life science.

At the level of microorganisms, particularly bacteria, the almost exponential growth of genomic databases in recent years facilitates global or targeted studies on a growing number of species, including exploration of molecular mechanisms, study of the regulatory systems and the interactants which intervene in cells with respect to their surrounding environment, and investigation of physiological state of cells under given conditions. Considering that a pathogen such as *Listeria monocytogenes* is able to adapt and live in natural environments even sometimes hostile for other microorganisms, to form biofilms on biotic and abiotic surfaces, to internalize and reproduce inside the host's cells causing listeriosis, it is particularly crucial to investigate the different cellular compartments and analyze their specific content in proteins. This enables gaining of knowledge in a dynamic range of analyzed proteins and targeting of proteins in their functional compartment. Therefore, strategies to extract and analyze subproteomes have developed considerably in recent years to explore the different subcellular proteomes which in Gram-positive bacteria include (1) intracellular or cytoplasmic proteins, (2) membrane associated proteins, (3) cell-wall-associated proteins, and (4) proteins secreted into the extracellular milieu. However, if the extraction and separation of intracellular and extracellular proteins can be achieved easily and efficiently by classical 2-DE procedures, it remains much more challenging for cell surface subproteomes due to the intrinsic properties of cell envelope associated proteins. Indeed, multitransmembrane proteins are generally highly hydrophobic and are almost impossible to solubilize and, on the other hand, proteins noncovalently or *a fortiori* covalently attached to the cell wall peptidoglycan are very difficult to extract and require specific and laborious treatments not always directly compatible with classical 2-DE separation. So, different strategies have been developed to tackle the difficulties in analysis of these cell envelope subproteomes. These strategies can associate different protocols of protein extraction with different techniques of separation and mass spectrometry [1]. Several studies have attempted to extract membrane associated proteins of Gram-positive bacteria by combining protocols described for Gram-negative or eukaryotes organisms [2]. Thus, the extraction procedures could include enzymatic treatment, fractionation of broken cells by centrifugation, use of chemical agents such as zwitterionic detergents for solubilization of hydrophobic proteins [3], solvents for delipidation [4], or protein extraction and separation [5]. The characterization of the cell wall proteome was performed by extracting proteins with the sequential action of two salts at high concentration [6] or by combining mechanical breakage with fractionation by centrifugation and treatment with high concentrations of SDS at high temperature [7, 8]. An alternative approach consists of "shaving" the bacterial surface with a specific protease (such as trypsin) to cleave surface-exposed proteins and to separate and analyze the peptide hydrolyzate by 2-D liquid chromatography coupled to tandem mass spectrometry (2-D LC MS/MS), also named "shotgun proteomics." This alternative approach is described in the following methodologies. It overcomes some limitations of gel-based 2-DE by identifying some proteins with a high number of transmembrane spanning regions or LPXTG motif which anchors covalently to the cell wall. These very attractive emergent technologies are continuously evolving, they provide a significant gain of time, allow the analysis of a larger dynamic range and the identification of supplementary proteins by the use of treatments that are not always compatible with classical 2-DE. However, contrary to 2-DE, shotgun proteomics does not allow a comprehensive view of a subproteome, separating thousands of complete proteins at once with high resolution and visualizing shifts due to posttranslational modifications. In addition, it uses high-tech and very expensive technologies (nano-LC, mass spectrometers) which are not accessible to all laboratories and microbiologists, but generally located on technological platforms and used by specialist engineers. Thus, 2-DE is an invaluable and irreplaceable technique for protein profiling. The combination of two or more complementary proteomic strategies would be one of the keys to generate valuable information on *L. monocytogenes* physiology, the role of cell envelope and extracellularly secreted proteins in pathogenic processes, bacterial communication, sensing of/exchange with and adaptation to its environment, motility, adhesion on, and colonization of biotic or abiotic surfaces. The methodologies described hereafter for subproteome extraction and separation can be easily and rapidly implemented in research laboratories to give answers or clarify the many questions that may arise in the minds of microbiologists.

2   Materials

All solutions are prepared using ultrapure water (obtained by purifying deionized water to attain 18 MΩ cm at 25 °C) and analytical grade reagents specifically dedicated to proteomics and stored apart from other laboratory chemicals at room temperature (unless indicated otherwise). Wear protective laboratory clothing such as powder-free gloves, closed overalls, and hair nets without cotton (for example in polypropylene) to protect yourself and to prevent protein contamination, particularly from skin and hair keratins.

2.1   Sample Preparation

1. TE buffer: 20 mM Tris–HCl, pH 7.5, 5 mM EDTA, 5 mM MgCl$_2$. Weigh 242 mg Tris, 186 mg EDTA, and 48 mg MgCl$_2$ and transfer in a graduated glass beaker containing 90 mL of water. Mix under magnetic stirring and adjust the pH with HCl. Remove the magnetic stir bar and make up to 100 mL. Store at 4 °C. Prepare the same TE buffer but adjusting the final pH to 9.0.

2. PMSF (phenylmethylsulfonyl fluoride) solution: 20 mM in ethanol 100 %. Weigh 34.84 mg PMSF and dissolve in 10 mL pure ethanol (*see* **Note 1**). Store in the dark at –20 °C.

3. Sodium deoxycholate solution: 2 % solution in water. Weigh 200 mg NA deoxycholate and dissolve in 10 mL of water (*see* **Note 2**).

4. Trichloroacetic acid (TCA) solution: 50 % TCA in water. Weigh 50 g TCA and dissolve in 80 mL of water by magnetic stirring. Remove the magnetic stir bar and make up to 100 mL with water (*see* **Note 3**).

5. DNase I and RNase A solutions in sterile water. Dissolve DNase I and RNase A powders in the required volumes of sterile water (variable according to the packaging and the suppliers) to obtain final solutions of 5 U/μL and 7.5 U/μL, respectively. Store at –20 °C.

6. Extraction buffer: 8 M urea, 4 M thiourea, 4 % CHAPS, and 4 mM TBP (*see* **Note 4**). For a final volume of 100 mL, weigh

48.05 g urea and transfer to a graduated glass beaker. Slowly add water by magnetic stirring with gentle heating (on a hotplate or in an oven at temperature < 30°C to avoid carbamylation) to completely dissolve the urea. Add 30.45 g thiourea and proceed in the same way to completely dissolve it before adding 4 g CHAPS and 2 mL TBP. Remove the magnetic stir bar and make up to 100 mL with water. Finish homogenizing the extraction buffer on a magnetic stirrer. Aliquot in 2 mL microtubes and store at 4 °C. Before use, remove the required amount from the refrigerator, and heat the solution (<30 °C) to redissolve the precipitate.

7. Tris buffer: Tris 40 mM, pH 8.5. Weigh 482 mg Tris, transfer in a graduated glass beaker containing 95 mL of water. Mix under magnetic stirring and adjust the pH with HCl. Remove the magnetic stir bar and make up to 100 mL. Store at 4 °C.

8. Delipidation solution: Tri-*n*-butylphosphate/acetone/methanol solution (1/12/1). Mix 7 mL of tri-*n*-butylphosphate, 84 mL acetone, and 7 mL methanol and store at 4 °C (*see* **Note 5**).

2.2 Isoelectric Focusing (IEF)

1. Sample solubilization buffer (=IEF buffer): 5 M urea, 2 M thiourea, 2 % CHAPS, 10 mM Tris–HCl, and traces of bromophenol blue (BPB) in 50 % trifluoroethanol (TFE) (*see* **Note 6**). For a final volume of 100 mL, weigh 30.03 g urea, transfer to a graduated glass beaker containing 80 mL of a 50 % TFE solution and dissolve by magnetic stirring with gentle heating (<30 °C) as indicated above for the extraction buffer. Add 15.22 g thiourea and let dissolve before adding 2 g CHAPS, a few grains of BPB and 10 mL of a 100 mM tris solution in 50 % trifluoroethanol. Remove the magnetic stir bar and make up to 100 mL with the 50 % TFE solution. Finish homogenizing the IEF buffer on a magnetic stirrer. Aliquot in 2 mL microtubes the IEF buffer and store at –20 °C. Before use, thaw the required amount and heat the solution (<30 °C) to redissolve the precipitate.

2. Rehydration buffer: add TBP (or DTT, final concentration 50 mM) and carrier ampholytes to IEF buffer just before use at final concentrations of 2 mM and 0.3 %, respectively. The pH range of carrier ampholytes must be consistent with the IPG strips used (*see* **Note 7**).

2.3 IPG Strip Equilibration

1. Buffer for the basic equilibration solution and for the resolving gel of the SDS polyacrylamide gel electrophoresis (SDS-PAGE): 1.5 M Tris–HCl, pH 8.8. Pour 600 mL of water at room temperature in a graduated glass beaker. Weigh 181.7 g Tris, transfer to the glass beaker and place it on a magnetic stirrer. Add water to a volume of 900 mL and mix with a magnetic stir bar. Adjust the pH with HCl to reach pH 8.8 (use a high to

low concentration of HCl from the starting pH to the required pH). Remove the magnetic stir bar, make up to 1 L with water and control the final pH. Store at 4 °C.

2. Basic solution for equilibration buffer I and II: prepare a 120 mL solution for six 18 cm IPG strips equilibration. Weigh 43.2 g urea (6 M final) and 2.4 g SDS (2 % final) (*see* **Note 8**), transfer to a graduated glass beaker. Add 4 mL of 1.5 M Tris–HCl, pH 8.8 buffer, 36 mL of glycerol, and about 50 mL of water (to reach the final volume). Dissolve the components of the solution by magnetic stirring with gentle heating (<30 °C). The final basic equilibration solution should be composed of 6 M urea, 2 % SDS, 50 mM Tris–HCl, pH 8.8 and 30 % glycerol.

3. Equilibration buffer I: Dissolve 1.2 g DTT (dithiothreitol) or add 2 mM TBP to 60 mL of basic equilibration solution for a final concentration of 2 % DTT.

4. Equilibration buffer II: Dissolve 1.5 g iodoacetamide to 60 mL of basic equilibration solution for a final concentration of 2.5 % iodoacetamide (*see* **Note 9**) and add 0.01 % of BPB.

2.4 SDS-PolyAcrylamide Gel Electrophoresis (SDS-PAGE)

1. Resolving gel buffer: 1.5 M Tris–HCl, pH 8.8. *See* point 1 in paragraph 2.3.

2. Acrylamide/N,N'-methylene-bis-acrylamide   (crosslinker) solution. Use ready-to-use 40 % Acrylamide/Bis solution (37.5:1; 2.6 % C, Bio-Rad, Hercules, CA, USA) and stored at 4 °C in the dark.

3. SDS solution: 10 % solution in water (*see* **Note 8**). Dissolve 10 g SDS in 60 mL water with gentle stirring to minimize the bubbles made by SDS. Complete with water to 100 mL.

4. Ammonium persulfate: 10 % solution in water (*see* **Note 10**). Weigh 200 mg ammonium persulfate and dissolve in a final volume of 2 mL of water.

5. TEMED (N,N,N',N'-tetramethylethylenediamine). Ready-to-use solution, store at 4 °C in the dark (*see* **Note 11**).

6. Preparation of a 12.5 % acrylamide gel. For a final volume of 100 mL (necessary for two 190 mm (length)×185 mm (width)×1 mm (thickness) gels), mix 25 mL of resolving gel buffer, 31.25 mL of 40 % acrylamide/Bis solution, and 42.15 mL of water. Degas the solution under vacuum for 10 min. Add 1 mL of 10 % SDS solution, 0.6 mL of 10 % ammonium persulfate solution, and 60 μL of TEMED. Let the final solution well homogenize for 1 min by moderate stirring before loading the gel.

7. Agarose solution: 1 % solution. Weigh 100 mg of low melting agarose and dissolve in 10 mL SDS-PAGE running buffer by

magnetic stirring with gentle heating (≤50 °C). Aliquot in microtubes and store at 4 °C. Melt the quantity required each time.

8. SDS-PAGE running buffer: 25 mM Tris–HCl, pH 8.3, 0.192 M glycine, 0.1 % SDS. Prepare 10× stock buffer (0.25 M Tris–HCl, 1.92 M glycine, 1 % SDS). Weigh 30.3 g Tris and 144 g glycine, dissolve by magnetic stirring in a graduated glass beaker containing 900 mL of water. Add 10 mL of 10 % SDS solution and complete to 1 L with water. Do not adjust the pH with acid or base. Store at room temperature. Before use, dilute 100 mL of 10× stock buffer to 900 mL of water.

2.5 Protein Staining with Brilliant Blue G

1. Fixing solution: 30 % ethanol, 2 % (v/v) phosphoric acid in water. An 85 % concentrated phosphoric acid is used. For 1 L fixing solution, add 20 mL of 85 % phosphoric acid.

2. Equilibration solution: 18 % ethanol, 2 % (v/v) phosphoric acid, and 15 % ammonium persulfate in water. For 1 L of solution, place 500 mL of water in a flask with magnetic stirring. Add 20 mL of 85 % phosphoric acid, then 150 g of ammonium sulfate. Allow to dissolve, transfer into a graduated cylinder, and adjust to 800 mL with water. Add 20 mL additional water, retransfer into the flask with stirring, and add 180 mL ethanol while stirring.

3. Brilliant Blue G (or colloidal Coomassie Blue G-250) 2 % solution: dissolve 2 g of pure Brilliant Blue in 100 mL of hot water with stirring. Wait about 30 min for complete dissolution.

3 Methods

Carry out all procedures at room temperature unless otherwise specified. Wear protective laboratory clothing as specified above.

3.1 Sample Preparation

The growth conditions (inoculum, culture medium, temperature, pH, agitation, etc.), the population and the growth phase of the bacterial culture at the time of its recovery must be perfectly controlled and mastered in order to reproduce the same experience for biological replicates. This is the same for all the extraction and protein separation methodologies which must be optimized and standardized to carry out technical replicates from the protein extracts. On average, it is necessary to perform three biological replicates with 2–4 technical replicates for each protein extract to ensure good reproducibility and repeatability of results [9]. Hereafter are described the methodologies to obtain different *L. monocytogenes* subproteomes as summarized in Fig. 1.

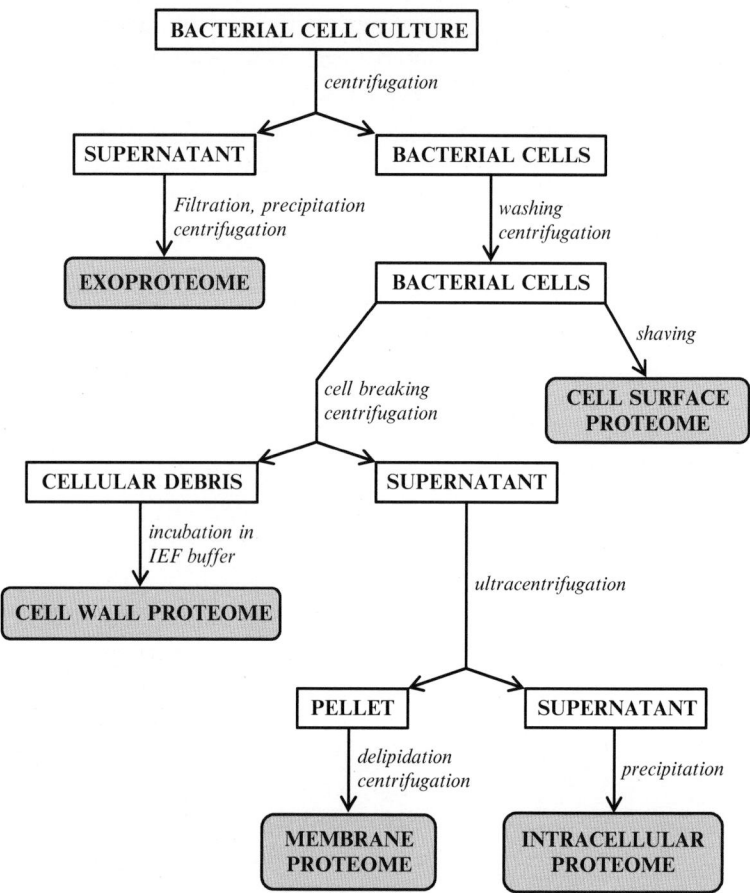

Fig. 1 Workflow for subproteomic analysis of *L. monocytogenes*

1. The first step, whatever the subproteome of interest, consists in separating the bacterial cells from the culture supernatant. The volume of culture treated is dependent on the biomass yield, the efficiency of protein extraction and the final quantity of proteins needed for the separation procedures. The procedures described hereafter have been used successfully in our laboratory in various studies on different strains of *L. monocytogenes* [10–12].

2. Harvest bacterial cells (usually 250 mL cultures) by centrifugation at $7,500 \times g$ for 15 min in a refrigerated centrifuge (at 4 °C).

3. Carefully remove the supernatant and filter through a 0.2 μm membrane either with syringe or bottle-top filter according to the volume of culture recovered (*see* **Note 12**). Maintain the supernatant in a cold environment with an ice bath.

4. Add 0.2 mM PMSF and 0.2 mg/mL Na deoxycholate. For 100 mL supernatant, add 1 mL of 20 mM PMSF solution and 1 mL of 2 % Na deoxycholate solution. Incubate for 30 min on ice.

5. Precipitate proteins by adding 10 % TCA and leave to incubate overnight at 4 °C. For 100 mL supernatant, gradually add 25 mL of 50 % TCA solution with manual agitation (*see* **Note 13**).

6. Centrifuge at 20,300×g for 30 min at 4 °C. Eliminate the supernatant and wash the pellet with ice-cold acetone (*see* **Note 14**). Repeat the washing step with ice-cold acetone at least twice, breaking up the pellet each time.

7. Aspirate the final supernatant and remove residual acetone by air drying or under a slow stream of nitrogen. Do not bring the pellet to complete dryness as this will make it difficult to bring proteins back into solution.

8. Solubilize the exoproteome protein sample into the IEF buffer (*see* **Note 15**).

9. Quantify protein concentration with a Bradford protein assay (Bio-Rad protein assay) and bovine serum albumin as the standard.

3.1.2 Intracellular Proteome

1. Recover the bacterial cell pellet from the culture centrifugation. Wash the pellet twice with 1 mL TE buffer at pH 7.5 followed by centrifugation (7,500×g, 15 min, 4 °C).

2. Resuspend the bacterial cell pellet in 500 µL to 1 mL TE buffer at pH 9.0.

3. Break bacterial cells with a cell disrupter (One shot cell disrupter, 1–8 mL, 2.7 kBar max, Constant Systems Ltd, Daventry, UK) by applying 2.5 kBar pressure.

4. Add 1 % DNase I solution and 1 % RNase A solution and incubate at room temperature for 30 min.

5. Add an equivalent volume of extraction buffer for a final concentration of 4 M urea, 2 M thiourea, 2 % CHAPS, and 2 mM TBP. Incubate for 30 min on ice with intermittent manual agitation.

6. Centrifuge the sample at 4,500×g for 10 min at 4 °C. Recover the supernatant and keep the pellet containing cell debris.

7. Ultracentrifuge the supernatant at 200,000×g for 30 min at 4 °C. Collect the supernatant corresponding to the intracellular proteome and keep the pellet containing cellular membranes.

8. Quantify protein concentration in the supernatant with a Bradford protein assay and bovine serum albumin as the standard.

9. Precipitate intracellular proteins with at least three volumes of ice-cold acetone and leave the sample for at least 2 h (up to overnight if possible) at −20 °C.

10. Pellet the proteins by centrifugation at $13,000 \times g$ for 40 min at 4 °C. Remove acetone and air-dry the pellet.

11. Dissolve the sample of intracellular proteins in the IEF buffer to obtain a final concentration of 5 μg/μL proteins and store at –20 °C.

3.1.3 Cell Envelope Subproteomes

1. The pellet containing cell debris (*see* Subheading 3.1.2, **step 6**) is washed three times with glacial absolute ethanol followed by centrifugation ($13,000 \times g$, 15 min, 4 °C).

2. Remove the supernatant and air-dry the pellet.

3. Resuspend the bacterial cell debris in a small volume of IEF buffer containing 7 M urea instead of 5 M and 2 % ASB14 instead of 2 % CHAPS (*see* **Note 16**) and keep for 1 h at ambient temperature, vortexing regularly.

4. Centrifuge at $13,000 \times g$ for 15 min before quantifying protein concentration with a Bradford protein assay and bovine serum albumin as the standard. Store this fraction enriched in cell wall proteins at –20 °C.

5. Recover the pellet containing cellular membranes from ultracentrifugation (*see* Subheading 3.1.2, **step 7**), wash twice with 40 mM Tris buffer pH 8.5 followed by ultracentrifugation ($200,000 \times g$, 30 min, 4 °C) and resuspend in 500 μL of Tris buffer.

6. Add 7 mL of ice-cold delipidation solution to the 500 μL of sample for a final acetone concentration of 80 % [4]. Incubate for 90 min at 4 °C, vortexing every 10 min.

7. Centrifuge the precipitate at $7,500 \times g$ for 15 min at 4 °C, wash twice with 1 mL delipidation solution and then air-dry.

8. Dissolve the delipidated membrane protein fraction in a small volume of IEF buffer containing 7 M urea and 2 % ASB14 (*see* **Note 16**) and keep for 1 h at ambient temperature, vortexing regularly.

9. Quantify protein concentration with a Bradford protein assay and bovine serum albumin as the standard. Store this fraction enriched in membrane proteins at –20 °C.

3.1.4 Complementary Method for Cell Surface Subproteome Analysis

Due to the difficulty in extracting and solubilizing cell-envelope proteins, alternative methods have been developed over the last years to explore and validate the presence of cell envelope proteins such as hydrophobic transmembrane or cell wall covalently linked proteins, generally inaccessible by classical 2-DE workflow, through the identification of at least one of their peptides. One of these alternative methods, named "shaving" of cell surface proteins (*see* Olaya-Abril et al. for review, [13]) applied on *L. monocytogenes* in our laboratory and on other Gram-positive bacteria such as *Bacillus subtilis* [14], *Staphylococcus aureus* [15, 16], *Enterococcus faecalis* [17], or *Streptococcus sp.* [18, 19], is briefly described hereafter.

1. Harvest bacterial cells from the culture medium by low speed centrifugation ($1,000 \times g$, 15 min, 4 °C) to minimize cell lysis. Gently wash the bacterial cell pellet twice with 2 mL of an ice-cold washing buffer containing 20 mM Tris–HCl, pH 7.6, 150 mM NaCl.

2. Resuspend the cell pellet in 1 mL of shaving buffer containing 20 mM Tris–HCl, pH 7.6, 150 mM NaCl, 10 mM $CaCl_2$, 1 M L-arabinose.

3. Add the enzyme for cell surface shaving, generally trypsin (1–10 μg/mL) or proteinase K (1–8 μg/mL) and incubate with gentle shaking at 37 °C for 15 min to 2 h (*see* **Note 17**).

4. Centrifuge bacterial cells ($1,000 \times g$, 15 min, 4 °C) and recover the supernatant containing the cleaved peptides. Filter the supernatant through a 0.2 μm membrane.

5. Add 0.6 μg/mL trypsin and incubate overnight at 37 °C to complete trypsic digestion of the peptides.

6. Purify and concentrate the extracted peptides on a Sep-Pak C18 Plus Light cartridge (Waters). Elute peptides with 500 μL of a 80 % acetonitrile (ACN) solution.

7. Evaporate the sample with a vacuum concentrator and add 20 μL solvent containing H_2O/ACN (98/2), 0.06 % trifluoroacetic acid.

8. Separate the peptides obtained from the cell wall shaving by nano-liquid chromatography on a PepMapC18 column (75 μm diameter, 150 mm length, 100 Å porosity) with a 4–70 % gradient of solvent containing ACN/H_2O (95/5), 0.5 % formic acid for 30 min at a flow rate of 300 nL/min. The eluted peptides are analyzed by direct coupling of the nano-LC with an electrospray ionization-ion trap-tandem mass spectrometer (ESI-IT-MS/MS) as for the exoproteome analysis (*see* **Note 15**).

3.2 Isoelectric Focusing (IEF)

1. Add an appropriate volume of rehydration buffer to the protein sample to reach the total volume necessary for Immobilized pH Gradients (IPG) rehydration according to the strip length used (*see* **Note 18**).

2. Remove the necessary number of IPG strips from –20 °C freezer and allow them to equilibrate at room temperature for a few minutes.

3. Place the strip holder on a flat surface. Distribute the appropriate volume of the protein sample–rehydration buffer mixture evenly in a channel of strip holder, either into the reswelling/equilibration tray (passive rehydration) or into the focusing tray (active rehydration) (*see* **Note 19**). Record the channel number with the sample name.

4. Remove the plastic protective cover from the IPG strip starting at the positive end. Place the IPG strip in the reswelling holder with the gel side down, starting at the anode (positive end), and laying the strip down to the cathode (negative end). Gently move the IPG strip back and forth in order to spread out the rehydration mixture and remove air bubbles underneath.

5. Completely cover each IPG strip by slowly adding mineral oil from one end and moving the pipette along the length of the IPG strip until the opposite side (*see* **Note 20**). The amount is dependent on the strip holder used.

6. Cover the strip holder with the lid and allow the IPG strip to completely absorb the sample during overnight incubation (at least 12 h).

7. After rehydration is complete, remove the lid from the strip holder.

8. When using a reswelling tray, transfer the IPG strips into the focusing tray. Wet (but not soak) two paper wicks per IPG strip with water (about 10 μL per wick) and place them at each end of the focusing tray to cover the electrode wire. By using the sloped end of the channel, carefully lift the end of the IPG strip with forceps, then hold it vertically for about 10 s and blot the tip of the IPG strip on a filter paper to allow the mineral oil to drain. Transfer the IPG strips to the corresponding channels into the focusing tray maintaining the gel side down and the correct polarity.

9. When using the focusing tray directly, insert wicks in contact with the electrodes. Wet (but not soak) two paper wicks per IPG strip with water (about 10 μL per wick). Carefully lift the end of the strip with forceps and insert the wet paper wick between the IPG strip and the electrode. Operate the same at the other end of the strip.

10. Overlay completely each IPG strip with fresh mineral oil and ensure that there is no air bubbles trapped beneath. Cover the focusing tray with the lid, transfer it in the IEF unit, and close the cover of the apparatus.

11. For the running conditions, follow the guidelines recommended by the IEF unit suppliers. Generally, the current should not exceed 50 μA per IPG strip (allow to minimize protein carbamylation reactions in urea sample buffer) and the temperature should be set at 20 °C. The total volt-hours delivered at each IPG strip is dependent on its length and pH range and the final voltage that can be applied is much higher if the strip is long (from 4,000 V for 7 cm IPG strips to 10,000 V for 17–24 cm IPG strips). Different focusing conditions can be chosen with linear rapid or slow voltage ramping mode or stepwise voltage programming mode with continuous voltage

changes. Anyway, it is important to start at low voltage for a few hours (50–200 V) to allow ionic constituents moving at the extremity of the IPG strip before protein focusing occurs (*see* **Note 21**).

12. When the migration is completed, remove the IPG strips from the focusing tray with forceps. Wash them rapidly with water and hold them vertically for about 10 s onto a piece of filter paper to allow mineral oil and water to drain. Transfer the IPG strips with the gel side up into a clean reswelling/equilibration tray in accordance with the channel number for each protein sample. Cover the tray and move onto the equilibration step.

If the equilibration step is not performed immediately, recover the IPG strips without draining mineral oil and place them in screw cap tubes or plastic bags for storing at –20 °C. Note the sample reference or number on each package.

3.3 IPG Strip Equilibration

1. When the IPG strips are stored at –20 °C, remove them from the freezer and let them thaw on the lab bench at ambient temperature for 10–15 min. Wash the IPG strips rapidly with water and drain them as previously described. The most convenient way to equilibrate several IPG strips at once is to use a disposable rehydration/equilibration tray.

2. Add the equilibration buffer I with DTT (or TBP) to each channel containing an IPG strip in order to overwhelm them. The volume needed is dependent on the length of the IPG strips and the tray used (from 3 mL for 7 cm IPG strips to 10 mL for 24 cm IPG strips). Incubate for 15 min on an agitation table under low speed.

3. Discard the used equilibration buffer I by carefully pouring the liquid until the tray is vertical. Dab the end of the channels with filter paper to remove the last drops.

4. As above in **step 2**, add the equilibration buffer II with iodoacetamide and BPB, incubate for 15 min more.

5. Discard the used equilibration buffer II as described in **step 3** and proceed to the next step in which the first two following points have been prepared before.

3.4 SDS-PolyAcrylamide Gel Electrophoresis (SDS-PAGE)

1. Load the 12.5 % acrylamide gel solution within assembled glass plates placed in a casting chamber. Leave space at the top for IPG strip and running buffer (about 1 cm).

2. Overlay very carefully the top of the gel with a solution containing 20 % ethanol and traces of BPB (*see* **Note 22**). After 1 h polymerization, replace the ethanol solution with 0.25× SDS-PAGE running buffer. Seal with stretch film to avoid evaporation and allow polymerization to continue overnight at low temperature (10–12 °C).

3. Remove the overlay solution from top of the gel and wash with 1× SDS-PAGE running buffer. Completely remove the buffer by inserting a filter paper between the two glass plates inclined to the side.

4. Apply the equilibrated IPG strips to the top of the second-dimension gel between the two glass plates making sure that the gel is lying in contact with the gel (*see* **Note 23**).

5. Overlay the IPG strips with 1 % molten agarose solution and eliminate air bubbles with a syringe needle if necessary. Wait a few minutes to allow agarose solidify (*see* **Note 24**).

6. Add 1× SDS-PAGE running buffer to the lower chamber of the electrophoresis apparatus then move the gels inside and add running buffer in the upper chamber to cover the top of the gels. Place the lid onto the electrophoresis apparatus to secure the device and allow electric current application.

7. Run the gels at 15 mA constant per gel (*see* **Note 25**) until BPB dye enters between 1 and 2 cm into the resolving gel (about 1 h) and then continue at 40 mA constant per gel until the BPB bands reach the bottom of the gel (4–6 h). In this condition, the voltage is not limiting but generally requires 100–400 V (*see* **Note 26**).

3.5 Protein Staining with Brilliant Blue G

1. After electrophoresis, proteins are stained according to procedure described by Neuhoff et al. [20, 21]. Perform all steps with gentle agitation on an orbital or back and forth shaker. Fix the gels with three successive baths of 30 min each in the fixing solution freshly prepared (about 300 mL per 190 mm × 185 mm × 1 mm gel).

2. Rinse 3 × 20 min in 2 % phosphoric acid in water.

3. Incubate the gels for 30 min in the equilibration solution, freshly prepared.

4. Add 1 % (v/v) of the Brilliant Blue G solution (3 mL for 300 mL of equilibration solution) to the equilibration solution. Let the stain proceed for 24–72 h.

5. If necessary, destain the background with water.

6. The stained proteins can be more easily visualized by placing the gel on a light table.

4   Notes

1. PMSF is an irreversible inhibitor of serine proteases and some cysteine proteases. It is a cytotoxic chemical which should be handled only inside a fume hood. Stock solution is prepared in ethanol due to its rapid degradation in water. Other protease inhibitors with similar specificity but less toxic

than PMSF can be used such as AEBSF (4-(2-Aminoethyl) benzene sulfonyl fluoride hydrochloride), marketed under the name of Pefabloc.

2. Sodium deoxycholate is an ionic detergent commonly used for disruption and dissociation of protein interactions. Here, it is use as a co-precipitant helping protein precipitation after addition of the stronger acid TCA.

3. TCA is a corrosive chemical, handle with care in a fume hood. The product is hygroscopic, keep it in a cool, dry and ventilated area in a tightly sealed container.

4. Urea and thiourea are chaotrope agents which disrupt hydrogen bonds and allow the proteins to unfold. CHAPS (3-((3-Cholamidopropyl)dimethylammonio)-1-propanesulfonate) is a sulfobetaine-type zwitterionic detergent which helps in protein solubility and minimizes protein aggregation. TBP (tributylphosphine) or DTT (dithiothreitol) are uncharged and negatively charged reducing agents, respectively, allowing the reduction of and preventing oxidative crosslinking through disulfide bonds. TBP is usually packaged in small volumes at a concentration of 200 mM.

5. Tri-n-butylphosphate is a lipophilic solvent which extracts lipids by forming micelles and improves the efficiency of acetone/methanol delipidation while retaining the acetone:methanol protein precipitation benefits of concentrating and desalting proteins. This step promotes a more complete delipidation that may reduce the intensity of the precipitate's hydrophobic interaction with aqueous buffers favoring more complete protein solubilization in aqueous buffers and thus greater two-dimensional gel electrophoresis spot resolution [4].

6. BPB is included in trace amounts to view the rehydration of the IPG strips. TFE is used as a co-solvent which affects the three-dimensional structure of proteins helping in their denaturation and solubilization. It is also used to improve extraction and focusing of membrane proteins spots in the IEF gels [5].

7. Carrier ampholytes help to keep proteins soluble. Add pH carrier ampholytes adapted to the IPG strips used. For example, pH 3–10 carrier ampholytes for pH 3–10 NonLinear or Linear IPG strips and a 50/50 mix of pH 4–6 and pH 5–7 carrier ampholytes for pH 4–7 IPG strips.

8. Wear a mask and work under a fume hood when weighing SDS.

9. Iodoacetamide is an alkylating agent that binds covalently with the thiol group of cysteine and thus avoids the reformation of disulfide bonds. BPB is used here to view the ions front (migration front) with which it migrates during SDS-PAGE.

10. Ammonium persulfate is a strong oxidizing agent used along with TEMED to catalyze the polymerization of acrylamide. It is better to prepare this solution freshly each time.

11. TEMED is a free radical stabilizer. Storing at 4 °C reduces its unpleasant smell.

12. If the volume of culture recovered is too important, it is recommended to concentrate by filtration with centrifugation at $3,500 \times g$ in a refrigerated centrifuge (4 °C) through a polyethersulfone membrane with a molecular weight cutoff at 5 kDa.

13. TCA is an efficient protein precipitant particularly suitable for large volumes; it allows instantaneous inactivation and precipitation of all the proteins including proteases while excluding small ionic molecules (salts, nucleotides, phospholipids, etc.) and polysaccharides which can result in poor focusing towards the anode and horizontal streaking, respectively. Other precipitation procedures can be used with ice-cold acetone (at least three volumes for one volume of sample) or with a solution of TCA in acetone (final concentration of 10 % TCA in the sample).

14. Acetone is an organic solvent commonly used to precipitate proteins. Use acetone-resistant centrifuge tubes and a glass pipette to add it. It is necessary to wash the pellet extensively (at least three times) with acetone (10–15 mL for an initial volume of 100 mL culture medium) to remove residual TCA which can acidify the sample and may cause degradation of proteins and highly disturb the IEF patterns. The pH can be controlled by placing a drop on a pH indicator strip.

15. An alternative method consists of the direct analysis of the exoproteome by liquid chromatography coupled to tandem mass spectrometry (LC-MS/MS) instead of 2-DE. In this case, the protein sample is solubilized in a buffer compatible with mass spectrometry consisting of 8 M urea and 2 M thiourea. By this approach, the exoproteome sample is treated with trypsin overnight at 37 °C and the resulting peptides are separated on a PepmapC18 column (75 μm × 150 mm) and analyzed by direct coupling with an electrospray ionization-ion trap-tandem mass spectrometer (ESI-IT-MS/MS).

16. The specific volume of IEF buffer to be added is dependent of the quantity of proteins usually recovered, which is generally quite small. ASB-14 (AmidoSulfoBetaine-14, tetradecanoyl-amidopropyl dimethyl ammonio propane sulfonate) is one of the new sulfobetaine zwitterionic detergents which improve protein solubilization and, in general, give better pattern and quality in 2-DE than DTT, particularly for membrane and cell wall proteins [22–25].

17. It is essential to avoid/limit bacterial cell lysis during shaving. Consequently, the parameters for shaving may be different

according to the growth parameters (T °C, culture medium, growth phase, etc.) of bacterial cells making them more or less susceptible to lysis. Cell lysis can be checked by the presence of nucleic acids in the culture medium.

18. Total volume of protein sample and rehydration buffer per IPG strip

IPG strip length (cm)	Total volume per strip (µL)
7	125
11	200
13	230
17	300
18	320
24	450

The quantity of proteins separated by 2-DE followed with Brilliant Blue G staining may vary between 400 and 800 µg.

19. Using a reswelling/equilibration disposable tray or cassette prevents cross-contamination of protein samples but necessitates more handling. The device used here is the IEF Protean® IEF cell (Bio-Rad).

20. When adding mineral oil avoid sample evaporation and urea crystallization.

21. The electrical conductivity of the IPG strip gel changes with time during the IEF run, particularly at the beginning where the current can be relatively high due to the presence of a large number of charge carriers (proteins, ampholytes, salts). So, low voltage prevents overheating during this first phase of IEF run. Then, the current decreases gradually as these molecules move toward their isoelectric points (pIs) or the electrodes (salts) and their charge decreases. At this step of the IEF run, it is necessary to apply progressively higher voltage allowing proteins to reach their pI. Towards the end of the IEF run, the current drops to nearly zero as the current carriers have stopped their migration in the gel.

If the ionic strength of the sample solution is too high, high voltage cannot be applied without high current and risk of overheating. To overcome this problem, it is possible to renew the paper wicks during (or at the end of) the first low voltage step to eliminate part of the ionic constituents accumulated at the electrodes.

Example of stepwise focusing condition for 18-cm 3–10 nonlinear or 4–7 IPG strips:

Voltage (V)	Gradient	Time (h)	Volt-hours (Vh)
50	Rapid	4–8[a]	200–400
200	Rapid	2	400
1,000	Linear	2	1,200
1,000		1	1,000
8,000	Linear	4	18,000
8,000		5	40,000
Total			60,800–61,000

[a]The duration of this step can be adjusted for time experimental requirements.

22. Be careful to avoid mixing of the overlayer solution with the gel solution. The overlay prevents contact with atmospheric oxygen which inhibits acrylamide polymerization and helps to level the resolving gel solution resulting in a plane and regular level of polymerization.

23. Proceed very carefully to positioning the IPG strips. Hold the IPG strip at each extremity with forceps, dip it in the SDS-PAGE running buffer and place it onto the long plate with the plastic backing against the plate. Slide the strip between the plates using a spatula to push against the plastic backing, preferably at each end of the strip to prevent damage of the gel. Push the IPG strip until there is direct contact with the acrylamide gel along its entire length.

24. The agarose solution seals the IPG strip in place and avoids shifting during the addition of the SDS-PAGE running buffer. Traces of BPB can be added in the agarose solution to view the front of migration when it is not included in the equilibration buffer II.

25. It is important to start the SDS-PAGE at low voltage to give time to the SDS front to resolubilize the proteins focalized at their pI while sweeping across the IPG strips [26]. The temperature of the running buffer in the lower chamber must be controlled to maintain a uniform temperature during the run which improves gel to gel reproducibility, especially when ambient temperature fluctuates in the laboratory. Generally, we use a circuit of cooling water at 18 °C.

26. For overnight migration, a constant current of 15 mA can be applied for the duration of migration but limited to 40 V for 1 h and then to 150 V for the rest of time (about 15 h).

Acknowledgment

The author thanks I. Chafsey for her valuable technical expertise in proteomics approaches and its constantly renewed investment in the training of students to the different proteomics methodologies. The author also thanks C. Chambon, engineer on the proteomics component of the Metabolism Exploration Platform (PFEMcp) from INRA Clermont-Ferrand, who brings his skills to perform all our analysis by mass spectrometry with a remarkable efficiency and dedication and a constant concern to improve the methods and to optimize the analytical performances. Finally, I thank all my students who have contributed to the development and improvement of these techniques in the lab.

References

1. Cordwell SJ (2006) Technologies for bacterial surface proteomics. Curr Opin Microbiol 9:320–329

2. Santoni V, Molloy M, Rabilloud T (2000) Membrane proteins and proteomics: un amour impossible? Electrophoresis 21:1054–1070

3. Santoni V, Kiefer S, Desclaux D, Masson F, Rabilloud T (2000) Membrane proteomics: use of additive main effects with multiplicative interaction model to classify plasma membrane proteins according to their solubility and electrophoretic properties. Electrophoresis 21:3329–3344

4. Mastro R, Hall M (1999) Protein delipidation and precipitation by tri-n-butylphosphate, acetone, and methanol treatment for isoelectric focusing and two-dimensional gel electrophoresis. Anal Biochem 273:313–315

5. Deshusses JMP, Burgess JA, Scherl A, Wenger Y, Walter N, Converset V, Paesano S, Corthals GL, Hochstrasser DF, Sanchez J-C (2003) Exploitation of specific properties of trifluoroethanol for extraction and separation of membrane proteins. Proteomics 3:1418–1424

6. Schaumburg J, Diekmann O, Hagendorff P, Bergmann S, Rohde M, Hammerschmidt S, Jansch L, Wehland J, Karst U (2004) The cell wall subproteome of *Listeria monocytogenes*. Proteomics 4:2991–3006

7. Pucciarelli MG, Calvo E, Sabet C, Bierne H, Cossart P, Garcia del Portillo F (2005) Identification of substrates of the *Listeria monocytogenes* sortases A and B by a non-gel proteomic analysis. Proteomics 5:4808–4817

8. Calvo E, Pucciarelli MG, Bierne H, Cossart P, Albar JP, Garcia del Portillo F (2005) Analysis of the *Listeria* cell wall proteome by two-dimensional nanoliquid chromatography coupled to mass spectrometry. Proteomics 5:433–443

9. Folio P, Chavant P, Chafsey I, Belkorchia A, Chambon C, Hebraud M (2004) Two-dimensional electrophoresis database of *Listeria monocytogenes* EGDe proteome and proteomic analysis of mid-log and stationary growth phase cells. Proteomics 4: 3187–3201

10. Dumas E, Meunier B, Berdague J-L, Chambon C, Desvaux M, Hebraud M (2008) Comparative analysis of extracellular and intracellular proteomes of twelve *Listeria monocytogenes* strains: differential protein expression is correlated to serovar. Appl Environ Microbiol 74:7399–7409

11. Dumas E, Desvaux M, Chambon C, Hebraud M (2009) Insight into the core and variant exoproteomes of *Listeria monocytogenes* species by comparative subproteomic analysis. Proteomics 9:3136–3155

12. Renier S, Chambon C, Viala D, Chagnot C, Hebraud M, Desvaux M (2013) Exoproteomic analysis of the SecA2-dependent secretion in *Listeria monocytogenes* EGDe. J Proteomics 80:183–195

13. Olaya-Abril A, Jiménez-Munguía I, Gómez-Gascón L, Rodríguez-Ortega MJ (2013) Surfomics: shaving live organisms for a fast proteomic identification of surface proteins. J Proteomics 97:164–176

14. Tjalsma H, Lambooy L, Hermans PW, Swinkels KW (2008) Shedding & shaving: disclosure of proteomic expressions on a bacterial face. Proteomics 8:1416–1428

15. Solis N, Larsen MR, Cordwell SJ (2010) Improved accuracy of cell surface shaving proteomics in *Staphylococcus aureus* using a false-positive control. Proteomics 10:2037–2049

16. Dreisbach A, Hempel K, Buist G, Hecker M, Becher D, van Dijl JM (2010) Profiling the surfacome of *Staphylococcus aureus*. Proteomics 10:3082–3096

17. Bohle LA, Riaz T, Egge-Jacobsen W, Skaugen M, Busk OL, Eijsink VGH, Mathiesen G (2011) Identification of surface proteins in *Enterococcus faecalis* V583. BMC Genomics 12:135

18. Olaya-Abril A, Gómez-Gascón L, Jiménez-Munguía I, Obando I, Rodríguez-Ortega MJ (2012) Another turn of the screw in shaving Gram-positive bacteria: optimization of proteomics surface protein identification in Streptococcus pneumoniae. J Proteomics 75: 3733–3746

19. Wei Z, Fu Q, Liu X, Xiao P, Lu Z, Chen Y (2013) Identification of *Streptococcus equi* ssp. *zooepidemicus* surface associated proteins by enzymatic shaving. Vet Microbiol 159: 519–525

20. Neuhoff V, Stamm R, Eibl H (1985) Clear background and highly sensitive protein staining with coomassie blue dyes in polyacrylamide gels: a systematic analysis. Electrophoresis 6:427–448

21. Neuhoff V, Arold N, Taube D, Ehrhardt W (1988) Improved staining of proteins in polyacrylamide gels including isoelectric focusing gels with clear background at nanogram sensitivity using Coomassie Brilliant Blue G-250 and R-250. Electrophoresis 9:255–262

22. Chevallet M, Santoni V, Poinas A, Rouquié D, Fuchs A, Kieffer S, Rossignol M, Lunardi J, Garin J, Rabilloud T (1998) New zwitterionic detergents improve the analysis of membrane proteins by two-dimensional electrophoresis. Electrophoresis 19:1901–1909

23. Tastet C, Charmont S, Chevallet M, Luche S, Rabilloud T (2003) Structure-efficiency relationships of zwitterionic detergents as protein solubilizers in two-dimensional electrophoresis. Proteomics 3:111–121

24. Planchon S, Chambon C, Desvaux M, Chafsey I, Leroy S, Talon R, Hebraud M (2007) Proteomic analysis of cell envelope from *Staphylococcus xylosus* C2a, a coagulase-negative *Staphylococcus*. J Res Proteome 6:3566–3580

25. Planchon S, Desvaux M, Chafsey I, Chambon C, Leroy S, Hebraud M, Talon R (2009) Comparative subproteome analyses of plaktonic and sessile *Staphylococcus xylosus* C2a: new insight in cell physiology of a coagulase-negative *Staphylococcus* in biofilm. J Res Proteome 8: 1797–1809

26. Rabilloud T, Lelong C (2011) Two-dimensional gel electrophoresis in proteomics: a tutorial. J Proteomics 74:1829–1841

Chapter 11

The Listeria Cell Wall and Associated Carbohydrate Polymers

Marcel R. Eugster and Martin J. Loessner

Abstract

Understanding molecular interactions of bacteria with their environment requires the purification and characterization of cell wall components. Here, we describe detailed experimental methods for the extraction, purification, and analysis of wall teichoic acids (WTA), which assume important roles as major constituents of Gram-positive cell walls, such as mediating interaction with cell wall-associated proteins, eukaryotic host cells, and bacteriophages. Specifically, we present a procedure for compositional WTA characterization to study large diversity of carbohydrate substitution on *Listeria monocytogenes* WTA. This protocol may also be used and adapted to analyze WTA from other bacteria.

Key words *Listeria monocytogenes*, Bacterial cell wall, Peptidoglycan, Wall teichoic acid (WTA), Acid hydrolysis, Mass spectrometry, ESI-MS/MS

1 Introduction

The Gram-positive bacterial cell wall comprises a mesh-like dynamic structure outside the cytoplasmic membrane and is responsible for a defined cell shape and bacterial cell integrity against high turgor [1]. As a scaffold for anchoring proteins and secondary cell wall polymers such as teichoic acids (TA) and additional glycopolymers, it is involved in cell growth, division, and a variety of other biological functions [2–4].

A major and most abundant class of these cell surface molecules of Gram-positive bacteria is the TA, which are either embedded in the membrane as lipoteichoic acids (LTA) or covalently anchored to the peptidoglycan as wall teichoic acids (WTA). Both polymers participate in various interactions with proteins, bacteriophages, and other microorganisms [4], but they differ in synthesis and chemical structure [5]. LTA display less structural variety, whereas the composition of WTA seems highly diverse [3].

As a general feature, WTA are anionic carbohydrate polymers, frequently featuring backbones of glycerol or ribitol phosphate

Kieran Jordan et al. (eds.), *Listeria monocytogenes: Methods and Protocols*, Methods in Molecular Biology, vol. 1157, DOI 10.1007/978-1-4939-0703-8_11, © Springer Science+Business Media New York 2014

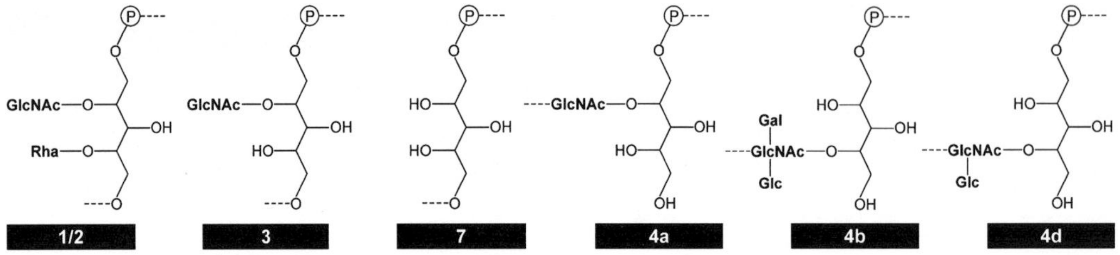

Fig. 1 Structures of the major WTA types of *L. monocytogenes* (modified from Eugster et al. 2011). WTA subtypes are marked in *black boxes*. Schematic structures represent the linear ribitol phosphate repeating units and their glycosidic substituents [6–8, 13]. *GlcNAc* N-acetylglucosamine, *Rha* rhamnose, *Gal* galactose, *Glc* glucose, *P* phosphate

repeating units, and can be modified by D-alanine and/or various sugar substituents. The composition of WTA is highly variable and may be species or even strain specific [3]. In *Listeria*, WTA diversity is mediated by a variety of sugar substitutions on a ribitol phosphate backbone. *Listeria monocytogenes* WTA of serovar 1/2 and 3 bear *N*-acetylglucosamine (GlcNAc) and/or rhamnose (Rha) linked to the C2 or the C4 positions of the ribitol molecule (Fig. 1). In contrast, serovar 4 (and SV 5 and SV 6 of *L. ivanovii* and *L. innocua*, respectively) strains contain GlcNAc moieties incorporated into the chain with glycosidic decoration (glucose and/or galactose) on the backbone residue [6–8]. The nature of these WTA subtypes determines the function of the polymer on the cell surface. *Listeria* WTA have been reported to interact with phages [9] and associated endolysins [10]. Defects in WTA synthesis pathways influence cell growth, bacterial shape, and viability of *Listeria* [10, 11]. Considering the role of these processes, the importance of structural information about WTA becomes very evident.

In this chapter, we describe established procedures for purifying peptidoglycan material and extraction of the WTA molecules. We also present a straightforward procedure for the identification and analysis of WTA carbohydrate substituents. However, note that functional analysis of WTA interactions or structural investigations by NMR spectroscopy are out of the scope of this protocol.

2   Materials

Prepare all solutions and liquid media at room temperature, using analytical grade reagents and ultrapure water (double-distilled water [ddH$_2$O]) or Milli-Q purified water, and store them in the fridge at 4 °C unless otherwise indicated. Solutions containing SDS should be stored at room temperature to avoid precipitation, whereas enzymes (DNase, RNase, Proteinase K) are stored in the freezer (−20 °C).

2.1 Bacterial Culture	1. *L. monocytogenes* strain, frozen stock.

2. Brain–heart infusion (BHI) broth: 37 g/L BHI in water, pH 7.4. Weigh 37 g of BHI and transfer it to a 1-L graduated cylinder (*see* **Note 1**). Add water to a volume of 950 mL. Mix using a magnetic stir bar, and adjust the pH to 7.4 (*see* **Note 2**). Make up to final volume of 1 L with ddH₂O. Sterilize by autoclaving (*see* **Note 3**).

3. BHI agar plates: Add 14 g/L agar to BHI. Mix the powder into the BHI broth prior to autoclaving (*see* **Note 4**). Prepare BHI agar plates by pouring 25 mL of autoclaved BHI agar per standard petri dish.

4. 1-liter conical flaks or 1-L Schott™ bottles for *Listeria* culture (*see* **Note 5**).

5. Shaker incubator for 30 °C.

6. Spectrophotometer.

7. Boling water bath or steamer (suitable for 1-L conical flaks or Schott™ bottles).

8. Refrigerated centrifuge suitable for large-volume centrifuge bottles (i.e., 250 mL or more), associated rotors.

9. SM buffer: 50 mM Tris–HCl, 100 mM NaCl, 8 mM $MgSO_4$, pH 7.4. Prepare a 1 L solution as in previous steps.

2.2   Isolation and Purification of L. monocytogenes Cell Walls

1. Bacterial cells from *L. monocytogenes* (i.e., pellet after harvesting by centrifugation ($7,000 \times g$, 10 min, 4 °C) resuspended in 25 mL of SM buffer).

2. FRENCH® Press cell disruptor (including 40,000 psi standard FRENCH® Press cell) or alternative machine used for physical disruption of bacteria.

3. Centrifuge and associated rotor.

4. SM buffer (*see* above).

5. DNase stock solution: 2 mg/mL in SM buffer. Store aliquots at −20 °C.

6. RNase stock solution: 20 mg/mL in SM buffer. Store aliquots at −20 °C.

7. Proteinase K stock solution: 20 mg/mL in SM buffer, 50 % (v/v) glycerol. Store aliquots at −20 °C.

8. Sodium dodecyl sulfate (SDS) solution (stock): 10 % (w/v) in distilled water. Dissolve 10 g of SDS in a total volume of 90 mL of distilled water. Make up to 100 mL with distilled water, and pass through a 0.22-μm filter. Store at room temperature (*see* **Note 6**).

9. 50-mL Falcon™ tubes.

10. Refrigerated high- and low-speed centrifuges and rotors.

2.3 Extraction of L. monocytogenes WTA

1. Purified peptidoglycan from *L. monocytogenes* (i.e., resuspended in distilled water).

2. Extraction buffer: 25 mM glycine–HCl, pH 2.5 (*see* **Note 2**). Sterilize by filtration through a 0.22-µm filter, and store at –20 °C.

3. High-speed centrifuge and rotor.

2.4 Purification of WTA

1. Extracted WTA dissolved in column running buffer 1.

2. Column running buffer 1: Tris–HCl, pH 7.5 (*see* **Note 2**). Sterilize by filtration through a filter with 0.22 µm in size, and store at 4 °C (*see* **Note 7**).

3. Column running buffer 2: Tris–HCl, 1 M NaCl, pH 7.5 (*see* **Note 2**). Sterilize by filtration through a 0.22-µm filter, and store at 4 °C (*see* **Note 7**).

4. Storage buffer: 20 % (v/v) ethanol in distilled water (*see* **Note 8**).

5. Vacuum filtration system (0.22 µm).

6. Bath sonicator.

7. Liquid chromatography system (e.g., ÄKTA™ FPLC) including an online UV detector and a fraction collector.

8. 5-mL HiTrap DEAE FF column (GE Healthcare, Glattbrugg, Switzerland). Use 20 % (v/v) ethanol as storage buffer (*see* **Note 8**).

9. Dialysis membrane (Spectra/Por, 1,000 Da MWCO).

10. High-speed centrifuge and rotor.

11. Freeze dryer.

2.5 Phosphate Determination for WTA Quantification

1. NANOCOLOR *NanOx* Metal (Macherey-Nagel, Oensingen, Switzerland). Used for oxidative decomposition of samples.

2. Phosphate test kit (Merck, Zug, Switzerland).

3. Glass reaction tubes with screw caps.

4. Heating block (up to 120 °C).

5. Phosphate standard solution (Merck): Aqueous solution of 1,000 mg/L PO_4 (KH_2PO_4) ready for use.

6. Spectrophotometer.

2.6 Acid Hydrolysis and Mass Spectrometry Analysis of WTA

1. Purified WTA from *L. monocytogenes* (i.e., lyophilized WTA after extraction).

2. Hydrofluoric acid (HF), ACS reagent, 48 % (Sigma-Aldrich, Buchs SG, Switzerland) (*see important safety* **Note 9**).

3. 1.5-mL Eppendorf tubes.

4. CaO filter system.

5. Acid-resistant freeze dryer.

6. Neoprene or nitrile gloves (*see* **Note 9**).

7. Safety goggles.

8. Polyethylene bag for items contaminated by hydrofluoric acid.

9. Calcium gluconate gel (*see* **Note 10**).

10. Fume hood.

11. Mass spectrometry buffer: 0.2 % formic acid in 50 % acetonitrile.

12. Mass spectrometer.

3 Methods

Carry out all procedures on ice or at 4 °C if possible. Keep solutions and preparations on ice unless otherwise specified. All procedures are based on previously published protocols [9, 10, 12, 13].

The basic procedure given below is for WTA purification from about 500 mL of total bacterial culture, which yields a few milligrams of purified polymer. It may be modified depending on the amount needed for subsequent assays. Up to 1 week may be required to prepare purified WTA.

3.1 Bacterial Culture

All bacterial growth steps should be carried out under aerobic conditions (*see* **Note 5**).

1. Streak the *L. monocytogenes* strain from a −80 °C glycerol stock onto a BHI agar plate. Grow bacteria at 30 °C overnight.

2. Pick freshly plated single *L. monocytogenes* colonies, and inoculate 50 mL of BHI broth in a 250-mL conical flask with aeration (*see* **Note 11**).

3. Incubate the culture overnight at 30 °C with shaking at 120 rpm.

4. The next day, transfer 10 mL of the overnight culture to a 1-L conical flask containing 500 mL of BHI (*see* **Note 11**). Take out a 1-mL aliquot of BHI broth prior to adding the overnight culture. Keep it as a reference to blank the spectrophotometer for optical density (OD) measurements. Incubate the culture at 30 °C with shaking at the same rpm as the overnight culture.

5. Monitor bacterial growth using the spectrophotometer. Every 30 min, transfer a 1-mL aliquot to a cuvette and measure the optical density of the bacterial suspension at 600 nm (OD_{600}). Grow the culture until an OD_{600} of approx. 0.6 (late exponential phase).

6. Place the culture flasks in a boiling water bath (or steamer) for 1 h to inactivate pathogenic *Listeria monocytogenes* and putative cell wall-degrading enzymes such as autolysins (*see* **Note 12**).

7. Transfer the culture to centrifuge bottles and centrifuge at 7,000×*g* for 10 min at 4 °C. Discard the supernatant.

8. Resuspend the cell pellet in SM buffer, and bring it to a total volume of 30 mL. Transfer the bacterial suspension to a 50-mL Falcon™ tube.

9. Store the bacterial suspension for at least 2 h at –80 °C (*see* **Note 13**).

3.2 Isolation and Purification of L. monocytogenes Cell Walls

Disruption of bacteria using the FRENCH® Press can be performed at room temperature, but the FRENCH® Press 40K cell should be cooled at 4 °C.

1. After thawing the cell suspension, lyse the bacteria by four successive passes through a FRENCH® Press cell at 2,200 psi, using an ice-cold 40K cell. Collect the suspension in a 50-mL Falcon™ tube, and keep it on ice.

2. Remove unbroken cells by centrifugation at 1,400×*g* for 5 min at 4 °C. Carefully transfer the supernatant into a centrifuge tube suitable for higher rpm (*see* **Note 14**).

3. Centrifuge the suspension at 30,000×*g* for 40 min at 4 °C (*see* **Note 15**). Discard the supernatant, and resuspend the pellet in cold distilled water. Repeat this washing step three times, before making a final resuspension in 7 mL of SM buffer per gram wet cell pellet (*see* **Note 16**).

4. Next, add 50 μl of DNase and 5 μl of RNase stock solution per gram wet cell pellet to the suspension, and gently shake the tube at room temperature for 2 h.

5. Add 5 μl of 20 mg/mL proteinase K per gram cell wall (wet weight) suspension, and digest at room temperature for 2 h with shaking.

6. Transfer the suspension to a 50-mL Falcon™ tube, and mix the sample with 2/3 volumes of boiling SDS solution to a final concentration of 4 % (w/v) SDS. Boil the mixture at 95 °C for 30 min (*see* **Note 17**).

7. Cool the tube at room temperature for 5–20 min (*see* **Note 18**).

8. Centrifuge at 30,000×*g* for 30 min at 25 °C. Discard the supernatant. Resuspend the pellet in 20 mL of prewarmed (50 °C) Milli-Q water (*see* **Note 19**).

9. Repeat the last step up to 4–6 times to completely remove residual SDS. Finally, resuspend the peptidoglycan fraction in approx. 5 mL of ultrapure water (*see* **Note 20**).

10. Lyophilize the prepared peptidoglycan fraction (*see* **Note 21**).

11. Weigh the lyophilized peptidoglycan, and dissolve it in ultrapure water to 10 mg/mL. Store purified cell walls in 1-mL aliquots at –20 °C.

**3.3 Extraction of
L. monocytogenes WTA**

Purified insoluble peptidoglycan is used for the extraction of WTA.

1. Transfer approx. 1–3 mL of thawed insoluble peptidoglycan stock (10 mg/mL) to an Oakridge centrifuge tube, and centrifuge cell walls at 30,000 × g for 15 min at 4 °C.

2. Carefully remove supernatant using a pipet tip or a Pasteur pipette. Resuspend the pellet in 3 mL of extraction buffer by pipetting and vortexing.

3. Incubate at 100 °C for 10 min to extract the WTA from the peptidoglycan.

4. Centrifuge at 30,000 × g for 15 min at 4 °C. Carefully collect the aqueous WTA fraction (supernatant) in a Falcon™ tube, and keep it on ice. Resuspend the remaining pellet in 3 mL of fresh extraction buffer by vortexing.

5. The extraction procedure is repeated twice (**steps 3** and **4**). Pool the supernatants in the same collection tube, and store the extractions on ice.

6. Dialyze the combined supernatants containing extracted WTA against ultrapure water to remove the extraction buffer. Change the dialysis medium (ddH$_2$O) 4–5 times (*see* **Note 22**).

7. Transfer the dialyzed fraction to a tube, and freeze-dry the extracted WTA by lyophilization (*see* **Note 21**).

8. Weigh the lyophilized crude WTA extract, and dissolve it in 1 mL of column running buffer 1. Store the samples at –20 °C.

**3.4 Purification
of WTA**

Chromatography steps are performed at room temperature, but fractions are collected on ice. All steps are described for an ÄKTA™ FPLC system and a HiTrap DEAE FF column. Equivalent systems may be used instead.

1. Filter the sample through a 0.22-µm syringe unit before loading the sample loop with 1 mL of crude WTA extract.

2. It is recommended to use a flow rate of 0.5–1 mL/min and UV detection at 205/212 nm and 280 nm (*see* **Note 23**). Equilibrate a 5-mL HiTrap DEAE FF column with 25 mL of column running buffer 1.

3. Load the crude WTA extract onto the column.

4. After the sample is loaded, rinse the column with 3 column volumes of running buffer 1.

5. Elute the sample with a linear gradient in 20 column volumes of 0–1 NaCl, i.e., from 0 to 100 % of running buffer 2.

6. Collect all eluates in 1-mL fractions, and determine the WTA content by phosphate quantification (*see* Subheading 3.5).

7. Combine the phosphate-containing WTA fractions and dialyze against ultrapure water to remove extraction buffer. Change the dialysis medium (ddH$_2$O) 4–5 times (*see* **Note 22**).

8. Freeze-dry the dialyzed WTA sample (*see* **Note 21**).

9. Weigh the purified WTA and dissolve in water (10 mg/mL). Store samples at −20 °C.

3.5 Phosphate Determination for WTA Quantification

Decomposition of WTA and determination of phosphate are performed according to the manufacturer's protocols (*see* Subheading 2.5).

1. For each sample, pipet 4.95 mL of distilled water and 50 μL of the sample into an empty glass reaction tube (dilution of 1:100). Prepare a reference sample required for the measurement according to the same protocol but using 5 mL of distilled water (*see* **Note 24**).

2. Add 1 level orange measuring spoon of *NanOx* Metal decomposition reagent, close the tube, and vortex.

3. Place the reaction tube into a heating block and heat at 120 °C for 60 min. Shake lightly after 15 min.

4. Remove the reaction tube from the heating block, turn it upside down once, and let cool down the solution to room temperature for about 10 min (*see* **Note 25**).

5. Carefully open the test tube, and add 1 microspoon *NanOx* Metal neutralization reagent. Close the lid, and shake thoroughly for 1 min (*see* **Note 26**).

6. For determination of phosphate, add five drops of reagent PO_4-1 and mix by vortexing.

7. Next, add 1 level blue microspoon of reagent PO_4-2 and shake vigorously until the reagent is completely dissolved. Let it stand at room temperature for 5 min.

8. For measurements in the spectrophotometer at 690 nm, transfer 1 mL of the reaction mixture into a cuvette. First, zero the spectrophotometer using the blank before recording the samples.

3.6 Acid Hydrolysis and Mass Spectrometry Analysis of WTA

To analyze WTA by mass spectrometry, it is useful to hydrolyze the high-molecular-weight WTA polymers. Mass spectrometry analysis of WTA monomers is then performed employing ESI-MS/MS.

Warning: HF is a highly toxic compound! Special safety precautions are necessary when using this chemical. Perform all the steps very carefully. Be sure that you are using appropriate protective equipment (*see* **Notes 9** and **10**).

1. Prepare approx. 1 mg of lyophilized WTA sample in a 1.5-mL (Eppendorf type) plastic reaction tube (weigh the empty tube prior to adding the sample); that is, transfer 100 μl of purified WTA (10 mg/mL in H_2O) to the tube and freeze-dry the sample.

2. Precool a 200-µl aliquot of HF in a plastic reaction tube on ice for 5 min (*see* **Note 27**).

3. Transfer 200 µl of cooled (0 °C) HF to the freeze-dried WTA in the reaction tube. Incubate the reaction mix for 16 h on ice. Then, freeze it in liquid nitrogen.

4. Place the tube in a plastic desiccator. Apply a vacuum to remove excess HF. The HF gas should be passed through a filter containing CaO to neutralize the vapors.

5. Add 100 µl of ice-cold ultrapure water, dissolve the sample for 5 min, and lyophilize it.

6. After freeze-drying, weigh the tube to determine the amount of monomeric WTA. Dissolve the sample in 10 µl of ultrapure water, and store it at −20 °C.

7. For positive-ion mass spectrometry, dissolve the monomeric WTA samples in 0.2 % formic acid in a mixture of 50 % acetonitrile/50 % water. Inject the sample into an ESI-MS/MS mass spectrometer. Individual settings depend on the instrument used.

4   Notes

1. Having water (about 100 mL) at the bottom of the 1-L graduated cylinder before adding BHI helps to dissolve the powder more easily.

2. Concentrated HCl (12 N) and NaOH (10 M) and appropriate dilution series (e.g., 4 and 1 M) are used to adjust the pH to the desired values.

3. Standard parameters for a complete sterilization of buffers and media by autoclaving are 121 °C for 15 min.

4. It is helpful to heat the broth (including the agar) by boiling and mix it to disperse the agar before autoclaving. Premelting avoids ending up with a layer of solid agar at the bottom of the flask.

5. Instead of using conical flasks, you can also use 1-L Schott™ bottles for *Listeria* cultures. For anaerobic conditions, inoculate BHI broth with bacteria and return the lid of the Schott™ bottle with a twist. Baffled flasks are not required.

6. Prewarm ddH$_2$O at 37 °C to dissolve SDS in water. The solution should always be stored at room temperature. At low temperature the SDS may precipitate. The stock solution is stable for at least 6 months. If some precipitation occurs, simply warm up the solution for complete dissolution, which can conveniently be done in a microwave oven.

7. Store the buffer at 4 °C, and prewarm it at room temperature before use. Filter it through a 0.22-µm filter prior to use.

8. Chromatography columns are rinsed with ddH$_2$O and 5 column volumes of 20 % (v/v) ethanol at 1 mL/min for storage to prevent microbial growth. We use parafilm to seal columns and the tubing in addition to the supplied stoppers, and store them at room temperature.

9. HF is one of the most dangerous acids known and needs to be handled with utmost care! It can readily penetrate the skin, causing destruction of deep tissue layers due to high affinity for calcium. If HF is not rapidly neutralized by binding the fluoride ion, tissue destruction may result in limb loss or death. If spills occur, immediately flush the acid from the affected skin area with water, and apply a calcium gluconate gel (*see* **Note 10**). Avoid contact with glass, metals, and water. Read the safety instruction sheet before use. Do not use latex gloves but neoprene or nitrile gloves instead. Always handle HF inside a properly functioning chemical fume hood!

10. Calcium gluconate gel as part of an HF exposure first-aid kit must be available and located in the laboratory working area. Calcium works by combining with HF to form insoluble calcium fluoride, preventing calcium extraction from bones and tissue. Use clean nitrile gloves to apply the gel.

11. The BHI broth should be prewarmed; that is, prepare the autoclaved media the day before and place the flasks in a 30 °C incubator. If needed, add antibiotics just prior to adding bacteria.

12. Centrifugation steps may also be carried out before heat-killing the bacteria. We prefer to heat treat the cultures before further manipulation to inactivate pathogenic bacteria as early as possible. To ensure that no viable *Listeria* cells remain, plate some of the suspension on an agar plate and incubate the plate at 30 °C overnight. Even for nonpathogenic bacteria, this step should not be omitted, because heating also inactivates the activity of cell wall-associated peptidoglycan hydrolases known as autolysins.

13. This step aids in pre-damaging bacterial cell walls for cell disruption.

14. Repeating the centrifugation step at 1,400×g (i.e., transfer supernatant to a fresh 50-mL tube) removes any remaining intact cells.

15. The pellet containing membranes and peptidoglycan is difficult to resuspend. Carefully dissolve the pellet in the buffer by using a 10-mL pipette and pipetting resuspended pieces up and down until broken cell walls are evenly suspended.

16. To determine the amount of wet cell walls in the Oakridge centrifuge tube, measure the weight of the empty tube before adding the suspension for centrifugation steps. Higher bacterial

culture volumes require more SM buffer and splitting the mixture into several Falcon™ tubes.

17. It is recommended to use a small magnetic stir bar inside the Falcon™ tube to ensure optimal mixing. This step removes membranes and contaminating proteins non-covalently bound to the peptidoglycan.

18. Keep the mixture at room temperature to prevent precipitation of SDS (*see* **Note 6**).

19. Pipetting the SDS solution up and down often results in foaming. To minimize this effect do not shake solutions.

20. The peptidoglycan pellet should appear translucent or whitish. If it appears brownish at the bottom of the tube, this may indicate the presence of intact bacteria cells. In this case, gently pipette the cell walls up and down to evenly resuspend the milky-white part of the pellet. Repeat this step two to three times before freeze-drying.

21. To quantify the total recovery of freeze-dried samples, first weigh the empty tube before adding the sample. Subtract the weight of the empty device to calculate the weight of the resulting dried peptidoglycan.

22. It is recommended to use 100:1 buffer-to-sample volume ratio and 4–5 changes over the period of 10 h. Perform the last buffer change prior to leaving overnight.

23. WTA polymers show low UV absorption. Fractions can be detected in the range of 205–212 nm but should be confirmed using the phosphate test kit.

24. The reference blank should contain the same solution except the sample substance. For analytical quality measurements, it is essential to use standard phosphate solutions to obtain a calibration curve. Ideally, sequential dilutions of solutions should cover the test kit measuring range.

25. The reaction mix must be clear and colorless.

26. Use pH test strips to check the pH value after neutralization. It should be between 3 and 7; otherwise, add additional neutralization reagent.

27. Acid hydrolysis using HF should be carried out on ice for best reaction specificity, since the carbohydrate substituents are usually not affected by HF treatment at 0 °C.

Acknowledgements

Dr. Serge Chesnov (FGCZ, University of Zurich, Switzerland) is gratefully acknowledged for his support in WTA analysis by mass spectrometry.

References

1. Vollmer W, Blanot D, de Pedro MA (2008) Peptidoglycan structure and architecture. FEMS Microbiol Rev 32:149–167

2. Navarre WW, Schneewind O (1999) Surface proteins of Gram-positive bacteria and mechanisms of their targeting to the cell wall envelope. Microbiol Mol Biol Rev 63:174–229

3. Kohler T, Xia G, Kulauzovic E, Peschel A, Otto H, Patrick JB, Mark von I (2009) Teichoic acids, lipoteichoic acids and related cell wall glycopolymers of Gram-positive bacteria. In: Moran A, Holst O, Brennan P, von Itzstein M (eds) Microbial Glycobiology: Structures, Relevance and Applications. San Diego: Elsevier p. 75–91

4. Weidenmaier C, Peschel A (2008) Teichoic acids and related cell-wall glycopolymers in Gram-positive physiology and host interactions. Nat Rev Microbiol 6:276–287

5. Neuhaus FC, Baddiley J (2003) A continuum of anionic charge: structures and functions of D-alanyl-teichoic acids in Gram-positive bacteria. Microbiol Mol Biol Rev 67:686–723

6. Fiedler F (1988) Biochemistry of the cell surface of Listeria strains: a locating general view. Infection 16:92–97

7. Uchikawa K, Sekikawa I, Azuma I (1986) Structural studies on teichoic acids in cell walls of several serotypes of Listeria monocytogenes. J Biochem 99:315–327

8. Fujii H, Kamisango K, Nagaoka M, Uchikawa K, Sekikawa I, Yamamoto K, Azuma I (1985) Structural study on teichoic acids of Listeria monocytogenes types 4a and 4d. J Biochem 97:883–891

9. Wendlinger G, Loessner MJ, Scherer S (1996) Bacteriophage receptors on Listeria monocytogenes cells are the N-acetylglucosamine and rhamnose substituents of teichoic acids or the peptidoglycan itself. Microbiology 142:985–992

10. Eugster MR, Haug MC, Huwiler SG, Loessner MJ (2011) The cell wall binding domain of Listeria bacteriophage endolysin PlyP35 recognizes terminal GlcNAc residues in cell wall teichoic acid. Mol Microbiol 81:1419–1432

11. Dubail I, Bigot A, Lazarevic V, Soldo B, Euphrasie D, Dupuis M, Charbit A (2006) Identification of an essential gene of Listeria monocytogenes involved in teichoic acid biogenesis. J Bacteriol 188:6580–6591

12. Eugster MR, Loessner MJ (2011) Rapid analysis of Listeria monocytogenes cell wall teichoic acid carbohydrates by ESI-MS/MS. PLoS One 6:e21500

13. Fiedler F, Seger J, Schrettenbrunner A, Seeliger HPR (1984) The biochemistry of murein and cell-wall teichoic-acids in the genus Listeria. Syst Appl Microbiol 5:360–376

Use of Bacteriophage Cell Wall-Binding Proteins for Rapid Diagnostics of Listeria

Mathias Schmelcher and Martin J. Loessner

Abstract

Diagnostic protocols for food-borne bacterial pathogens such as *Listeria* need to be sensitive, specific, rapid, and inexpensive. Conventional culture methods are hampered by lengthy enrichment and incubation steps. Bacteriophage-derived high-affinity binding molecules (cell wall-binding domains, CBDs) specific for *Listeria* cells have recently been introduced as tools for detection and differentiation of this pathogen in foods. When coupled with magnetic separation, these proteins offer advantages in sensitivity and speed compared to the standard diagnostic methods. Furthermore, fusion of CBDs to differently colored fluorescent reporter proteins enables differentiation of *Listeria* strains in mixed cultures. This chapter provides protocols for detection of *Listeria* in food by CBD-based magnetic separation and subsequent multiplexed identification of strains of different serotypes with reporter-CBD fusion proteins.

Key words *Listeria monocytogenes*, Phage cell wall-binding domains, Diagnostics

1 Introduction

Listeriosis is exclusively transmitted via contaminated food [1]. Since *Listeria* often occurs in low numbers within foods and is frequently accompanied by other organisms of the background flora, high sensitivity and selectivity are important requirements for diagnostic protocols. In addition are other relevant criteria such as low costs, rapidity, and ease of use. While the officially adopted culture methods for detection of *Listeria* (e.g., IDF 143A:1995 and ISO 11290-1) are still considered as gold standard, they are hampered by lengthy enrichment and incubation periods. The recently introduced cell wall-binding domain-based magnetic separation (CBD-MS) assay employs paramagnetic beads coated with recombinant CBD proteins derived from *Listeria* bacteriophage cell wall hydrolases, the endolysins [2]. They have been used for specific immobilization and subsequent separation of *Listeria* cells from food samples and were shown to be superior to the standard protocols both in terms of sensitivity and time requirement [2]. Endolysins are phage-encoded

Kieran Jordan et al. (eds.), *Listeria monocytogenes: Methods and Protocols*, Methods in Molecular Biology, vol. 1157,
DOI 10.1007/978-1-4939-0703-8_12, © Springer Science+Business Media New York 2014

enzymes that are produced inside an infected host cell at the end of the lytic cycle of the phage and degrade the bacterial peptidoglycan from within, which results in cell lysis and liberation of progeny phage particles. As Gram-positive organisms lack an outer membrane shielding them from the outside, endolysins can also access their substrate in the Gram-positive cell wall from without when added as purified proteins, making them promising agents for control and detection of pathogens [3, 4]. Like all endolysins from a Gram-positive background, *Listeria* phage endolysins described to date show a modular architecture consisting of N-terminal enzymatically active domains (EAD), which catalyze peptidoglycan degradation, and C-terminal CBDs. The latter have been shown to bind to their ligands in the cell wall (which can be the peptidoglycan itself or other cell wall molecules [5, 6]) with high affinity and specificity [7, 8] and therefore present ideal tools for the aforementioned magnetic separation procedure [2]. Furthermore, fusion of these affinity molecules to fluorescent reporter proteins such as the green fluorescent protein (GFP) or the *Discosoma* red fluorescent protein (dsRed) makes it possible to visualize cell wall localization by fluorescence microscopy and thereby elucidate binding properties and spectra of strains recognized by each CBD [7, 8]. As opposed to binding modules from phages of most other Gram-positive genera, *Listeria* phage CBDs exhibit specificity down to the serovar level and display characteristic binding patterns that coincide with the various serovar groups. This unique property enables differentiation of *Listeria* strains in mixed cultures by using a mixture of suitable CBDs, each fused to a differently colored fluorescent reporter [8] (Fig. 1).

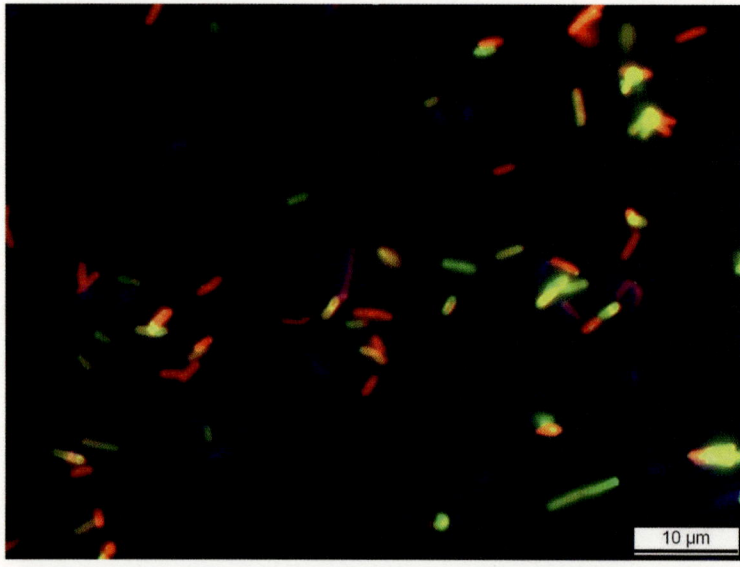

Fig. 1 Differentiation of several *Listeria* strains of different serovars in a mixed culture by fluorescence microscopy, using three differently colored FR-CBD fusion proteins

This chapter includes protocols for design, construction, recombinant production, and purification of fluorescent protein-CBD (FP-CBD) fusion proteins and their use in specific detection by CBD-MS and multiplexed differentiation of *Listeria* cells from food samples.

2 Materials

2.1 Materials Required for Construction of FR-CBD Fusion Constructs

The materials described in this section are only required if novel FR-CBD fusions are to be generated.

1. High-Fidelity PCR polymerase Master Mix (Phusion, Finnzymes/New England Biolabs).
2. Genomic DNA of bacteriophage of interest.
3. Plasmid constructs based on a pQE-30 backbone (Qiagen) bearing sequences encoding various differently colored fluorescent marker proteins inserted into BamHI and SacI restriction sites (omitting native stop codons and thereby allowing fusion with sequences inserted downstream), such as pHGFP, pHRS, and pHCFP [8]. If such plasmids are not available in the laboratory, they have to be constructed using standard molecular cloning techniques [9] prior to performing the cloning described in this protocol.
4. PCR purification, gel extraction, and plasmid preparation kits.
5. *Sac*I and *Sal*I restriction enzymes.
6. T4 ligase and ligase buffer.
7. Competent cells of *E. coli* strain XL1-Blue "MRF".
8. Luria Bertani (LB) broth: 10 g/L Tryptone (pancreatic digest of casein), 5 g/L yeast extract, 8 g/L NaCl, pH 7.4. For LB agar, add 14 g/L of agar.
9. Ampicillin stock solution: Dissolve 50 mg/mL of ampicillin in ultrapure water, filter-sterilize using 0.2 µM syringe filters, aliquot, and store at −20 °C.
10. Tetracycline stock solution: Dissolve 15 mg/mL of tetracycline in 70 % ethanol, filter-sterilize, aliquot, and store at −20 °C.

2.2 Materials Required for Protein Expression and Purification

1. *E. coli* XL1-Blue "MRF" strain harboring the desired FP-CBD construct in a pQE-30 plasmid backbone.
2. LB broth, ampicillin, and tetracycline stock solutions (*see* Subheading 2.1).
3. Modified LB broth for protein expression (LB-PE): 15 g/L Tryptone, 8 g/L yeast extract, 5 g/L NaCl, pH 7.8.
4. 20 % (~840 mM) isopropyl-β-D-thiogalactopyranoside (IPTG) stock solution: Dissolve 2 g IPTG in 10 mL ultrapure water, filter-sterilize (0.2 µM), aliquot, and store at −20 °C.

5. Buffer A: 50 mM Na_2HPO_4, 500 mM NaCl, 5 mM imidazole, 0.1 % Tween 20, pH 8.0. For 1 L of buffer, weigh 17.91 g of Na_2HPO_4 (dodecahydrate), 29.23 g NaCl, and 0.34 g imidazole; dissolve in 900 mL ultrapure water; adjust the pH to 8.0; add 1 mL of Tween 20; and fill up to 1 L total volume with water. Store at 4 °C.

6. Buffer B: 50 mM Na_2HPO_4, 500 mM NaCl, 250 mM imidazole, 0.1 % Tween 20, pH 8.0. Proceed as described for buffer A, but weigh 17.02 instead of 0.34 g imidazole per 1 L total volume.

7. Phosphate-buffered saline (PBS): 50 mM Na_2HPO_4, 120 mM NaCl, pH 8.0. For 1 L of buffer, weigh 17.91 g of Na_2HPO_4 (dodecahydrate) and 7.02 g NaCl, dissolve in 900 mL ultrapure water, adjust the pH to 8.0, and fill up to 1 L total volume with water. Autoclave and store at room temperature.

8. French press (SLM Aminco) equipped with a 20K or 40K pressure cell or any comparable device for cell disruption by high-shear mechanical processing (e.g., FPG12800 Pressure Cell Homogenizer, Stansted Fluid Power Ltd., or OneShot, Constant Cell Disruption Systems).

9. Ni-NTA Superflow resin (Qiagen) for purification of 6xHis-tagged recombinant proteins or equivalent product.

10. Standard laboratory overhead rotator.

11. Econo-Pac chromatography columns (empty) including upper bed supports, end caps, and tip closures (Bio-Rad) or equivalent products.

12. Nanodrop spectrophotometer (Thermo Scientific) or an equivalent product for determination of protein concentration.

13. Dialysis buffer: 50 mM Na_2HPO_4, 100 mM NaCl, 0.005 % Tween 20, pH 8.0. For 1 L of buffer, weigh 17.91 g of Na_2HPO_4 (dodecahydrate) and 5.85 g NaCl, dissolve in 900 mL ultrapure water, adjust the pH to 8.0, add 50 µl of Tween 20, and fill up to 1 L total volume with water. Store at 4 °C.

14. Dialysis membrane tubing (6,000–8,000 MWCO) with tubing clamps or an equivalent product for protein dialysis.

15. Acrodisc syringe filter with Supor membrane (0.2 µm pore size, low protein binding; Pall Corporation) or equivalent.

2.3 Materials Required for Cell Wall Binding Assay

1. TB medium: 20 g/L Tryptose, 1 g/L glucose, 5 g/L NaCl, 0.005 g/L of thiamine HCl, pH 7.4.

2. PBS-T buffer: PBS including 0.01 % Tween 20. For 1 L of buffer, proceed as described above for PBS and add 100 µl of Tween 20 before autoclaving.

3. Suitable *Listeria* strain (*see* **Note 17**).

4. Purified FP-CBD protein (*see* Subheading 3.2).

5. Fluorescence microscope with suitable excitation and emission filter sets (*see* **Note 20**).

2.4 Materials Required for Bead Coating

1. Lyophilized M-270 Epoxy Dynabeads (Invitrogen).

2. Purified FP-CBD protein (*see* Subheading 3.2).

3. Magnet for molecular and cell separation applications (for 1.5–2.0 mL tubes).

4. Diglyme (diethylene glycol-dimethyl ether).

5. Phosphate buffer (PB): 100 mM Na_2HPO_4, pH 7.4. For 1 L of buffer, weigh 35.82 g of Na_2HPO_4 (dodecahydrate), dissolve in 900 mL ultrapure water, adjust the pH to 7.4, and fill up to 1 L total volume with water.

6. Ammonium sulfate (3 M): Dissolve 39.6 g $(NH_4)_2SO_4$ in PB, and adjust volume to 100 mL.

7. Overhead rotator (*see* above).

8. Tris buffer: 1 M Tris, pH 7.4. Weigh 12.11 g Tris, dissolve in 90 mL ultrapure water, adjust pH to 7.4, and fill up with water to a total volume of 100 mL.

2.5 Materials Required for Recovery of Listeria from Food Samples by CBD-MS

1. Food sample to be tested (minimum 25 g of solid or 25 mL of liquid foods).

2. Stomacher 400 homogenizer (Seward, UK) or a similar laboratory blender, including suitable sterile plastic (Stomacher) bags.

3. Sterile polypropylene plastic bags.

4. TSB-ANC medium: 17 g/L peptone from casein, 3 g/L NaCl, 2.5 g/L K_2HPO_4, 2.5 g d(+) glucose, 6 g/L yeast extract. Dissolve in 1 L purified water, and adjust pH to 7.2. Aliquot in bottles (175 mL portions), and autoclave. Before use, add the following selective agents: acriflavine (dissolved in purified water) at a final concentration of 0.009 %; nalidixic acid (dissolved in 0.05 N NaOH) at 0.036 %; and cycloheximide (dissolved in 40 % EtOH) at 0.045 %.

5. Double-concentrated PBS-T buffer: 100 mM Na_2HPO_4, 240 mM NaCl, 0.02 % Tween 20, pH 8.0. For 1 L of buffer, weigh 35.82 g of Na_2HPO_4 (dodecahydrate) and 14.04 g NaCl, dissolve in 900 mL ultrapure water, adjust the pH to 8.0, and fill up to 1 L total volume with water. Add 200 µl of Tween 20, and autoclave.

6. M-270 Epoxy Dynabeads coated with a suitable FP-CBD fusion protein (*see* **Note 27**), resuspended in PBS.

7. Magnet for molecular and cell separation applications (for 1.5–2.0 mL tubes).

8. Oxford agar plates (*Listeria*-selective agar): Weigh 55.5 g of *Listeria*-selective agar base, Oxford formulation (Oxoid, UK), and additionally 3 g of agar; dissolve in 1 L of purified water; adjust the pH to 7.2; and autoclave. Allow to cool to 50 °C before aseptically adding two portions of Oxford-selective supplement (Oxoid, UK) that were previously dissolved in 5 mL of 70 % EtOH each. Mix and pour into sterile petri dishes.

2.6 Material Required for Differentiation of Listeria Isolates by Multiplex Binding Assay

1. Oxford agar plates with presumptive *Listeria* colonies from the CBD-MS assay (Subheading 3.5).

2. Collection of suitable FP-CBD fusion proteins (*see* Subheading 3.2).

3. PBS-T buffer (*see* Subheading 2.3).

4. Fluorescence microscope with suitable excitation and emission filter sets (*see* Subheading 2.3).

3   Methods

3.1 Construction of Novel FR-CBD Fusion Constructs

If novel FR-CBD fusion proteins are to be generated, follow the instructions in this paragraph. If existing constructs (such as those described in Schmelcher et al. 2010 [8]) are to be used, proceed directly to the paragraph in Subheading 3.2. Please note that this paragraph is based on the assumption that sequence information regarding the particular CBD to be cloned is available in public databases and the researchers have access to the encoding phage or genomic DNA thereof. It does not include instructions on how to isolate new phages, prepare phage genomic DNA, and sequence and annotate phage genomes. Researchers are encouraged to read the entire protocol before commencing construction of FR-CBD fusions described here.

1. Obtain the nucleotide and corresponding amino acid sequence of the putative *Listeria* phage endolysin harboring the CBD of interest by sequencing or from a database.

2. Identify the sequence region coding for the putative CBD and the linker region between the EAD and the CBD of the endolysin by bioinformatic analysis. As a first step, analyze the complete endolysin amino acid sequence by Standard Protein BLAST (using the BlastP algorithm available on http://blast.ncbi.nlm.nih.gov, searching against non-redundant protein sequences and applying the default settings) in order to approximate the functional domain borders (*see* **Note 1**). As a second step, predict the putative linker region between EAD and CBD from the amino acid sequence using the DomCut tool available on http://www.bork.embl.de/~suyama/domcut/ (*see* **Note 2**).

3. Design oligonucleotide primers for PCR amplification of the identified CBD coding region and cloning into *Sac*I/*Sal*I restriction sites of pHGFP (or another pQE-30-based plasmid bearing a fluorescent reporter gene of the researcher's choice). This includes one reverse primer introducing the *Sal*I site and complementary to the 3′ end of the target sequence and 3–4 forward primers introducing *Sac*I sites and complementary to sequences spanning the border region between EAD and CBD of the

target gene, thereby allowing generation of several versions of the CBD varying in length at the 5' end (*see* **Note 3**).

4. Amplify CBD sequences by PCR using a high-fidelity polymerase, the primers from **step 3**, and phage genomic DNA as template (*see* **Note 4**). If desired, generate FR-CBD fusion constructs employing standard molecular cloning techniques [9] and following the manufacturer's instruction for the use of kits and enzymes. In brief, purify PCR products using a commercial PCR purification kit. Digest both PCR products and FR-bearing plasmids with *Sac*I and *Sal*I restriction endonucleases and purify via PCR purification and gel extraction kits, respectively. Ligate purified fragments into plasmids using T4 ligase. Use the ligation reaction mixture to transform competent *E. coli* XL1-Blue "MRF" cells. Plate on LB agar supplemented with 100 µg/mL ampicillin and 30 µg/mL tetracycline (LB + Amp + Tet agar) for plasmid selection. Select individual colonies after overnight incubation at 37 °C, and check for the presence of plasmids bearing inserts of the expected size (using colony PCR, control digest of purified plasmid DNA, or any other suitable method). Verify sequence identity of positive clones by nucleotide sequencing (*see* **Note 5**).

3.2 Protein Expression and Purification

1. Inoculate 10 mL of LB + Amp + Tet in a test tube with a single colony from an *E. coli* strain harboring the FP-CBD construct of interest and incubate overnight at 37 °C in a shaker at 220 rpm. To provide a negative control for the cell wall binding assay (Subheading 3.3), inoculate a separate culture with an *E. coli* strain expressing the FP only (no fusion to CBD).

2. Use the overnight culture to inoculate 1 L of LB-PE + Amp + Tet in baffled shaker flasks (1:100 dilution) for each construct and incubate at 30 °C and 120 rpm until the optical density at 600 nm ($OD_{600 \, nm}$) reaches approximately 0.5 (*see* **Note 6**).

3. Induce protein expression by adding 0.5 M IPTG (add 600 µl of the 20 % stock solution to 1 L of culture), and continue shaking at 30 °C for 4 h. Afterwards, incubate overnight (12–18 h) shaking at 4 °C (*see* **Note 7**).

4. Harvest cells by centrifugation ($7,000 \times g$, 10 min, 4 °C), discard supernatant, and resuspend pellet in 10 mL buffer A per liter culture volume (*see* **Note 8**). Transfer to a 15 or a 50 mL Falcon tube, centrifuge again, discard supernatant, and freeze cell pellets at −80 °C.

5. Thaw pellets on ice and resuspend in 10 mL buffer A. Disrupt bacterial cells by using a French Press (SLM Aminco) equipped with a 20K or a 40K standard cell at 100 MPa or an equivalent device for cell disruption (*see* **Note 9**).

6. Centrifuge samples ($10,000 \times g$, 30 min, 4 °C) to remove cellular debris. In the meantime, prepare the Ni-NTA resin

required for purification of 6xHis-tagged recombinant proteins. For each protein to be purified, add 2 mL of Ni-NTA slurry (corresponding to 1 mL of resin) per 1 L original culture volume (*see* **step 2**) into a 15 mL Falcon tube. Allow the resin to settle at the bottom of the tube or gently centrifuge ($300 \times g$, 5 min, 4 °C). Gently remove (by pipetting) and discard supernatant (*see* **Note 10**).

7. Add supernatant from the centrifuged samples to Ni-NTA resin (*see* **Note 11**), and resuspend by inverting or gently vortexing. Incubate at 4 °C for 1 h under constant agitation using an overhead rotator.

8. Centrifuge the resin ($300 \times g$, 5 min, 4 °C), and gently remove and discard supernatant. Resuspend resin in 10 mL of buffer A.

9. Pack the chromatography column with resin, and let the column drain by gravity flow (*see* **Note 12**).

10. Wash the column with 25 column volumes (CV) of buffer A, followed by 5 CV of wash buffer (90 % buffer A, 10 % buffer B). Elute target proteins with buffer B, collecting ½ CV fractions in 1.5 mL tubes. Keep fractions on ice (*see* **Note 13**).

11. Determine protein concentration in each fraction by using a Nanodrop spectrophotometer or an equivalent method, and pool all fractions with satisfactory concentration (as a rule of thumb, use fractions with concentrations greater 1 mg/mL). Discard the remaining fractions (*see* **Note 14**).

12. Dialyze the purified protein (combined fractions) overnight at 4 °C against two changes of dialysis buffer (2 x 1.5 L) (*see* **Note 15**).

13. Filter-sterilize the dialyzed protein using a 0.2 μm syringe filter. Determine the protein concentration, and store the final preparation on ice. Alternatively, add sterile glycerol to a final concentration of 50 % (vol/vol) and store at –20 °C (*see* **Note 16**).

3.3 Cell Wall Binding Assay for Verification of FP-CBD Protein Functionality

1. Inoculate a culture of a suitable *Listeria* strain in 3 mL TB medium and incubate overnight at 30 °C without shaking (*see* **Note 17**).

2. Add 7 mL of fresh TB medium to the culture, and further incubate at 30 °C for 2.5 h in order to obtain an exponential growth-phase culture.

3. Harvest the cells by centrifugation ($7,000 \times g$, 10 min, 4 °C), resuspend in 1/10 volume of PBS-T, pH 8.0, and store on ice (*see* **Note 18**).

4. Dilute the FP-CBD protein in PBS-T to a concentration of ~0.25 mg/mL (*see* **Note 19**).

5. Mix 100 μl of cells and 50 μl of protein in a 1.5 mL microcentrifuge tube, and incubate at room temperature for 15 min. Set up a control experiment with FP not fused to a CBD (no cell wall localization is expected in this case).

6. Harvest the cells by centrifugation (maximum speed for 1 min in a conventional table-top centrifuge), wash twice in 1 mL of PBS-T by repeated resuspension and centrifugation, and finally resuspend the cells in 50 μl PBS-T.

7. Add 1–2 μl of the cell suspension on a glass slide, and cover with a cover slip. Determine cell wall binding of the CBD (visualized by fluorescent bacterial cells) by fluorescence microscopy employing suitable excitation and emission filter sets and a 100× objective (*see* **Note 20**).

3.4 Coating of Paramagnetic Beads with CBD Proteins

1. Suspend lyophilized M-270 Epoxy Dynabeads in diglyme to a final concentration of 30 mg/mL (*see* **Note 21**).

2. Dilute the FP-CBD protein in PB, pH 7.4, to a concentration of approximately 1 mg/mL.

3. Resuspend the beads well by vortexing for 1–2 min. Add a 400 μl aliquot of suspended beads to a 2.0 mL centrifuge tube. Place the tube on the magnet for 4 min, remove the supernatant (diglyme), and then remove the tube from the magnet.

4. Wash the beads with 2×1 mL PB by repeated resuspension, incubation in an overhead rotator (10 rpm, 10 min, room temperature), and application of the magnet, and finally remove and discard the supernatant, keeping only the beads in the tube.

5. Add 200 μl of FP-CBD protein diluted in PB to the beads, mix thoroughly, and then add 200 μl of 3 M ammonium sulfate (*see* **Note 22**). Seal tube with parafilm, and let stand for 5 min.

6. Incubate the mixture at 4 °C in an overhead rotator at 10 rpm for 24 h.

7. Wash the beads in 4×1 mL of Tris buffer for 10 min each time. Resuspend in 400 μl Tris buffer for storage at 4 °C (*see* **Note 23**).

8. Resuspend the beads in 400 μl PBS (corresponding to a concentration of 2×10^9 coated beads/mL) for the following detection assay.

3.5 Detection of Listeria in Food Samples by CBD-MS

1. Add a 25 mg portion of the food sample to be investigated into a sterile polypropylene plastic bag, and homogenize using a Stomacher laboratory blender (*see* **Note 24**). In case of liquid foods, add 25 mL to the bag and omit the homogenization step.

2. Add 175 mL of TSB-ANC medium to the homogenized food sample in the plastic bag and incubate at 30 °C for 24 h (*see* **Note 25**).

3. Remove a 100 μl sample of enrichment culture from the bag, and mix with 90 μl of double-concentrated PBS-T (*see* **Note 26**) and 10 μl of FP-CBD-coated Dynabeads (corresponding to 2×10^7 beads) in a 1.5 mL centrifuge tube (*see* **Note 27**).

4. Incubate the mixture in an overhead rotator at 10 rpm for 40 min at room temperature.

5. Place the tube on the magnet for 4 min for bead separation, remove and discard the supernatant, remove the tube from the magnet, and resuspend the beads in 100 μl PBS-T.

6. Plate resuspended beads on Oxford agar and incubate for 24 h at 37 °C.

7. Emerging *Listeria* colonies on the Oxford agar indicate the presence of *Listeria* in the food sample (*see* **Note 28**).

3.6 Differentiation of Recovered Listeria Cells by Multiplex Binding Assays Using FR-CBDs

1. Recover presumptive *Listeria* colonies from Oxford agar plates (*see* Subheading 3.5) and collectively resuspend in PBS-T buffer (*see* **Note 29**).

2. Dilute each FR-CBD fusion protein to be used for the multiplex assay in PBS-T to a concentration of 0.5 mg/mL (if two different proteins are used) or 0.75 mg/mL (if three proteins are used) (*see* **Note 30**).

3. Mix equal volumes of each FP-CBD protein, so that the final concentration of each protein in the mixture is 0.25 mg/mL. The total volume of the mixture required for each binding assay is 50 μl.

4. Mix 100 μl of resuspended colony material and 50 μl of FP-CBD mixture, incubate at room temperature for 15 min, and proceed as described above for the cell wall binding assay (Subheading 3.3).

5. *Listeria* strains of different serovars present in the cell suspension can be distinguished by the color of cells in fluorescence microscopy, depending on the combination of FP-CBD proteins selected (*see* **Note 31**).

4   Notes

1. *Listeria* phage endolysins described to date consist of one N-terminal EAD and one C-terminal CBD, with the two domains mostly connected by a flexible linker region. As the CBDs in many cases constitute unique sequences without homology to known conserved domains (that would show up in the conserved domain display of the BLAST graphic summary), approximate positions of domain borders may have to be deduced on the basis of BLAST hits with homology to the EAD domain only.

2. It should be noted that other algorithms, e.g., for secondary structure prediction, may be equally suited to identify the positions of functional domains and linker sequences. In general, prediction of a CBD region by bioinformatics methods does not guarantee its functionality in an FR-CBD fusion protein. Therefore, it is recommended to generate several fusion proteins comprising different length variants of the putative CBD region and to empirically identify the construct with the highest functionality.

3. Primers should be designed so that they contain the desired restriction site at their 5′ end (preceded by four additional nucleotides to ensure proper functioning of the restriction enzymes) [7] and allow in-frame fusion of the amplified CBD-fragments downstream of the fluorescent reporter gene in the pQE-30 backbone. The absence of internal *Sac*I and *Sal*I restriction sites within the target sequence should be verified, and if necessary, alternative restriction sites within the multiple cloning site of the recipient vector should be selected. The different length variants of the amplified fragments should be selected in a way that they differ by 10–30 amino acids in length; the shorter variants comprise the complete putative CBD with and without the linker region, and the longer variants extend into the C-terminal region of the EAD. The pQE-30 plasmid encodes a 6×His tag fused to the N-terminus of the recombinant protein, allowing purification by immobilized metal ion affinity chromatography using nickel–NTA resin.

4. If purified phage genomic DNA is not available, a complete phage preparation can alternatively serve as a template for PCR. As a rule of thumb, use 1 μl of a 10^{-2} dilution of a phage suspension containing 10^{11} pfu/mL and boil for 10 min before adding to the PCR reaction.

5. It is recommended to first create fusions of GFP with each length variant of the target PCR, produce and purify the GFP-CBD fusion proteins, and test them for functionality in binding assays as described below. As a second step, create fusions of other FR proteins of interest with only the CBD variant that has proven most functional in these preliminary binding assays.

6. To allow sufficient aeration of the culture, the culture volume in the flask should not exceed ¼ of the total flask volume (e.g., use four 1 L flasks with 250 mL culture each for a total culture volume of 1 L).

7. The 4 °C overnight incubation with shaking is recommended for GFP and GFP-derived fluorescent proteins to promote autocatalytic chromophore formation. Depending on the FP used, this step may be omitted. If the expression protocol described here yields low concentrations of soluble protein and protein is present as inclusion bodies inside the cells instead,

we recommend using standard LB broth instead of LB-PE and lowering the incubation temperature to 19 °C after induction. If producing a novel FP-CBD construct, we recommend monitoring the expression and purification procedure by taking samples of the culture during overexpression and of various fractions during protein purification (*see* below) and analyzing the samples by SDS-PAGE. During overexpression, take one 500 μl sample each at time points 0, 1, 2, and 4 h after induction and one sample after overnight incubation. Dilute all samples to the same $OD_{600\ nm}$ as measured for the $t=0$ h sample using LB-PE. Harvest cells from 500 μl of each diluted sample, discard supernatant, resuspend the pellet in 50 μl of PBS, and freeze at −20 °C until use.

8. To facilitate resuspension of pellets after centrifugation, spin the dry pellet (after removal of supernatant) at low speed ($500 \times g$, 2 min, 4 °C) with the centrifuge cup turned by 180° in order to loosen the firm pellet from the wall of the cup. After addition of buffer A, resuspend pellet by vortexing or pipetting up and down using a 10 mL pipette. Try to get the suspension as homogeneous as possible prior to cell disruption. Keep buffers and cell suspensions on ice at all times.

9. Please refer to the manufacturer's instructions on how to operate the particular cell disruption device available in your lab. The sample should be kept as cold as possible by placing pressure cells on ice between individual runs and collecting the processed sample in a tube/beaker (preferably glass) placed in ice water (best is to let the sample run down the iced side wall of the collection tube). If you do not have access to any high-shear-based mechanical processing device, alternative methods such as sonication or bead beating can be used. Depending on the method used, a smaller or a greater sample volume may be optimal, and more time may be required for complete cell disruption. Efficacy can be monitored by microscopy, comparing the concentration of intact cells in the sample before and after processing (use phase contrast and 100× magnification).

10. Gently swirl the Ni–NTA slurry before use and between pipetting steps to ensure homogeneous resuspension and to prevent settling and transfer of unequal amounts of the resin to the Falcon tubes.

11. Pour the supernatant from the centrifuge tube onto the resin, including any soft/viscous material sitting on top of the solid pellet, since this material may contain a high concentration of target proteins.

12. Close the column outlet using the included tip closure. Put column upright into a rack. Add resin resuspended in buffer A into the column, and allow the resin to settle for 20–30 min. Place the upper bed support on top of the resin bed by inserting

it into the column and pushing it down using the upper end of a 10 mL plastic pipette (remove the filter from the pipet first). Be careful when pushing the support down the inside of the column as it tends to flip sideways. The support should be sitting snugly on top of the resin bed without compressing the resin. Place the assembled column in a metal rack or anything of the like on top of a large collection container, and remove the tip closure, allowing the buffer to drain into the container until the liquid inside the column is level with the top of the resin bed. If a novel FP-CBD construct is being purified, take a sample for SDS-PAGE (5 μl each) of the flow-through (*see* **Note 7**). Avoid letting the resin run dry.

13. Keep buffers on ice during the entire purification procedure. One column volume corresponds to 1 mL when columns with 1 mL resin bed are used. During elution of target protein, add elution buffer in ½ CV (= 0.5 mL) increments and collect eight fractions at first. Measure the protein concentration in each fraction (e.g., using a spectrophotometer). If concentrations are still increasing up to the last fraction, more fractions can be collected. If a novel FP-CBD construct is being purified, take a sample for SDS-PAGE (5 μl each) of the 10 % buffer B wash fraction (*see* **Note 7**).

14. If the protein concentration is determined by spectrophotometry, measure the absorption at 280 nm and correct the value by taking into account the protein's molar extinction coefficient. If a novel FP-CBD construct is being purified, take a sample for SDS-PAGE (5 μl each) of the pooled eluate (*see* **Note 7**).

15. If using dialysis tubing, soak the tubing in ultrapure water for 15 min and then briefly in dialysis buffer. Close the tubing on one end using a tubing clamp. Carefully pipette the protein sample into the open end of the tubing (twist the wet tubing between two fingers to create an opening), and close the second end with a clamp. Avoid including large air bubbles. Place the dialysis tubing into a beaker with 1.5 L of cold dialysis buffer and gently agitate on a magnetic stirrer for 3 h at 4 °C. Change the buffer, and continue stirring overnight. After dialysis, open the tubing on one end and remove the sample using a pipette. If solubility problems are encountered (cloudy dialysate), try increasing the NaCl concentration of the dialysis buffer to 300 or 500 mM.

16. If a novel FP-CBD construct is being purified, take a sample for SDS-PAGE (5 μl each) of the final protein preparation (*see* **Note 7**). For SDS-PAGE, load 20–30 μl of the samples taken during overexpression and 2–3 μl of the samples taken during protein purification (including final preparation) onto the gel.

17. If a previously characterized FP-CBD protein (such as those described in [8]) is being used, use a *Listeria* strain from a

serovar group that is recognized by this particular CBD. In case of a novel, yet uncharacterized FP-CBD construct, it is recommended to perform the binding assay with a representative set of *Listeria* strains, including strains from all major species and serovars [7].

18. In case of a novel FP-CBD protein, the optimal pH for cell wall binding may have to be determined in a preliminary set of experiments.

19. The FP-CBD protein concentration required to yield completely saturated *Listeria* cell surfaces in the binding assay depends on the number of binding ligands present per average cell (mostly in the range of 10^4 to 10^5) and the bacterial concentration. The protein concentration of 0.25 mg/mL suggested here represents excess protein under normal circumstances.

20. Gently tap the cover slip with a pen or something of the like to spread the sample underneath the slip, and enable localization of cells in a single focal layer. Alternatively, cells can be fixed by quickly pulling the slide through a Bunsen burner flame. The filter set to be used for fluorescent microscopy depends on the spectral properties of the FP. We recommend using the following filter sets for selected FPs: excitation BP 450–490 nm, beam splitter FT 510 nm, emission LP 520 nm for green fluorescent protein (GFP); excitation G 365 nm, beam splitter 395 nm, and emission LP 420 nm for blue fluorescent protein (BFP); excitation BP 436/10 nm, beam splitter FT 460 nm, and emission LP 470 nm for cyan fluorescent protein (CFP); excitation BP 500/20 and emission BP 535/30 for yellow fluorescent protein (YFP); and excitation BP546/12 nm, beam splitter FT 560 nm, and emission BP 575–640 nm for dsRed.

21. When suspending the beads in an organic solvent such as diglyme, they are stable for at least 1 year when stored at 2–8 °C. 30 mg/mL corresponds to a bead concentration of 2×10^9 beads/mL. When working with the beads, keep them well suspended at all time (vortex before each pipetting step).

22. The volumes of beads, protein, and ammonium sulfate added to the reaction mixture may be adjusted as required. However, the ratio of the individual components should remain constant. A final ammonium sulfate concentration of 1.5 M is desired for the coating procedure. The FP portion of the FP-CBD fusion protein has no relevance for detection but functions as a spacer in the bead-coating procedure, decreasing the chance of CBD inactivation through immobilization on the bead surface.

23. This step blocks residual epoxy groups by formation of covalent bonds between the amino group of Tris and the activated epoxy groups on the bead surface. To prevent microbial growth, NaN_3 may be added at a concentration of 0.02 % before storage.

24. To facilitate homogenization of particularly dry food samples, up to half of the volume of the 175 mL TSB-ANC medium (*see* **step 2**) can be added to the sample prior to processing with the stomacher.

25. This step allows selective enrichment of *Listeria* cells present in the food sample. Incubation in TSB-ANC medium for 24 h at 30 °C has been shown to be sufficient for detection of *Listeria* concentrations as low as 0.1–1.0 CFU per gram of food [2].

26. Addition of PBS-T, pH 8.0 ensures optimum pH conditions for binding of previously described CBDs to the listerial cell wall [7]. Please note that pH dependence of CBD binding should be investigated for every novel CBD and buffers adjusted accordingly.

27. To ensure the recovery of *Listeria* cells of all serovars from the food sample, a CBD with a broad binding range (and ideally high binding affinity) has to be used, such as CBDP40 [8]. Alternatively, a 1:1 mixture of beads coated with CBD500 and CBD118, which feature complementary binding ranges and together recognize *Listeria* strains from all species and serovars, can be employed as previously described [2].

28. *Listeria* can be recognized from characteristic black zones formed around the colonies, which is due to hydrolysis of aesculin present in the Oxford agar and consequently formation of black iron phenolic compounds. Growth of Gram-negative bacteria is completely inhibited, and most Gram-positive bacteria are suppressed. Staphylococci may grow, however, without formation of a black halo. As an alternative to the culture-based method described here, recovered *Listeria* cells can also be quantified by real-time PCR techniques as recently demonstrated [10].

29. Colonies can be recovered by scraping them off the plate using an inoculation loop. If less than 100 CFU have emerged on the Oxford agar for a given food sample, recover all colonies. The volume of buffer required for resuspension depends on the mass of the cell material. This requires some experience. As a rule of thumb, a pellet resulting from centrifugation of 100 µl of the suspension should be similar in size to a pellet from 1 mL of an exponentially growing *Listeria* culture (*see* Subheading 3.3).

30. Mixtures of up to three different FP-CBD proteins have been used for multiplex binding assays. It should be noted that the excitation and emission spectra of the FPs should be non-overlapping to allow sufficient separation of signals by the filter sets employed for microscopy. Examples for suitable combinations include dsRed/BFP, dsRed/CFP, dsRed/GFP, dsRed/YFP, BFP/GFP, BFP/YFP, BFP/GFP/dsRed, and BFP/YFP/dsRed.

31. For binding properties of a set of previously characterized Listeria phage CBDs, please refer to Schmelcher et al. 2010 [8]. Addition of novel CBDs featuring other binding patterns to this existing set of tools will make the CBD-based differentiation system increasingly powerful. As an alternative to the multiplex setup described here, a collection of unique CBDs fused to the same FP could serve for CBD-based typing of *Listeria* isolates. In this case, the isolate would be analyzed by performing multiple binding assays (each with a different CBD) in parallel, similar to the use of bacteriophage sets in phage typing [11]. Compared to phage typing, however, CBD-based typing offers several advantages with respect to time required, reproducibility, and ease of use [8].

Acknowledgment

We thank Fritz Eichenseher for critical reading of the manuscript.

References

1. Vazquez-Boland JA, Kuhn M, Berche P, Chakraborty T, Domínguez-Bernal G, Goebel W, González-Zorn B, Wehland J, Kreft J (2001) *Listeria* pathogenesis and molecular virulence determinants. Clin Microbiol Rev 14:584–640

2. Kretzer JW, Lehmann R, Schmelcher M, Banz M, Kim KP, Korn C, Loessner MJ (2007) Use of high-affinity cell wall-binding domains of bacteriophage endolysins for immobilization and separation of bacterial cells. Appl Environ Microbiol 73:1992–2000

3. Schmelcher M, Donovan DM, Loessner MJ (2012) Bacteriophage endolysins as novel antimicrobials. Future Microbiol 7:1147–1171

4. Fischetti VA (2010) Bacteriophage endolysins: a novel anti-infective to control Gram-positive pathogens. Int J Med Microbiol 300:357–362

5. Eugster MR, Haug MC, Huwiler SG, Loessner MJ (2011) The cell wall binding domain of *Listeria* bacteriophage endolysin PlyP35 recognizes terminal GlcNAc residues in cell wall teichoic acid. Mol Microbiol 81:1419–1432

6. Eugster MR, Loessner MJ (2012) Wall teichoic acids restrict access of bacteriophage endolysin Ply118, Ply511, and PlyP40 cell wall binding domains to the *Listeria monocytogenes* peptidoglycan. J Bacteriol 194:6498–6506

7. Loessner MJ, Kramer K, Ebel F, Scherer S (2002) C-terminal domains of *Listeria monocytogenes* bacteriophage murein hydrolases determine specific recognition and high-affinity binding to bacterial cell wall carbohydrates. Mol Microbiol 44:335–349

8. Schmelcher M, Shabarova T, Eugster MR, Eichenseher F, Tchang VS, Banz M, Loessner MJ (2010) Rapid multiplex detection and differentiation of *Listeria* cells by use of fluorescent phage endolysin cell wall binding domains. Appl Environ Microbiol 76:5745–5756

9. Sambrook J, Fritsch EF, Maniatis T (1989) Molecular cloning: a laboratory manual, vol 2. Cold Spring Harbor Laboratory, Cold Spring Harbor, NY

10. Walcher G, Stessl B, Wagner M, Eichenseher F, Loessner MJ, Hein I (2010) Evaluation of paramagnetic beads coated with recombinant *Listeria* phage endolysin-derived cell-wall-binding domain proteins for separation of *Listeria monocytogenes* from raw milk in combination with culture-based and real-time polymerase chain reaction-based quantification. Foodborne Pathog Dis 7:1019–1024

11. Loessner MJ, Busse M (1990) Bacteriophage typing of *Listeria* species. Appl Environ Microbiol 56:1912–1918

Chapter 13

Virulence Characterization of *Listeria monocytogenes*

Swetha Reddy and Mark L. Lawrence

Abstract

Listeria monocytogenes, a foodborne intracellular pathogen, is responsible for about 300 deaths every year in the USA. It has the ability to escape host defense mechanisms and causes listeriosis more frequently in immunocompromised individuals. Virulence mechanisms in *L. monocytogenes* are highly regulated and tightly controlled. A number of virulence factors that play important roles in pathogenesis of listeriosis have been identified and characterized. This review highlights the power of comparative genomics and functional genomics in identifying genes and proteins involved in the infection process. These genes and proteins are potentially useful as biomarkers for detecting virulent *L. monocytogenes*. This review also focuses on developments in the in vivo and in vitro models used in characterization of listerial virulence.

Key words *Listeria monocytogenes*, Genomics, Virulence, Animal model, Cell culture, PCR, Biomarker

1 Introduction

Listeria monocytogenes is the second most common cause of death from foodborne infections in the USA, and it has the highest case fatality rate of all foodborne pathogens [1]. Listeriosis is characterized by bacteremia, meningoencephalitis, abortions, and fatalities. It causes around 300 deaths every year in the USA, and the listerial cantaloupe outbreak in 2011 is thought to be one of the deadliest foodborne outbreaks in recent times. It not only infects humans but also rabbits, horses, and guinea pigs. A unique feature of *L. monocytogenes* is that it can survive harsh environments such as low pH, low temperature, and high salt; it is a particular problem in refrigerated food. *L. monocytogenes* has the ability to infect both phagocytic and non-phagocytic cells. Another unique feature of *L. monocytogenes* is its ability to cross three host barriers: intestinal, blood–brain, and materno-fetal. For these reasons, *L. monocytogenes* is considered a model pathogen [2].

L. monocytogenes isolates are classified into four different lineages (I, II, III, and IV) of which most isolates (and disease cases) belong to lineages I and II. Based on somatic and flagellar antigens,

Kieran Jordan et al. (eds.), *Listeria monocytogenes: Methods and Protocols*, Methods in Molecular Biology, vol. 1157,
DOI 10.1007/978-1-4939-0703-8_13, © Springer Science+Business Media New York 2014

L. monocytogenes is divided into 12 serotypes (1/2a, 1/2b, 1/2c, 3a, 3b, 3c, 4a, 4b, 4c, 4d, 4e, and 7) [3]. These isolates vary in pathogenic potential; serotypes 4b, 1/2a, 1/2b, and 1/2c account for >98 % of listeriosis in humans [4–6] and are considered to have the highest risk for foodborne illness. Other serotypes (3a, 3b, 3c, 4c, 4d, and 4e) are pathogenic in the mouse model, yet they have much lower incidence in listeriosis [7]. Thus, there is need for methods for determining pathogenic potential of listerial isolates. In this review, we discuss methods for determining virulence of *L. monocytogenes*, including traditional in vivo and cell culture methods as well as newer molecular methods.

Virulence in *L. monocytogenes* is serotype and strain dependent. Characterization of strain virulence would not only help understand virulence mechanisms but also potentially help predict the risk of a strain causing listeriosis. Methods for determining strain virulence include in vivo (animal models), in vitro (cell culture assays), and molecular methods (PCR or protein biomarkers).

1.1 In Vivo Studies: Animal Models

L. monocytogenes naturally infects many animal species, but it is not always practically possible to use them as laboratory animal models. An appropriate listerial animal model should have comparable cell and tissue tropisms as humans as well as a similar physiology, immune response, and pathophysiology of infection. Other criteria that impact the selection of practical animal models are size and cost of the animal, husbandry requirements, ability to reproduce in captivity, and length of gestation period. Considering all these factors, mice have been widely used as animal models to study virulence in *L. monocytogenes* [8, 9].

Listerial virulence is usually assessed by determining bacterial concentrations in liver and spleen at specific time points after infection or evaluating the 50 % lethal dose (LD_{50}). Routes of infection include oral, nasal, intraperitoneal, intravenous, or subcutaneous routes. The oral route most closely mimics natural infection; however, mice do not naturally express a receptor for internalin A (InlA), which is important for listerial invasion of the intestine. Therefore, a transgenic mouse line was developed that expresses human E-cadherin, which is a species-specific ligand receptor for *L. monocytogenes* [10–12]. Murine E-cadherin does not bind to InlA and thus does not allow *L. monocytogenes* invasion. Transgenic mice expressing human E-cadherin helped in elucidating the role of InlA–cadherin interaction during invasion of intestinal epithelium [10–12].

Zebrafish (*Danio rerio*) has become a popular vertebrate model for infectious disease investigations. Zebrafish larvae have proven to be effective in evaluation of listerial virulence genes, showing similar patterns of infection as mice [13, 14]. Three invertebrate hosts, *Galleria mellonella* (wax moth), *Drosophila melanogaster* (fruit fly), and *Caenorhabditis elegans* (round worm), have also been evaluated as model organisms and show potential for evaluating listerial virulence [15–17].

1.2 Cell Culture Studies

L. monocytogenes has the ability to survive and replicate in host cells and also is capable of cell-to-cell spread. Several mammalian cell lines have been used in listerial pathogenesis studies, and some are useful for evaluating listerial virulence. Parameters that have been measured using cell lines include adherence, invasion, intracellular replication, cell-to-cell spread, and plaque formation. Examples of cell lines used include Caco-2 (human epithelial colorectal adeno-carcinoma cells), HT-29 (human colon adenocarcinoma cells), Vero (kidney epithelial cell line), Hep-G2 (hepatocytes), Henle 407 (human embryonic epithelial cell line), A549 (lung alveolar basal epithelial cells), HEK293 (embryonic kidney cells), THP-1 (monocytes), L2 (mouse fibroblasts), J774 (murine macrophage cell line), PtK2 (male rat kidney cell line), and LLC-PK$_1$ and PK$_{15}$ (pig kidney epithelial cell line) [18–21]. Myeloid dendritic cells have also been used to study in vitro suppression of T cell functions after *L. monocytogenes* infection [22]. Of these, the Caco-2 cell line has been most widely used to investigate intestinal adherence/invasion and intracellular replication of *L. monocytogenes* [23, 24]. This cell line is also useful for evaluating virulence of listerial strains.

1.3 Molecular Methods

Development of molecular methods for predicting listerial virulence is complicated by the complex pathogenesis of listeriosis. Multiple sets of virulence genes and regulatory cascades are required for *L. monocytogenes* to successfully invade, spread, evade host defenses, and replicate in multiple cell types (including professional phagocytes). Redundancy of virulence genes and overlapping functions further complicates identification of gene targets that can reliably distinguish virulent and avirulent strains. For this reason, accurate molecular prediction of listerial virulence is not simple and is likely to require the detection of multiple gene targets. Nevertheless, advances in genomics have enabled progress in the identification of effective virulence target genes. We will review the use of comparative genomics and gene/protein expression studies to identify potential virulence gene targets.

1.4 Comparative Genomics

Comparative genome sequencing investigates the differences in gene composition between virulent and avirulent *L. monocytogenes* strains. It not only gives information about the systematics of various genetic lineages but also identifies candidate genes responsible for listerial virulence. The complete genome sequence of *L. monocytogenes* was first released in 2001 and used for comparative purposes to identify virulence genes based on comparison with *L. innocua* [25]. Based on comparison of these genome sequences and comparative hybridizations, genes were identified that are present primarily in clinical or food listerial isolates and missing in environmental isolates [7]. Among these, *lmo2821* (encoding InlJ) was present in all the virulent isolates as determined by mouse virulence assay, and *lmo2470* (encoding an interalin-like protein) was present in all but one of the virulent isolates. Subsequently, sequences of several *Listeria* strains (both virulent and avirulent) were released, which resulted in better understanding of listerial genome organization [26, 27].

Comparative genomics of lineages I, II, and III revealed loss of surface proteins (responsible for virulence) in lineage III (serotype 4a) along with prophage insertions [28]. Among the surface protein genes identified as missing in lineage III were *lmo2821* (InlJ) and *lmo2470* (internalin-like protein), both of which were identified previously [7]. Comparing the genomes of 13 *Listeria* strains, 2,032 core and 2,918 accessory genes were identified [26]. A group of 22 "virulence-associated" genes were found to be variably present in these strains. Cluster analysis of these genes divided the *Listeria* into two clusters, and only the gene encoding internalin A (*inlA*) was consistently present in the cluster containing virulent isolates and missing in the second cluster. The gene encoding internalin C (*inlC*) was the only gene that was present in the four virulent strains in their analysis and missing in all the other strains.

Comparison of the serotype 4a *L. monocytogenes* isolate HCC23 genome (NC_011660) to the complete genomes of *L. innocua* strain CLIP1182 and pathogenic isolates EGD-e (serotype 1/2a) and F2365 (serotype 4b) allowed identification of 58 protein-coding genes that are present in EGD-e and F2365 but are missing in HCC23 and *L. innocua*. Some of the virulence-associated genes encode known virulence factors such as InlC, InlH, and InlJ. Interestingly, six PTS enzyme II transporters and cell surface proteins were identified that are present in only pathogenic isolates.

Based on these comparative genomics studies, *L. monocytogenes-specific* and serovar-specific marker genes have been used to develop diagnostic PCR methods [29, 30]. One multiplex method separates *L. monocytogenes* into five serovar groups: 1/2a-3a, 1/2b-3b-7, 1/2c-3c, 4b-4d-4e, and 4a-4c [29]. Another multiplex PCR method utilizes internalin genes *inlA* (for species-specific detection of *L. monocytogenes*), *inlJ*, and *inlC* for differentiating virulent and avirulent isolates [30]. Sixteen *L. monocytogenes* isolates that were pathogenic for mice were positive for either *inlC* or *inlJ* or both, and six nonpathogenic isolates were negative for both genes.

1.5 Functional Genomics Methods to Identify Virulence Genes

Pathogenicity is a complex process that involves coordination of virulence gene expression during different stages of infection. One target virulence gene can be under the influence of several regulators that coordinate to ensure the expression of a specific gene under favorable conditions. Several studies have been conducted to analyze changes in listerial gene expression during intracellular growth or in the host. Proteins that are differentially expressed by virulent isolates during infection could potentially be useful biomarkers for distinguishing virulent and avirulent *L. monocytogenes* isolates.

Genes involved in glycolysis and TCA cycle (downregulated) and chaperones (upregulated) were found to be regulated during the initiation of strain EGD-e infection in macrophages [31]. *L. monocytogenes* strain EGD-e upregulates the expression of *prfA-regulated* genes, genes required for utilizing alternate carbon sources, and chaperone genes within the epithelial cell line Caco-2 [32].

Strain EGD-e actively replicates within the spleen of infected mice, and it undergoes a shift from expression of metabolic genes to increased expression of virulence genes during this stage of active growth in the spleen [33]. EGD-e was also found to increase the expression of proteins involved in cell wall biosynthesis and some metabolic pathways following a 24-h infection in macrophages as identified through 2D-DIGE methods [34]. Whole-genome transcriptional profiling identified sigma B, sigma 54, and VirR regulons as genes upregulated during stress conditions and virulence [35–37].

Comparison of protein expression of three *L. monocytogenes* strains was conducted in response to phagocytosis by J774A.1 macrophages using multidimensional protein identification technology coupled with electrospray ionization tandem mass spectrometry [38]. A large number of proteins were identified that were differentially expressed in virulent strains and not in avirulent strain HCC23. For example, oxidative stress response proteins and protein chaperones are potentially important components for long-term listerial survival in macrophages. Furthermore, induction of central intermediary metabolism (glycolysis) may contribute to the ability to replicate intracellularly. An example of a differentially expressed protein that may be useful as a biomarker of listerial virulence is LMOf2365_2464 (accession #46908663), which had over 100-fold increased quantity at 3 h post-infection compared to pre-infection, and it had over 60-fold increased quantity at 5 h versus pre-infection. In EGD-e, the LMOf2365_2464 ortholog (accession #16804529) had over 60-fold increased quantity at 3 h post-infection versus pre-infection, and it had almost 40-fold increased quantity at 5 h versus pre-infection. This protein was not differentially expressed in avirulent strain HCC23 in macrophages.

2 Materials

2.1 Mouse Virulence Assay

1. Phosphate-buffered saline (PBS), pH 7.2: To prepare 1 L 0.1 M PBS (10×), mix 10.9 g of Na_2HPO_4 (anhydrous), 3.2 g of NaH_2PO_4 (anhydrous), and 90 g of NaCl. Adjust the pH to 7.2, and store it at room temperature. To make 1× PBS, mix 100 ml of 10× PBS in 900 ml water. Filter sterilize the buffer prior to use.

2. 0.05 % Triton X-100: Mix 500 μL Triton X-100 in 1,000 ml water and store at room temperature. Filter sterilize the solution prior to use.

3. Divide fifty A/J, 6–8-week-old female, into groups of five. Allow the mice to acclimatize to new environment for 1 week prior to infection.

4. Sterilize forceps and meshes for maceration of tissue, and prepare BHI agar plates.

5. The day before experimental infection, inoculate 25 ml BHI broth with wild-type and mutant strains. Grow the cultures till they reach an OD_{540} of 1.35 (usually takes 10–12 h).

2.2 Cell Culture Assay

1. Grow human colon carcinoma enterocyte-like epithelial cells (Caco-2) in *Eagle's* Minimum Essential Medium and 20 % fetal bovine serum at 37 °C and 5 % CO_2.

2. 0.1 % Triton X-100: Mix 10 ml Triton X-100 in 990 ml water and store at 4 °C. Filter sterilize the solution prior to use.

3. Two days prior to the assay, seed Caco-2 cells in 12-well plates with approximately 10^5 cells per plate. On the day of assay, wash cells with PBS, and add fresh prewarmed media to the wells.

4. On the day prior to infection, inoculate 10 ml BHI broth with wild-type and mutant strains. Grow the cultures till they reach an OD_{600} of 1.

3 Methods

3.1 Mouse Virulence Assay

1. On the day of infection, remove 1 ml of bacterial broth and pellet by centrifugation. Resuspend the pellet in 1 ml PBS. Perform serial dilutions to obtain a bacterial dose of 10^{-5}, 10^{-6}, 10^{-7}, and 10^{-8}. Spread these dilutions on BHI agar to quantify CFU/ml by plate counting.

2. Using humane animal handling techniques, intraperitoneally inoculate four groups of 6–8-week-old female A/J mice, five mice per group, with appropriate dilutions of bacteria [5–8].

3. Inoculate five mice with 100 μL of sterile saline, which serves as a negative control for the experiment. Do not inject any solution to a group of five mice. This serves as a second negative control.

4. Observe the mice two to three times a day to monitor morbidity and mortality. If any mice show unusual behavior, record it for future reference.

5. If mice become moribund, they should be humanely euthanized. Mice that are depressed but not moribund should be left alone. Depression may include decreased appetite and piloerection, but mice will be responsive to handling. Moribund mice are defined as those that are unresponsive to handling or are immobile and unable to eat or drink. Aseptically remove liver, spleen, and thymus from each mouse using a sterile forceps.

6. On day 6 after recording clinical signs, humanely euthanize all mice. Aseptically remove liver, spleen, and thymus from each mouse using a sterile forceps.

7. Add 1 ml of 0.05 % Triton X-100 to each tissue, and homogenize it using a sterile mesh. Dilute this tissue homogenate to 10^{-2} and spread on BHI agar. Incubate the plates at 37 °C for 48 h.

8. Count and record colonies on the plates. Use this data to calculate CFU/ml. Compare the data obtained for mice infected with wild-type and mutant strain, and perform a Student's *t*-test to determine if the difference observed is statistically significant.

3.2 Cell Culture Assay

1. Adjust overnight cultures of wild-type strain and mutant strain to an $OD_{600} = 1.0$, and then add approximately 10^6 bacteria to each well to yield a multiplicity of infection (MOI) of 10:1.

2. For adhesion experiments, briefly centrifuge plates for 45 s and incubate at 37 °C for 30 min. Wash cells five times with PBS and lyse using 500 μL of cold 0.1 % Triton X-100. Dilute the resulting suspensions and spread on BHI agar plates. Incubate plates at 37 °C for 48 h.

3. For invasion assay, incubate the plates for 2 h after infection. Wash cells twice with PBS, and add fresh media containing gentamicin (1 μg/ml) to kill extracellular bacteria. At 2, 4, 6, 8, 10, 12, 14, 16, 18, 20, 22, and 24 h post-infection, wash cells twice with PBS and lyse using 500 μL of cold 0.1 % Triton X-100 with sonication (Fisher Scientific Sonic Dismembrator Model 100, setting 3, 3 pulses, 5 s each). Dilute resulting suspension and spread on BHI agar plates. Incubate plates at 37 °C for 48 h.

4. Perform three independent assays to obtain consistent data. Record colony counts, and calculate CFU/ml.

5. Perform Student's *t*-test to determine significant differences in intracellular bacterial (wild type and mutant) quantities in Caco-2 cells.

3.3 Conclusions

As illustrated throughout this review, considerable progress has been made to identify virulence factors and understand their implication in listeriosis. The gold standard method for assessing listerial virulence is using the mouse model; however, newer fish and invertebrate host models have been developed to evaluate listerial virulence. Cell culture models are also useful for evaluating listerial virulence, and the Caco-2 cell line is one of the most widely used. Comparative genomics methods have been useful for the identification of virulence gene markers; in particular, an assay has been developed that utilizes internalin genes *inlJ* and *inlC* for predicting listerial virulence. Functional genomics assays have the potential for identifying useful protein biomarkers for identifying virulent *L. monocytogenes*. LMOf2365_2464 is an example of such a biomarker.

References

1. Barton Behravesh C, Jones TF, Vugia DJ, Long C, Marcus R, Smith K, Thomas S, Zansky S, Fullerton KE, Henao OL, Scallan E (2011) Deaths associated with bacterial pathogens transmitted commonly through food: foodborne diseases active surveillance network (FoodNet), 1996–2005. J Infect Dis 204:263–267

2. Hamon M, Bierne H, Cossart P (2006) *Listeria monocytogenes*: a multifaceted model. Nat Rev Microbiol 4:423–434

3. Seeliger HP (1984) Modern taxonomy of the *Listeria* group relationship to its pathogenicity. Clin Invest Med 7:217–221

4. Barbour AH, Rampling A, Hormaeche CE (2001) Variation in the infectivity of *Listeria monocytogenes* isolates following intragastric inoculation of mice. Infect Immun 69:4657–4660

5. Kim SH, Bakko MK, Knowles D, Borucki MK (2004) Oral inoculation of A/J mice for detection of invasiveness differences between *Listeria monocytogenes* epidemic and environmental strains. Infect Immun 72:4318–4321

6. Roche SM, Gracieux P, Albert I, Gouali M, Jacquet C, Martin PM, Velge P (2003) Experimental validation of low virulence in field strains of *Listeria monocytogenes*. Infect Immun 71:3429–3436

7. Liu D, Ainsworth AJ, Austin FW, Lawrence ML (2003) Characterization of virulent and avirulent *Listeria monocytogenes* strains by PCR amplification of putative transcriptional regulator and internalin genes. J Med Microbiol 52: 1065–1070

8. Lecuit M, Cossart P (2002) Genetically-modified-animal models for human infections: the *Listeria* paradigm. Trends Mol Med 8: 537–542

9. Lecuit M, Sonnenburg JL, Cossart P, Gordon JI (2007) Functional genomic studies of the intestinal response to a foodborne enteropathogen in a humanized gnotobiotic mouse model. J Biol Chem 282:15065–15072

10. Mengaud J, Ohayon H, Gounon P, Mege RM, Cossart P (1996) E-cadherin is the receptor for internalin, a surface protein required for entry of *L. monocytogenes* into epithelial cells. Cell 84:923–932

11. Lecuit M, Dramsi S, Gottardi C, Fedor-Chaiken M, Gumbiner B, Cossart P (1999) A single amino acid in E-cadherin responsible for host specificity towards the human pathogen *Listeria monocytogenes*. EMBO J 18:3956–3963

12. Lecuit M, Vandormael-Pournin S, Lefort J, Huerre M, Gounon P, Dupuy C, Babinet C, Cossart P (2001) A transgenic model for listeriosis: role of internalin in crossing the intestinal barrier. Science 292:1722–1725

13. Levraud JP, Disson O, Kissa K, Bonne I, Cossart P, Herbomel P, Lecuit M (2009) Real-time observation of *Listeria monocytogenes-phagocyte* interactions in living zebrafish larvae. Infect Immun 77:3651–3660

14. Coombes JL, Robey EA (2010) Dynamic imaging of host-pathogen interactions in vivo. Nat Rev Immunol 10:353–364

15. Chambers MC, Song KH, Schneider DS (2012) *Listeria monocytogenes* infection causes metabolic shifts in Drosophila melanogaster. PLoS One 7:e50679

16. Guha S, Klees M, Wang X, Li J, Dong Y, Cao M (2013) Influence of planktonic and sessile *Listeria monocytogenes* on Caenorhabditis elegans. Arch Microbiol 195:19–26

17. Joyce SA, Gahan CG (2010) Molecular pathogenesis of *Listeria monocytogenes* in the alternative model host Galleria mellonella. Microbiology 156:3456–3468

18. Bouwer HG, Bai A, Forman J, Gregory SH, Wing EJ, Barry RA, Hinrichs DJ (1998) *Listeria monocytogenes*-infected hepatocytes are targets of major histocompatibility complex class Ib-restricted antilisterial cytotoxic T lymphocytes. Infect Immun 66:2814–2817

19. Temm-Grove CJ, Jockusch BM, Rohde M, Niebuhr K, Chakraborty T, Wehland J (1994) Exploitation of microfilament proteins by *Listeria monocytogenes*: microvillus-like composition of the comet tails and vectorial spreading in polarized epithelial sheets. J Cell Sci 107(Pt 10):2951–2960

20. Bierne H, Travier L, Mahlakoiv T, Tailleux L, Subtil A, Lebreton A, Paliwal A, Gicquel B, Staeheli P, Lecuit M, Cossart P (2012) Activation of type III interferon genes by pathogenic bacteria in infected epithelial cells and mouse placenta. PLoS One 7:e39080

21. Yamada F, Ueda F, Ochiai Y, Mochizuki M, Shoji H, Ogawa-Goto K, Sata T, Ogasawara K, Fujima A, Hondo R (2006) Invasion assay of *Listeria monocytogenes* using Vero and Caco-2 cells. J Microbiol Methods 66:96–103

22. Popov A, Driesen J, Abdullah Z, Wickenhauser C, Beyer M, Debey-Pascher S, Saric T, Kummer S, Takikawa O, Domann E, Chakraborty T, Kronke M, Utermohlen O, Schultze JL (2008) Infection of myeloid dendritic cells with *Listeria monocytogenes* leads to the suppression of T cell function by multiple inhibitory mechanisms. J Immunol 181:4976–4988

23. Gaillard JL, Berche P, Mounier J, Richard S, Sansonetti P (1987) In vitro model of penetration and intracellular growth of *Listeria monocytogenes* in the human enterocyte-like cell line Caco-2. Infect Immun 55:2822–2829

24. Jaradat ZW, Bhunia AK (2003) Adhesion, invasion, and translocation characteristics of *Listeria monocytogenes* serotypes in Caco-2 cell and mouse models. Appl Environ Microbiol 69:3640–3645

25. Glaser P, Frangeul L, Buchrieser C, Rusniok C, Amend A, Baquero F, Berche P, Bloecker H, Brandt P, Chakraborty T, Charbit A, Chetouani F, Couve E, de Daruvar A, Dehoux P, Domann E, Dominguez-Bernal G, Duchaud E, Durant L, Dussurget O, Entian KD, Fsihi H, Garcia-del Portillo F, Garrido P, Gautier L, Goebel W, Gomez-Lopez N, Hain T, Hauf J, Jackson D, Jones LM, Kaerst U, Kreft J, Kuhn M, Kunst F, Kurapkat G, Madueno E, Maitournam A, Vicente JM, Ng E, Nedjari H, Nordsiek G, Novella S, de Pablos B, Perez-Diaz JC, Purcell R, Remmel B, Rose M, Schlueter T, Simoes N, Tierrez A, Vazquez-Boland JA, Voss H, Wehland J, Cossart P (2001) Comparative genomics of *Listeria* species. Science 294:849–852

26. den Bakker HC, Cummings CA, Ferreira V, Vatta P, Orsi RH, Degoricija L, Barker M, Petrauskene O, Furtado MR, Wiedmann M (2010) Comparative genomics of the bacterial genus *Listeria*: genome evolution is characterized by limited gene acquisition and limited gene loss. BMC Genomics 11:688

27. Buchrieser C, Rusniok C, Garrido P, Hain T, Scortti M, Lampidis R, Karst U, Chakraborty T, Cossart P, Kreft J, Vazquez-Boland JA, Goebel W, Glaser P (2011) Complete genome sequence of the animal pathogen *Listeria ivanovii*, which provides insights into host specificities and evolution of the genus *Listeria*. J Bacteriol 193:6787–6788

28. Hain T, Ghai R, Billion A, Kuenne CT, Steinweg C, Izar B, Mohamed W, Mraheil MA, Domann E, Schaffrath S, Karst U, Goesmann A, Oehm S, Puhler A, Merkl R, Vorwerk S, Glaser P, Garrido P, Rusniok C, Buchrieser C, Goebel W, Chakraborty T (2012) Comparative genomics and transcriptomics of lineages I, II, and III strains of *Listeria monocytogenes*. BMC Genomics 13:144

29. Doumith M, Buchrieser C, Glaser P, Jacquet C, Martin P (2004) Differentiation of the major *Listeria monocytogenes* serovars by multiplex PCR. J Clin Microbiol 42:3819–3822

30. Liu D, Lawrence ML, Austin FW, Ainsworth AJ (2007) A multiplex PCR for species- and virulence-specific determination of *Listeria monocytogenes*. J Microbiol Methods 71:133–140

31. Chatterjee SS, Hossain H, Otten S, Kuenne C, Kuchmina K, Machata S, Domann E, Chakraborty T, Hain T (2006) Intracellular gene expression profile of *Listeria monocytogenes*. Infect Immun 74:1323–1338

32. Joseph B, Przybilla K, Stuhler C, Schauer K, Slaghuis J, Fuchs TM, Goebel W (2006) Identification of *Listeria monocytogenes* genes contributing to intracellular replication by expression profiling and mutant screening. J Bacteriol 188:556–568

33. Camejo A, Buchrieser C, Couve E, Carvalho F, Reis O, Ferreira P, Sousa S, Cossart P, Cabanes D (2009) In vivo transcriptional profiling of *Listeria monocytogenes* and mutagenesis identify new virulence factors involved in infection. PLoS Pathog 5:e1000449

34. Van de Velde S, Delaive E, Dieu M, Carryn S, Van Bambeke F, Devreese B, Raes M, Tulkens PM (2009) Isolation and 2-D-DIGE proteomic analysis of intracellular and extracellular forms of *Listeria monocytogenes*. Proteomics 9:5484–5496

35. Hain T, Hossain H, Chatterjee SS, Machata S, Volk U, Wagner S, Brors B, Haas S, Kuenne CT, Billion A, Otten S, Pane-Farre J, Engelmann S, Chakraborty T (2008) Temporal transcriptomic analysis of the *Listeria monocytogenes* EGD-e sigmaB regulon. BMC Microbiol 8:20

36. Raengpradub S, Wiedmann M, Boor KJ (2008) Comparative analysis of the sigma B-dependent stress responses in *Listeria monocytogenes* and *Listeria innocua* strains exposed to selected stress conditions. Appl Environ Microbiol 74: 158–171

37. Arous S, Buchrieser C, Folio P, Glaser P, Namane A, Hebraud M, Hechard Y (2004) Global analysis of gene expression in an rpoN mutant of *Listeria monocytogenes*. Microbiology 150:1581–1590

38. Donaldson JR, Nanduri B, Pittman JR, Givaruangsawat S, Burgess SC, Lawrence ML (2011) Proteomic expression profiles of virulent and avirulent strains of *Listeria monocytogenes* isolated from macrophages. J Proteomics 74:1906–1917

Chapter 14

Internalization Assays for *Listeria monocytogenes*

Andreas Kühbacher, Pascale Cossart, and Javier Pizarro-Cerdá

Abstract

Listeria monocytogenes is a model intracellular pathogen that can invade the cytoplasm of host mammalian cells. Cellular invasion can be measured using standard techniques such as the classical gentamicin protection assay, based on the quantification of colony-forming units from lysates of infected cells. In addition, there are methods based on immunofluorescence microscopy which allow for assaying invasion in a medium- to high-throughput manner. In the following sections we detail two different assays that can be used alone or in combination to quantify the internalization of *L. monocytogenes* in host cells.

Key words Cell invasion, Invasion assay, Colony-forming unit, Fluorescence microscopy, Medium throughput

1 Introduction

Listeria monocytogenes is a facultative intracellular Gram-positive bacterium which invades a broad range of host cells by actin- and clathrin-dependent mechanisms triggered through interaction of the bacterial invasion internalin (InlA) with cellular E-cadherin and/or InlB with the hepatocyte growth factor receptor Met [1–4]. In epithelial cells, the lytic activity of the pore-forming toxin listeriolysin O (LLO) allows *L. monocytogenes* disruption of the vacuole in which the bacterium is entrapped after invasion and favors bacterial translocation to the host cell cytoplasm. Actin polymerization at the surface of cytoplasmic *L. monocytogenes* initiated by the bacterial membrane protein ActA leads to the formation of cytoskeletal structures known as "actin comet tails" which provide a force that enables bacteria to move throughout the cytoplasm of primary infected cells and to spread to neighboring cells [1]. In secondary infected cells, *L. monocytogenes* is initially located in a double-membrane vacuole that is lysed by the cooperative activities of the bacterial phospholipases PlcA and PlcB together with LLO, allowing spreading of bacteria to start a new infection cycle (Fig. 1).

Kieran Jordan et al. (eds.), *Listeria monocytogenes: Methods and Protocols*, Methods in Molecular Biology, vol. 1157, DOI 10.1007/978-1-4939-0703-8_14, © Springer Science+Business Media New York 2014

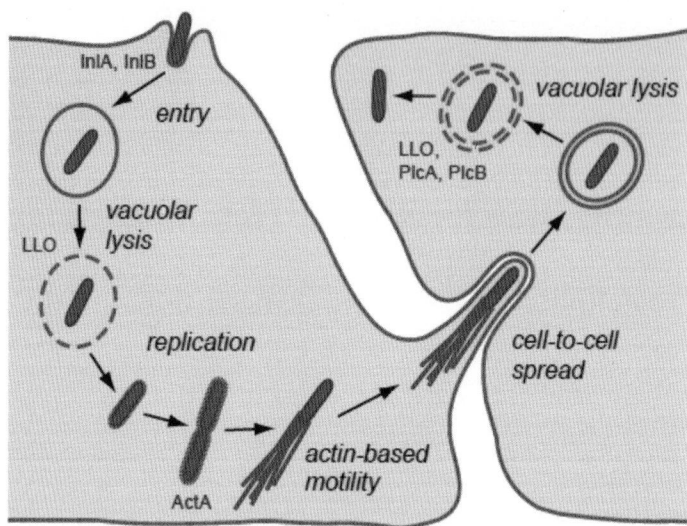

Fig. 1 *L. monocytogenes* intracellular cycle. *L. monocytogenes* promotes entry into target cells through interaction of the bacterial surface proteins InlA and InlB with their cellular receptors E-cadherin and Met, respectively. Rupture of the bacterial internalization vacuole is subsequently achieved by the secreted toxin LLO. Within the host cell cytoplasm, *L. monocytogenes* proliferates and induces actin polymerization by the surface protein ActA, favoring bacterial invasion of neighboring cells. The *L. monocytogenes*-containing vacuole in secondary infected cells is disrupted by the combined activity of LLO and the two bacterial phospholipases PlcA and PlcB. Reproduced from [2] (Copyright to Cold Spring Harbor Laboratory Press)

Different methods have been developed and improved over time to monitor the invasion of host cells by *L. monocytogenes* [5]. A common standard assay that is used for quantifying bacterial entry, intracellular survival, and replication is the gentamicin protection assay [6]: in that assay, cells are infected with bacteria for a given time, followed by treatment with the non-cell-permeable antibiotic gentamicin to kill extracellular *L. monocytogenes*; host cells are then lysed and lysates serially diluted and plated on agar plates. Quantification of bacterial colony-forming units (CFUs) allows determination of the fraction of invasive *L. monocytogenes* that were present inside host cells at the time of gentamicin addition and were therefore protected from the antibiotic treatment. The use of different incubation times after gentamicin addition allows the system to provide additional information about the entry efficiency or the ability of bacteria to survive and/or replicate within infected cells. Microscopic assessment of intracellular versus total bacterial populations is another method to specifically assay the entry step of *L. monocytogenes* into host cells [7]. This technique allows for a more sensitive measurement of bacterial invasion after very short internalization times since it does not require the use of gentamicin treatment after infection; moreover, this method is more robust towards cell number variation. In this chapter we focus on the use of these two different assays in a medium-throughput manner, allowing the screening of several perturbations simultaneously.

2 Materials

All assays are optimized for HeLa cells which are only permissive for the InlB/Met-dependent entry pathway. Cells are grown using standard cell culture procedures in the complete absence of antibiotics (to avoid residual bacterial killing by endocytosed antibiotics). Gentamicin protection assays are suited to investigate bacterial invasion using all strains from *L. monocytogenes*; as an example, we have used the strain EGDe in our experiments. Infections for microscopy-based assays in HeLa cells require more invasive strains such as the *L. monocytogenes* EGDe-PrfA* strain [8] used in our protocols, which harbors a point mutation in PrfA, the major transcriptional regulator of *L. monocytogenes* virulence genes, rendering this regulator constitutively active and therefore leading to an increase in the expression levels of the invasion protein InlB (*see* **Note 1**).

2.1 Gentamicin Protection Assay

1. Cell preservation medium: 50 % Dulbecco's modified Eagle medium (DMEM), 40 % heat-inactivated fetal calf serum (FCS), 10 % DMSO.

2. HeLa cells clone CCL2 (American Type Cell Collection): Kept in frozen stocks in cell preservation medium in liquid nitrogen.

3. Standard cell culture medium: DMEM supplemented with 10 % heat-inactivated FCS.

4. Infection medium: DMEM supplemented with 1 % heat-inactivated FCS.

5. Phosphate-buffered saline (PBS).

6. Trypsin/EDTA solution.

7. *L. monocytogenes* strain EGDe (our own laboratory collection): Kept in frozen glycerol stocks at −80 °C.

8. *L. monocytogenes* culture medium: Brain–heart infusion (BHI) as liquid medium or in agar plates.

9. 96-well cell culture plates.

10. Gentamicin.

11. Distilled water.

12. Multichannel pipette.

2.2 Differential Bacterial Staining Assay

1. Standard cell culture medium: DMEM, 10 % heat-inactivated FCS.

2. *L. monocytogenes* strain EGDe.PrfA*-green fluorescent protein (GFP).

3. Infection medium: DMEM supplemented with 1 % heat-inactivated FCS.

4. *L. monocytogenes* culture medium: BHI as a liquid medium or in the form of agar plates.

5. Dark 96-well cell culture plates for cell microscopy.

6. Fixation solution: PBS, 4 % paraformaldehyde (PFA).

7. PBS supplemented with 1 % bovine serum albumin (BSA).

8. PBS supplemented with 1 % BSA and 0.05 % saponin.

9. Anti-*L. monocytogenes* antibody.

10. Fluorescently labeled secondary antibodies.

11. DAPI or Hoechst, 1 mg/mL.

12. Fluorescently labeled phalloidin.

13. Fluorescence microscope equipped with a 10× or a 20× objective and a screening module.

14. Multichannel pipette.

15. Erythromycin.

3   Methods

3.1   Gentamicin Protection Assay

We present the assay in a 96-well plate format, but it can be easily downscaled to a 24-well plate format if required.

1. At least 3 days before the infection and starting from a frozen glycerol stock, streak *L. monocytogenes* strain EGDe on a BHI agar plate and grow for 48 h at 37 °C: individual bacterial colonies should appear on the plate which can then be stored at 4 °C for 1 month (while stored, seal the plate with parafilm to avoid agar desiccation).

2. At least 2 days before the infection, thaw HeLa cells by resuspending a frozen cellular aliquot in 10 mL of cold standard cell culture medium and centrifuge during 5 min at $200 \times g$ in a 15 mL polystyrene tube. Discard the supernatant, resuspend the cellular pellet in 12 mL of warm standard cell culture medium, add the cells to a 75 cm² flask, and transfer to a cell incubator with a humidified 5 % CO_2—containing atmosphere at 37 °C.

3. The day before infection, trypsinize a flask of confluent HeLa cells, prepare a cell suspension of 100,000 cells per mL in standard cell culture medium, and plate 100 µl of this suspension per well in a 96-well cell culture plate (*see* **Notes 2** and **3**). Do not use the outer wells of the plate for seeding cells, and fill them instead with PBS or cell culture medium (*see* **Note 4**).

4. Move the plate quickly back and forth to uniformly distribute cells, and let the cells settle down for 10 min at room temperature (*see* **Note 5**).

5. Transfer the plate to an incubator with a humidified 5 % CO_2—containing atmosphere at 37 °C where cells will be allowed to attach and spread overnight.

6. The day before the infection prepare a liquid culture of *L. monocytogenes* by inoculating a bacterial colony from the BHI agar plate to 5 mL of BHI liquid medium contained within a 15 mL polystyrene tube. Let the culture grow overnight at 37 °C in a shaker.

7. The actual day of the infection, take 1 mL of the overnight *L. monocytogenes* culture, centrifuge during 2 min at $10,600 \times g$ in a tabletop centrifuge, discard the supernatant (containing the secreted cytotoxin listeriolysin O), and resuspend the pellet in 1 mL of PBS (*see* **Note 6**).

8. Repeat the washing steps three more times.

9. Read the bacterial optical density at 600 nm, and estimate the number of bacteria (OD = 1 is equivalent of 1×10^9 bacteria/mL).

10. Prepare the inoculum by diluting bacteria in the appropriate volume of infection medium (*see* **Notes 7** and **8**).

11. Aspirate the medium from HeLa cells, and add 100 µl of the inoculum to each well (*see* **Note 9**).

12. Transfer the inoculated plate to an incubator with a humidified 5 % CO_2 atmosphere at 37 °C, and allow *L. monocytogenes* to invade the cells for 1 h.

13. During this incubation time, plate a dilution of the inoculum to precisely establish the number of bacteria that were used for infection. For this, using four wells from a 96-well plate row, put 90 µl of distilled water in each well and make a first 1:10 dilution of the inoculum by adding 10 µl of the inoculum to the first well of the row; homogenize and take 10 µl of this dilution that will be diluted in 90 µl of distilled water in the second well of the row; repeat two additional 1:10 dilutions, and plate 10 µl of dilutions 2, 3, and 4 on the top of a 10 cm BHI agar plate in triplicate using a multichannel pipette. Plate six lysates per 10 cm plate by pipetting the drops on top of the plate, and then tilt the plate to let them run down to make lines of lysate (Fig. 2) (*see* **Note 10**).

14. After the cellular infection has been completed for 1 h, wash the plate once with 100 µl per well of pre-warmed standard cell culture medium supplemented with 40 µg/mL gentamicin and replace by 100 µl of the same gentamicin-containing medium.

15. Transfer the 96-well plate back to the incubator, and keep it at 37 °C in 5 % CO_2 humidified atmosphere for another hour for assaying entry (if intracellular survival or replication is assayed, start the experiment by preparing and infecting two identical

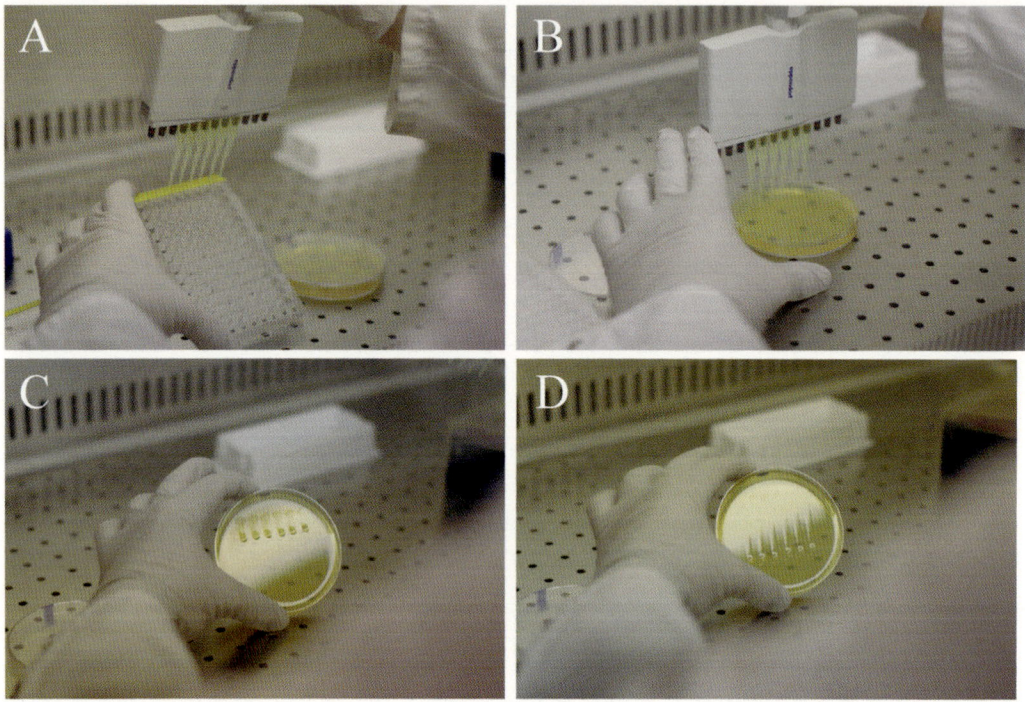

Fig. 2 Pipetting *L. monocytogenes* CFUs for the gentamicin protection assay. (**a**) After cellular lysis with distilled water, 10 µl containing resuspended *L. monocytogenes* CFUs from six independent wells (using a 96-well plate) is aspirated using a multichannel pipette. (**b**) The 10 µl is deposited at the border of 10 cm BHI agar plate. (**c**) The agar plate is tilted to allow the water drops containing the CFUs to run along the plate. (**d**) The drops are allowed to make a water path along the agar, after which the plate is allowed to dry and incubated at 37 °C

96-well plates, of which one should be kept in the incubator for 1 h and the other one for 4 h after addition of gentamicin) (*see* **Note 11**).

16. After the killing of the extracellular bacteria is completed by the gentamicin treatment, wash the cells once with 100 µl of standard cell culture medium without gentamicin per well, remove the medium, and lyse the cells by adding 100 µl of distilled water per well.

17. To completely disrupt the HeLa cells, pipette the water 5–10 times up and down using a multichannel pipette and scrape the cells by pressing the pipette tips on the well bottom and rotating while pipetting (*see* **Note 12**).

18. Make a 1:10 dilution in a new 96-well plate by adding 10 µl lysate to 90 µl distilled water per well. Dilute the lysates two more times to obtain 1:100 and 1:1,000 dilutions.

19. Plate 10 µl of each lysate and dilution on a 10 cm BHI agar plate: six lysates can be plated on one plate as described in **step 13** and Fig. 2.

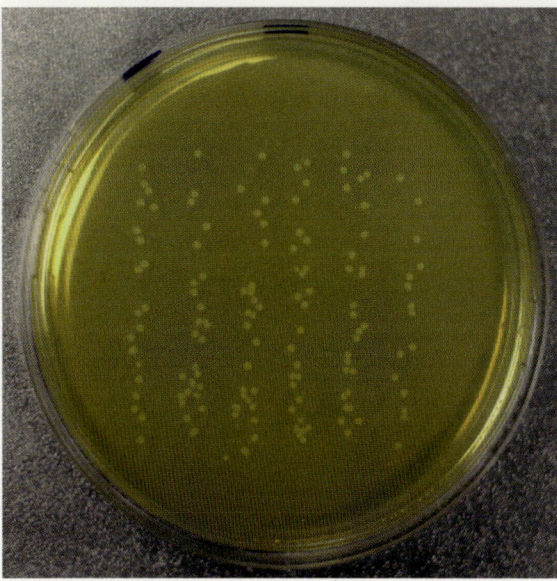

Fig. 3 Fully grown *L. monocytogenes* CFUs. 24 h after plating and growth at 37 °C, the individual *L. monocytogenes* CFUs can be counted to estimate bacterial entry into host cells

20. Let the lysates dry by opening the BHI agar plates under the microbiological hood for a few minutes.

21. Put the lid back on the BHI agar plates, and transfer them to an incubator at 37 °C to let bacterial colonies grow overnight.

22. The day after the infection, count colonies of each row for the dilution where colonies are well separated from each other (Fig. 3) and calculate entry as the ratio of CFU of lysates and inoculum. For quantifying intracellular survival or replication, calculate then the ratio of the late and early time points (*see* **Note 13**).

3.2 Differential Bacterial Staining

As with the previous protocol, we present this particular assay in a 96-well plate format with the goal of performing medium- to high-throughput assays with image acquisition using a microscope equipped with an automatic plate-reading system. The procedure can be upscaled to 384-well plates for the study of small interfering RNA libraries (*see* **Note 3**) or downscaled to a 24-well plate format in which cover slips are added in each well for individual treatment of each condition.

1. HeLa cells and *L. monocytogenes* EGDe.PrfA*-GFP are grown as described in **steps 1** and **2** of Subheading 3.1 of the gentamicin protection assay protocol (bacterial agar plates and liquid medium should be supplemented with 5 μg/mL of erythromycin to select for GFP-expressing clones).

2. Prepare HeLa cells for the experiment using similar procedures as the ones described in **steps 3–5** of Subheading 3.1, plating the cells in dark microscopy 96-well plates (instead of transparent plates).

3. Prepare the *L. monocytogenes* EGDe.PrfA*-GFP strain as described in Subheading 3.1, **steps 6–10**.

4. Infect cells as described in Subheading 3.1, **step 11**.

5. Centrifuge the plate for 5 min at $200 \times g$ at room temperature to bring *L. monocytogenes* in close contact to HeLa cells in order to synchronize infection.

6. Transfer the plate to an incubator with a humidified 5 % CO_2 atmosphere at 37 °C, and let bacteria enter cells for 30 min (*see* **Note 14**).

7. Aspirate infection medium, and wash cells once with 100 μl of standard cell culture medium per well.

8. Fix cells by adding 100 μl of PBS supplemented with 4 % PFA to each well and incubate at 4 °C for 15 min.

9. Remove the fixative, wash cells three times with 100 μl PBS per well, and keep cells in 100 μl PBS per well.

10. Prepare the primary antibody solution by diluting a rabbit-derived anti-*L. monocytogenes* serum 1:300 in PBS supplemented with 1 % bovine serum albumin, and add 30 μl of primary antibody to each well after removing the PBS.

11. Incubate for 15 min at room temperature to label extracellular *L. monocytogenes*.

12. Discard the primary antibody solution and wash six times with 100 μl of PBS per well.

13. Dilute the secondary AlexaFluor 546-coupled anti-rabbi antibody (1:300) in PBS, and add 30 μl of this secondary antibody solution to each well after removing the PBS.

14. Incubate for 30 min at room temperature in the dark to protect the fluorescent probe.

15. Discard the secondary antibody solution and wash six times with 100 μl of PBS per well.

16. Dilute DAPI solution (1:1,500) and phalloidin Dy 647 (1:150) in PBS and 0.05 % saponin, and add 30 μl of this staining solution to each well after removing the PBS.

17. Incubate for 30 min at room temperature in the dark to protect the fluorescent probes.

18. Discard the secondary staining solution and wash four times with 100 μl of PBS per well. And keep the cells in 100 μl of PBS per well.

19. Acquire images in the 350 nm (DAPI), 488 nm (total *L. monocytogenes*), 546 nm (extracellular *L. monocytogenes*), and 647 nm (actin) channels using a 20× objective mounted on an automated microscope.

20. Entry can be quantified as a ratio of intracellular *L. monocytogenes* (GFP-positive bacteria) minus extracellular *L. monocytogenes* (AlexaFluor 546-positive bacteria) divided by total cell-associated bacteria (*see* **Note 15**). In addition, the total number of bacteria per cell can give indications on the ability of bacteria to bind to cells. Bacteria can be counted manually or by using appropriate image analysis software.

4   Notes

1. The protocols that we present in this chapter are optimized for HeLa cells (human uterus) which are invaded by *L. monocytogenes* only via the InlB/Met pathway. Polarized epithelial cells such as Caco-2 (human intestine), LoVo (human colon), and Jeg3 or Bewo (human trophoblasts) can be additionally invaded via the InlA/E–cadherin pathway and are therefore more sensitive to *L. monocytogenes* infection; our protocols can be applied to these cell lines but the multiplicity of infection (MOI) should be reduced to account for the greater number of bacteria that will invade these host cells. These cell lines are particularly relevant for the microscopical study of bacterial internalization using poorly invasive strains of *L. monocytogenes* such as the EGDe strain.

2. The number of plated wells will vary according to the number of different conditions that will be analyzed: we typically seed cells in triplicate (three independent wells) for each condition.

3. All assays described here can be used in combination with small interfering RNA (siRNA)-mediated knockdown of host cell factors to study the involvement of a given host cell protein in the *L. monocytogenes* infection process. If cells are transfected with siRNA 72 h before infection, the cell number plated per well should be reduced to 5,400 cells for 96-well plates (or to 600 cells for 384-well plates). Special attention should be made to ensure that cell death due to siRNA treatment is not perturbing the final cell number counts.

4. Outer wells are exposed to more temperature variations and evaporation than inner wells, and results proceeding from cells seeded in these wells differ consistently from those of cells seeded in more central wells. We therefore avoid using the outer wells for cell seeding, and we fill them instead with a solution that will serve as a temperature and evaporation buffer for inner wells.

5. Differences in infection levels dependent on the well position on a plate can cause important artifacts. There are several possibilities to reduce these effects: for example, letting cells settle down after seeding for 10 min at room temperature may also reduce plate effects. Keeping plates on aluminum blocks in the incubator ensures more equal temperature distribution across the plate, and sealing plates with parafilm prevents evaporation of medium from edge and especially corner wells. In addition, a random distribution of controls and perturbation conditions on the plate can minimize a systematic accumulation of plate effects causing artifacts.

6. LLO is a cytotoxin that can induce cell death especially if bacteria are not well washed before infection, if the MOI is too high, or if there is too much time passing between washing bacteria and infecting cells. LLO can especially cause problems when highly invasive strains like EGDe.PrfA* are used. Therefore, it is important to check under a microscope if cells detach during the experiment. This is especially important for classical gentamicin protection assay since the total number of bacteria in the whole-cell lysate is used as the measure for infection, meaning that differences in cell number can significantly influence the outcome of the experiment.

7. The specific MOI of the inoculum needs to be adjusted according to the *L. monocytogenes* strain that is examined in the assay: when using the poorly invasive strain EGDe we prepare an inoculum containing between 1×10^7 and 2×10^7 bacteria per mL (the final MOI = 50–100 estimated for 20,000 cells per well). When using more invasive *L. monocytogenes* strains such as EGDe.PrfA* the MOI can be decreased by a factor of 5.

8. We use usually 1 % FCS in infection media; however, in some cases residual FCS can influence cellular signaling in a way that it interferes with the effect that one would like to assay. In these cases it is recommended not to use FCS in the infection medium.

9. Since the results will strictly depend on the number of cells present in individual wells, it is critical not to increase variation by removing cells throughout the different washing steps that the experiment requires. Therefore, addition of new medium should be performed gently to avoid removal of attached cells by sheer mechanical forces.

10. In our experience, the drops of lysate run straight down on tilted BHI agar plates when the agar is fresh with an even surface, meaning that the agar should not be too dry on the surface. We noticed that some dilutions run down better than others. Running down of the lysates can be improved by slightly shaking the tilted plate vertically against the bench.

Nevertheless, we recommend preparing spare agar plates for the experiment in case certain plates need to be repeated.

11. For the study of survival/replication of *L. monocytogenes* within host cells, the incubation time after addition of gentamicin can be varied between 1 and 5 h after infection (after that length of time, in addition to entry into primary infected cells there may be bacterial cell-to-cell spread to neighboring cells). For a more precise quantification of replication, an individual plate can actually be used to screen each individual time point after gentamicin addition.

12. HeLa cells are easily disrupted using the proposed protocol; however, other cell lines could require additional rounds of pipetting for complete disruption and microscopical verification of cell lysis using transmitted white light is suggested.

13. Example of calculations for a theoretical experiment:

    CFU (input) dilution 4: 10

    CFU (output 1 + 1 h) dilution 2: 12

    CFU (output 1 + 4 h) dilution 3: 9:

    $$\% \text{ of invasion (of inoculum)} = 100 \times \frac{\text{CFU (output)} \times \text{dilution (output)}}{\text{CFU (input)} \times \text{dilution (input)}}$$

    $$\% \text{ of invasion after } 1 + 1 \text{ h} = 100 \times \frac{12 \times 100}{10 \times 10,000} = 1.2$$

    $$\% \text{ of invasion after } 1 + 4 \text{ h} = 100 \times \frac{9 \times 1,000}{10 \times 10,000} = 9$$

    $$\text{Fold change of intracellular bacteria within 3 h} = \frac{\% \text{ invasion } (1 + 4 \text{ h})}{\% \text{ invasion } (1 + 1 \text{ h})}$$

    $$\text{Fold change of intracellular bacteria within 3 h} = \frac{9}{1.2} = 7.5$$

14. For microscopical assessment of invasion, cells can be infected for time periods as short as 5 min or as long as 4 h (the actual time limits will depend on the specific cell lines and the bacterial strains used). A time course for entry can be done using this technique.

15. For differential bacteria staining, it is possible to use non-GFP-expressing bacteria: do a first staining of extracellular bacteria without saponin in the staining solution and wash thoroughly 3–4 times with PBS and 1 % BSA followed by a second staining with primary and secondary antibody against total *L. monocytogenes*, using staining solution containing 0.05 % saponin for

cell permeabilization. We do this for small-scale experiments on cover slips; however, for large-scale experiments the use of GFP expression to visualize total bacteria can simplify the experiment and make it more robust towards cross-staining between total and extracellular bacteria.

Acknowledgments

Research in the P. Cossart laboratory is supported by the Pasteur Institute, the Institut National de la Santé et de la Recherche Médicale, the Institut National de la Recherche Agronomique, ERC Advanced Grant (233348), the Agence Nationale de la Recherche (Grant MIE-SignRupVac), the Louis-Jeantet Foundation, and the Fondation Le Roch Les Mousquetaires. A.K. is a recipient of a scholarship from the Pasteur-Paris University International Doctoral Program/Institut Carnot Maladies Infectieuses.

References

1. Cossart P (2011) Illuminating the landscape of host-pathogen interactions with the bacterium *Listeria monocytogenes*. Proc Natl Acad Sci U S A 108:19484–19491

2. Pizarro-Cerdá J, Kühbacher A, Cossart P (2012) Mechanisms of entry of *Listeria monocytogenes* in mammalian cells. Cold Spring Harb Perspect Med. 2: a010009

3. Bonazzi M, Kühbacher A, Toledo-Arana A et al (2012) A common clathrin-mediated machinery co-ordinates cell-cell adhesion and bacterial internalization. Traffic 13: 1653–1666

4. Kühbacher A, Dambournet D, Echard A et al (2012) Phosphatidylinositol 5-phosphatase oculocerebrorenal syndrome of Lowe protein (OCRL) controls actin dynamics during early steps of *Listeria monocytogenes* infection. J Biol Chem 287:13128–13136

5. Pizarro-Cerdá J, Lecuit M, Cossart P (2002) Measuring and analysing invasion of mammalian cells by bacterial pathogens: the *Listeria monocytogenes* system. In: Sansonetti PJ, Zychlinsky A (eds) Methods in microbiology, vol 31. Academic, London, pp 161–177

6. Vaudaux P, Waldvogel FA (1979) Gentamicin antibacterial activity in the presence of human polymorphonuclear leukocytes. Antimicrob Agents Chemother 16:743–749

7. Pizarro-Cerdá J, Cossart P (2006) *Listeria monocytogenes*: techniques to analyse bacterial infection in vitro. In: Celis J (ed) Cell biology: a laboratory handbook, 3rd edn. Academic, London, pp 407–415, vol 2, chapter 52

8. Becavin C, Bouchier C, Lechat P, Archambaud C, Creno S, Wu Z-F, Kuhbacher A, Brisse S, Garcia-del-Portillo F, Lecuit M, Portnoy D, Pizarro-Cerdá J, Mozer I, Bierne H, Cossart P (2014) Comparison of widely used *Listeria monocytogenes* strains EGD, 10403 S, and EGD-e highlights genomic variations underlying differences in pathogenicity. mBio 5: e00969-14

Part III

Strain Manipulation

Chapter 15

Extraction and Analysis of Plasmid DNA from *Listeria monocytogenes*

Aidan Casey and Olivia McAuliffe

Abstract

A plasmid preparation is a method used to extract and purify plasmid DNA. Methods developed to purify plasmid DNA from bacteria generally involve harvesting and alkaline lysis of the bacteria and precipitation of chromosomal DNA and protein, followed by purification of the plasmid DNA. Here, we describe the mini-preparation of plasmid DNA by a rapid small-scale method, adapted for *Listeria monocytogenes*. The quality of plasmid DNA isolated using this method is sufficient for analytical purposes but may be upscaled for further downstream analysis. Electrophoretic separation of the resultant lysate allows conclusions to be made on the presence, number, copy number, and size of the plasmids in the analyzed bacterial strains.

Key words Plasmid, DNA, Purification, Alkaline lysis, Gel electrophoresis

1 Introduction

Plasmids naturally exist in virtually all bacterial species. These small, circular, double-stranded DNA molecules are distinct from the cell's chromosomal DNA, replicate independently of the genome, and are stably inherited. Plasmids generally only account for a small fraction of the bacterial genome, and their lengths can vary in size from a few kilobases to many hundred kilobases [1]. Many copies of plasmids of lower molecular weight may exist in a cell, whereas fewer copies of larger plasmids are found. Plasmids are of significant importance in the prokaryotic world as they often provide the bacterial cell with a genetic advantage [2], such as antibiotic resistance or other traits [3, 4]. In addition, plasmids are a key vector for horizontal gene transfer in bacteria through conjugation [1, 5]. Given their importance for conferring adaptive traits to the host organism, interest in studying and sequencing plasmids is growing.

For plasmid isolation, bacterial cells are generally grown to late logarithmic or early stationary phase. Lysozyme is used to weaken the cell wall, thus achieving easier enzymatic lysis. Like chromosomes,

Kieran Jordan et al. (eds.), *Listeria monocytogenes: Methods and Protocols*, Methods in Molecular Biology, vol. 1157, DOI 10.1007/978-1-4939-0703-8_15, © Springer Science+Business Media New York 2014

plasmids generally exist in a covalently closed circular (supercoiled) form, and most isolation procedures are based on this fact. After gentle alkaline lysis, plasmids can be found in linear, open circle or multiple-supercoiled form which allows for easier discrimination. Since DNA is very sensitive to mechanical stress, all mixing steps during and after cell lysis should be performed carefully by inverting the tubes to prevent shearing. At this point, it is necessary to remove all intracellular macromolecules in order to enrich and purify the plasmid DNA. Adding to sodium acetate allows the circular DNA to renature, while sheared cellular DNA remains denatured as ssDNA. This ssDNA is then precipitated along with other large molecules, which allow the separation of the contaminants from the plasmid DNA. Phenol–chloroform extractions are used in this instance, followed by ethanol precipitation of the DNA and RNase treatment. Analysis of plasmid DNA is achieved by horizontal electrophoresis at 4 °C at low voltage. This mini-preparation method generally yields between 5 and 10 μg of relatively pure plasmid DNA which is generally sufficient for a number of uses, such as strain characterization, in vitro transcription or translation, and restriction digestion. For other downstream applications such as ligation, transformation, or sequencing, scale-up and additional purification techniques may be required which will yield higher concentrations of plasmid DNA.

2   Materials

2.1   Sucrose (25 %) Containing 30 mg/mL Lysozyme

To 50 mL water, add 12.5 g sucrose to give a 25 % sucrose solution. The resulting solution is autoclaved at 121 °C for 15 min. Add 1.5 g of lysozyme to the autoclaved solution and stir thoroughly to dissolve, yielding a 25 % sucrose solution containing 30 mg/mL lysozyme. Once the lysozyme is fully dissolved in solution, sterilize the solution once more by passing it through a 0.45 μm filter into a sterile container.

2.2   Alkaline SDS Solution

To 200 mL of H_2O, 1.6 g of NaOH is added, stirring to dissolve, to yield a concentration of NaOH of 0.2 N (*see* **Note 1**). The solution is then autoclaved at 121 °C for 15 min. To the autoclaved solution, add 6 g of sodium dodecyl sulphate (SDS) to give a final concentration of 3 % w/v SDS in 0.2 N NaOH.

2.3   Sodium Acetate: 3 M (pH 4.8)

To 100 mL of H_2O, add 24.61 g of anhydrous sodium acetate to give a concentration of 3 M sodium acetate in solution. Adjust the pH of the solution to 4.8 by adding a small quantity of concentrated hydrochloric acid (*see* **Note 2**). Once the pH has been adjusted to the desired level, sterilize the solution by passing it through a 0.45 μm filter. Put the solution on ice prior to use for the plasmid isolation protocol.

2.4 Ammonium Acetate (7.5 M)	To 50 mL of H_2O, add 28.91 g of ammonium acetate powder in order to make up a solution with a final concentration of 7.5 M. Once acid has been fully dissolved, sterilize the solution by passing it through a 0.45 μm filter.

2.5 TAE Buffer (1×)	Firstly, make up 50× TAE buffer by weighing out into a large conical flask 242 g of Tris (Trizma buffer from Sigma), 57.1 mL of glacial acetic acid, and 18.6 g of EDTA. Make up to 1 L with distilled water, stirring to dissolve. Transfer into five separate 200 mL bottles. The buffer is then autoclaved at 121 °C for 15 min. Dilute 50× TAE buffer down to 1× by adding 20 mL of 50× buffer to 980 mL of sterile water. The final concentrations are as follows: For 50×: Tris—2 M, glacial acetic acid—1 M, EDTA—50 mM. For 1×: Tris—40 mM, glacial acetic acid—20 mM, EDTA—1 mM. Once 1 L of 1× TAE buffer is made up, add 20 μL of ethidium bromide and mix by inverting (*see* **Note 3**).

3   Methods

3.1 Preparation of Bacterial Cultures for Plasmid Isolation Protocol	1. Picking a colony from a previously streaked agar plate containing the desired strain of *Listeria monocytogenes*, resuspend the bacteria in 10 mL of enrichment broth (*see* **Note 4**) and incubate overnight at 37 °C. 2. Following the resulting overnight incubation, transfer 100 μL of the bacterial culture into 5 mL of sterile media and incubate for 4–5 h at 37 °C. 3. Once sufficiently turbid, centrifuge the culture at $2,500 \times g$ for 10 min, and discard the supernatant. The remaining bacterial pellet is then used in the subsequent plasmid isolation protocol (*see* **Note 5**).

3.2 Isolation of Plasmids from Listeria monocytogenes Strains	1. Resuspend each bacterial pellet in 200 μL of the 25 % sucrose containing 30 mg/mL lysozyme solution (*see* **Note 6**), and for each, transfer the 200 μL solution to a sterile 1.5 mL Eppendorf tube. 2. Incubate Eppendorf tubes at 37 °C for 20–25 min. 3. To each Eppendorf, add 400 μL of the alkaline SDS solution (*see* **Note 7**), and mix thoroughly by inverting the tube several times. 4. Incubate Eppendorf tubes at room temperature for 7 min. 5. Vortex each sample for 30 s (*see* **Note 8**). 6. To each sample, add 300 μL of ice-cold 3 M sodium acetate. Mix samples immediately by inversion, before centrifuging the Eppendorf tubes at $18,500 \times g$ for 15 min and at a temperature of 4 °C.

7. Transfer 700–800 μL of the resulting supernatant from each tube to another sterile 1.5 mL Eppendorf tube, and discard the pellet.

8. To each of these, add 650 μL of isopropanol (room temperature).

9. Centrifuge these samples once more at $18,500 \times g$ for 15 min and at a temperature of 4 °C.

10. Once centrifugation is complete, remove all liquid from each of the tubes (*see* **Note 9**), and resuspend the remaining pellet in 320 μL of sdH$_2$O.

11. When pellet is completely dissolved, add 200 μL of 7.5 M ammonium acetate to each tube and mix by inversion.

12. To each tube, add 350 μL of phenol/chloroform, and invert tubes.

13. Centrifuge samples at $18,500 \times g$ for 5 min and at room temperature.

14. Transfer the upper phase (approximately 300 μL) from each tube into a sterile Eppendorf tube, and to each, add 1 mL of 100 % ethanol (stored at a temperature of –20 °C).

15. Immediately centrifuge samples at $18,500 \times g$ for 15 min and at a temperature of 4 °C.

16. Following centrifugation, pour off the 100 % ethanol, and wash the remaining pellets with 200 μL of 70 % ethanol (*see* **Note 10**).

17. Completely remove the ethanol from the tubes, and dissolve each pellet in 40 μL of H$_2$O containing 0.1 mg/mL of RNase.

18. Store samples at 4 °C for subsequent running on an agarose gel.

3.3 Assessment of Quality of Isolated Plasmid DNA from Listeria monocytogenes

1. Make up a 0.7 % agarose gel by weighing out 1.4 g of agarose and suspending it in 200 mL of 1× TAE buffer (containing ethidium bromide).

2. Microwave for 5–6 min to fully dissolve agarose, and allow it to temper at room temperature on worktop bench, before pouring into a previously set up gel cast unit. Insert a 10-well comb immediately without introducing air bubbles.

3. Allow the gel to set and solidify for 10–15 min.

4. Prepare each of the plasmid DNA samples for running on the gel by adding 10 μL of plasmid DNA to 10 μL of 2× loading dye (final concentration of 1× loading dye).

5. Transfer the gel into the running unit, and carefully remove the comb.

6. Pipette the 20 μL solutions for each sample into each of the wells.

7. Into a well at either end of the loaded samples, add 4 μL of a 1 kb ladder.

8. Run the gel overnight at a temperature of 4 °C and at a voltage of 45 V.

9. Take an image of gel under UV light in order to visually confirm the isolation of plasmid DNA. Successfully isolated plasmid DNA should appear in one, or several, clear and concise bands on the agarose gel.

4   Notes

1. SDS powder is harmful to the respiratory system if inhaled. When weighing out powder, ensure that necessary precautions (such as weighing powder inside a fume hood) are taken for safety.

2. Addition of concentrated hydrochloric acid to reagents when adjusting the pH should be done in a laminar flow hood as fumes from concentrated hydrochloric acid may be harmful.

3. Ethidium bromide is mutagenic and carcinogenic and should be handled with care at all times. Protective gloves should be worn when working with this chemical.

4. Enrichment broth such as tryptic soy broth or brain–heart infusion broth is sufficient for growth of the *Listeria monocytogenes* strains in preparation for the plasmid isolation protocol.

5. It is also recommended that a bacterial strain be grown in tandem, of which the plasmid content is already known, to use as a positive control for the experiment. If possible, use a strain with known plasmid sizes, and which have a combination of large and small plasmids, so that this strain may also be used as a reference ladder when run out on an agarose gel.

6. Following addition of lysozyme solution to bacteria, the time of incubation required to achieve lysis of the cells depends on the nature of the bacteria of interest. For bacteria such as *Lactococcus*, the incubation time required is generally between 10 and 15 min. For *Listeria*, these times need to be increased to ensure that full lysis is achieved.

7. Alkaline SDS solution should be tempered at 40 °C prior to use, as storage of the solution at cooler temperatures can cause the SDS to precipitate out of solution.

8. Vortexing of the samples is optional. While vortexing of samples is not generally recommended at any point during the protocol, in this instance, it should have the effect of shearing genomic DNA while not affecting the closed plasmid DNA.

9. After the 15-min centrifugation in isopropanol, resulting samples should consist of a clear supernatant with a white pellet at

the bottom of the tube. Ensure that all of the supernatant has been removed before resuspending the pellet in dsH₂O. If necessary, spin the tubes several times on a small bench-top centrifuge and remove small quantities of isopropanol with 20 µL tips. The more isopropanol that remains in the tube, the more difficult the sample will be to resuspend in water. Dissolving of the pellet in dsH₂O can take anywhere from 2 to 10 min, and incubating the tubes at 37 °C may help the pellets dissolve faster.

10. When washing the pellets with 70 %, be careful to ensure that the pellet does not become dislodged from the bottom of the tube.

References

1. Smillie C et al (2010) Mobility of plasmids. Microbiol Mol Biol Rev 74:434–452

2. Ratani SS et al (2012) Heavy metal and disinfectant resistance of *Listeria monocytogenes* from foods and food processing plants. Appl Environ Microbiol 78:6938–6945

3. Kuenne C et al (2010) Comparative analysis of plasmids in the genus Listeria. PLoS One 5:e12511

4. Poyart-Salmeron C et al (1990) Transferable plasmid-mediated antibiotic resistance in Listeria monocytogenes. Lancet 335:1422–1426

5. Harrison E, Brockhurst MA (2012) Plasmid-mediated horizontal gene transfer is a coevolutionary process. Trends Microbiol 20:262–267

Chapter 16

Generation of Nonpolar Deletion Mutants in *Listeria monocytogenes* Using the "SOEing" Method

Kathrin Rychli, Caitriona M. Guinane, Karen Daly, Colin Hill, and Paul D. Cotter

Abstract

The ability to manipulate chromosomally encoded genes is a fundamental biological tool for the analysis of gene function. Here, we provide in greater depth protocols for the creation of nonpolar unlabeled gene deletions in *Listeria* (*L.*) *monocytogenes* that are facilitated by the splicing overlap extension PCR technique. For mutagenesis in *L. monocytogenes*, we describe two different plasmid-based approaches, which facilitate the introduction of this spliced amplicon in place of the corresponding segment of chromosomal DNA: the pKSV7 system and the pORI280/pVE6007 system.

Key words Nonpolar deletion mutants, *Listeria monocytogenes*, SOEing method, pKSV7 vector, pORI280/pVE6007 system

1 Introduction

The ability to manipulate chromosomally encoded genes is a fundamental biological tool for the analysis of gene function, which is of immense importance when studying microbial genetics. In *Listeria* (*L.*) *monocytogenes* a number of strategies and vectors have been used to create chromosomal mutations [1–3]. Here, we describe in greater depth protocols for the creation of nonpolar unlabeled gene deletions in *L. monocytogenes* that are facilitated by the splicing overlap extension PCR (SOEing-PCR) technique, developed by Horton et al. [4]. For SOEing-PCRs, two segments of DNA, which flank the region that is targeted for mutagenesis, are amplified by initial PCRs. The oligonucleotides employed for these PCRs are designed such that overlapping stretches are incorporated into these primers to facilitate subsequent splicing. These amplicons are purified, mixed, and used as a template for another PCR to create this spliced amplicon [5]. For mutagenesis in *L. monocytogenes*, we describe two different plasmid-based approaches,

Kieran Jordan et al. (eds.), *Listeria monocytogenes: Methods and Protocols*, Methods in Molecular Biology, vol. 1157, DOI 10.1007/978-1-4939-0703-8_16, © Springer Science+Business Media New York 2014

which facilitate the introduction of this spliced amplicon in place of the corresponding segment of chromosomal DNA: the pKSV7 system (variant A) [6, 7] and the pORI280/pVE6007 system (variant B), previously documented by Monk et al. [3].

pKSV7, first generated for use in *Bacillus subtilis*, is a temperature-sensitive 6.9 kbp shuttle vector conferring both ampicillin and chloramphenicol resistance [8]. The pORI vectors, which have originally been developed for use in *Lactococcus lactis* [9], do not contain an origin of replication (repA⁻) and therefore require the replication ignition protein RepA to be provided *in trans*. Therefore, an *E. coli* strain containing a chromosomal integration of the *repA* gene (e.g., *E. coli* EC101) is used for cloning and plasmid propagation and a Rep⁺ helper plasmid (e.g., pVE6007) is required for the process of mutagenesis in a Gram-positive host. The 5.3 kb pORI280 vector contains an erythromycin resistance and constitutively expressed *lacZ* genes, which easily allows the selection of colonies containing the plasmid on plates containing 5-bromo-4-chloro-indolyl-β-D-galactopyranoside (X-gal). The RepA⁺ helper plasmid pVE6007 confers chloramphenicol resistance and also has a temperature-sensitive (Ts) origin of replication [10]. The Ts nature of the plasmids pKSV7 and pVE6007 ensures that these vectors can replicate in the cytoplasm at permissive (lower) temperatures, but upon temperature increase these vectors, or in the case of the pORI280 vector that requires the presence of pVE6007, can no longer replicate. In the majority of instances the host strain becomes sensitive to the relevant antibiotic markers. However, a minority of cells will retain resistance to chloramphenicol or erythromycin as a consequence of the natural integration of pKSV7 or pORI280, respectively, into the chromosome through a single crossover homologous recombination event. Subsequent sequential subculturing of the strain at appropriate temperatures ultimately results in the development of subpopulations of cells in which the plasmids are excised through a second crossover event. This will either result in a reversion to the genotype of the original wild-type gene or in the incorporation of the SOEing fragment to generate the targeted deletion mutant.

The protocol below describes the creation of the SOEing-PCR product that facilitates the subsequent generation of an in-frame deletion (Fig. 1). This involves the amplification of two initial PCR products SoeAB and SoeCD. These products are then mixed in a 1:1 ratio and used as a template to create a spliced SoeAD product. We also provide the protocols for the construction of the plasmid variants and the method for chromosomal integration, excision, and mutant selection through the use of the pKSV7 (Fig. 2) and pORI280/pVE6007 (Fig. 3) systems.

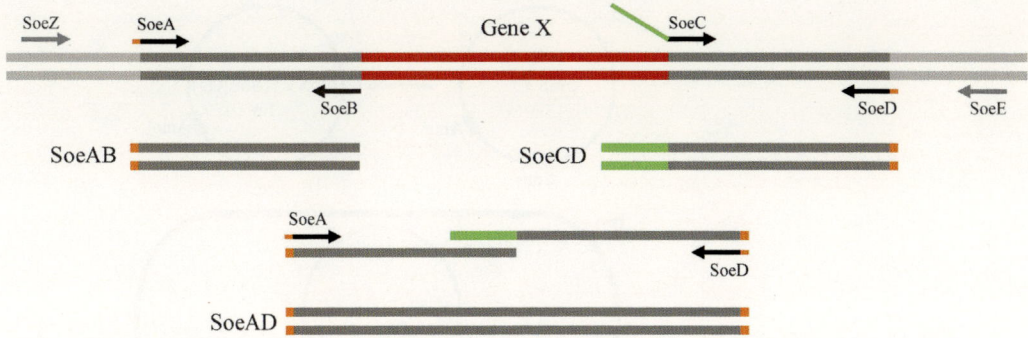

Fig. 1 The SOEing-PCR involves the two initials PCR SoeAB and SoeCD and a third PCR, which uses the PCR products SoeAB and SoeCD as a template resulting in the SoeAD PCR product. *Red* the gene targeted for deletion, *orange* restriction enzyme site, *green* overlapping gene segment

2 Materials

2.1 SOE PCR

Keep all components for the PCR reaction on ice.

1. DNA template: Isolate chromosomal DNA using, e.g., soil isolation kit (Mol BIO Laboratories, Carlsbad, CA, USA).

2. VENT-PCR: VENT DNA Polymerase, 10× ThermoPol reaction buffer, 100 mM MgSO$_4$ (all New England BioLabs, Ipswich, MA, USA), 10 mM deoxynucleotide (dNTP) solution mix (Bioline, London, UK), PCR-grade H$_2$O (Sigma Aldrich, St. Louis, MO, USA).

3. Soe PCR primers (Microsynth, Balgach, Switzerland).

4. 1.5 % agarose gel: Prepare 1× Tris–borate–EDTA (TBE) buffer from a 10× TBE stock solution with ddH$_2$O, melt 1–1.5 g agarose in 100 mL 1× TBE buffer using a microwave, and add 2 μL 10 mg/mL ethidium bromide (all Sigma Aldrich).

5. QIAquick PCR purification kit (QIAGEN, Hilden, Germany).

6. QIAquick gel extraction kit (QIAGEN).

2.2 Cloning of SoeAD into Vector

2.2.1 Variant A (Cloning of SoeAD into pKSV7)

1. Luria broth (LB) and LB agar (both Oxoid, Cambridge, UK) supplemented with 50 μL/mL ampicillin (LB-Amp^{50}, Sigma-Aldrich) for cultivation of *E. coli* DH5α pKSV7 transformants.

2. QIAprep spin miniprep kit (QIAGEN).

3. For digestion: Restriction enzymes and the appropriate buffers (Roche, Basel, Switzerland), QIAquick PCR purification kit.

4. For ligation: T4-DNA ligase, 10× ligation buffer (Roche).

5. For microdialysis: MF-Millipore™ membrane filter, VSWP, 0.025 μm pore size (Millipore, Billerica, MA, USA); sterile H$_2$O, petri dish.

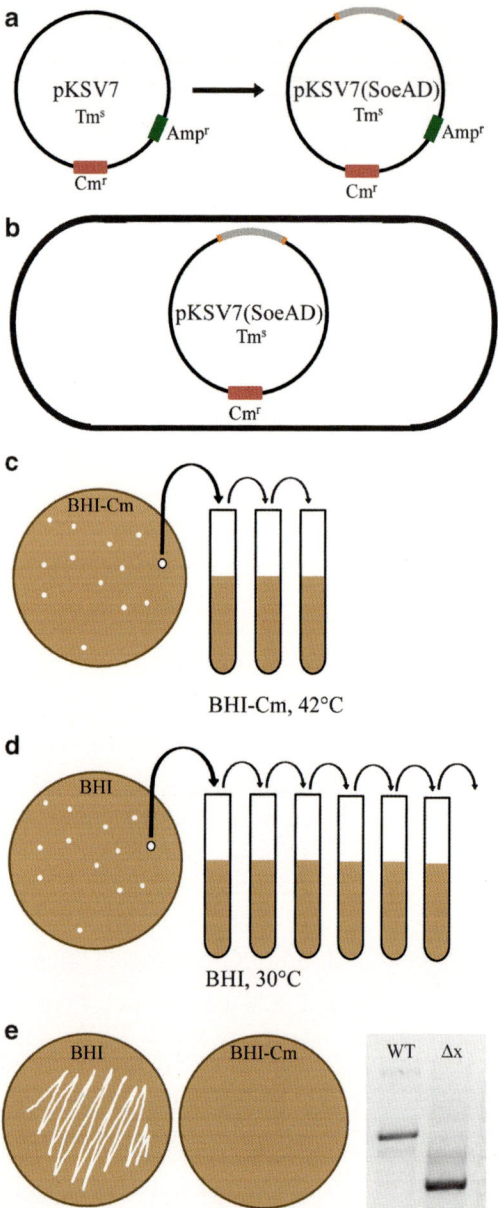

Fig. 2 Gene deletion protocol using the pKSV7. In *step A* the SoeAD-PCR product was cloned into pKSV7, which is a temperature-sensitive (Tm^s) plasmid conferring ampicillin (Amp^r) and chloramphenicol (Cm^r) resistance, generating pKSV7(SoeAD). In *step B*, pKSV7(SoeAD) was transformed into the recipient electrocompetent *L. monocytogenes* strain and selected on BHI agar containing 10 μg/mL chloramphenicol (BHI-Cm). In *step C*, chromosomal integration of the plasmid was induced by serial passage of a transformant in BHI-Cm broth 42 °C. In *step D*, plasmid excision and curing were induced by continuous passage in BHI at 30 °C. In *step E*, putative mutants were selected for chloramphenicol sensitivity by streak colonies in replica onto BHI and BHI-Cm-agar. Cm-sensitive colonies were screened by colony PCR with SoeE and SoeZ primers. *WT* wild type, Δ*x* deletion mutant

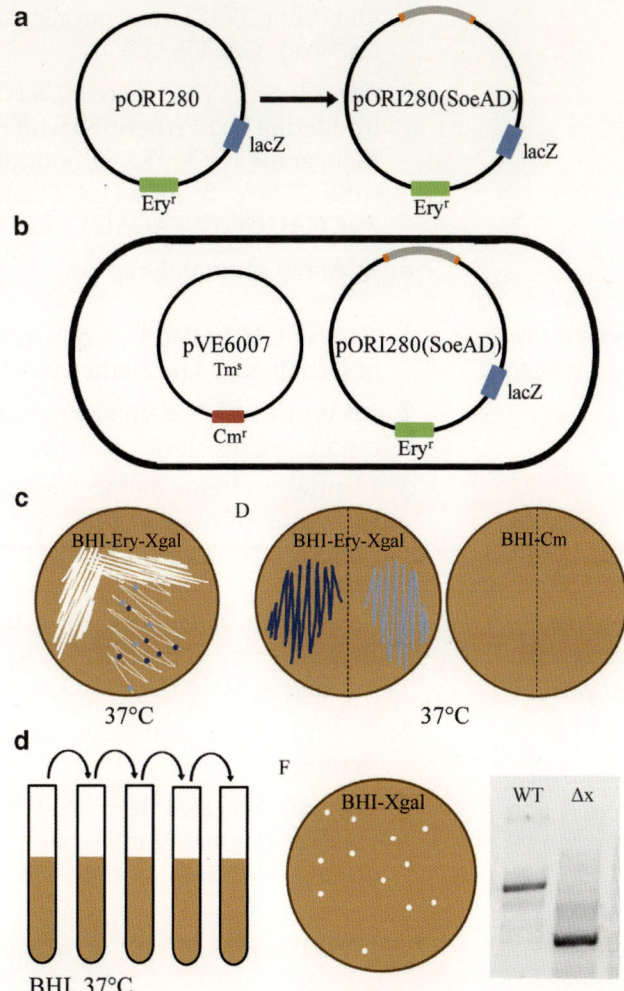

Fig. 3 Gene deletion protocol using the pORI280/pVE6007 system, adapted from Monk et al. [3]. In *step A* the SoeAD-PCR product was cloned into pORI280, which confers erythromycin resistance (Ery^r) and a constitutively expressed *lacZ* gene, generating pORI280(SoeAD). In *step B*, pORI280(SoeAD) and pVE6007, which are temperature-sensitive (Tm^s) Rep^+ helper plasmid conferring chloramphenicol resistance (Cm^r), were transformed into the recipient electrocompetent *L. monocytogenes* strain. Transformants were selected on BHI agar containing 5 µg/mL erythromycin and 100 µg/mL X-gal (BHI-Ery-Xgal). In *step C*, the loss of pVE6007 and the chromosomal integration of pORI280(SoeAD) were induced by complex streaking a blue colony on BHI-Ery-Xgal and incubation at 37 °C. In *step D* the loss of pVE6007 was confirmed by streaking a light and dark blue colony in replica onto BHI Ery-Xgal and BHI-Cm-agar. In *step E*, plasmid excision and curing were induced by continuous passage in BHI at 37 °C. In *step F*, putative mutants were selected by plating on BHI-Xgal. White colonies were screened by colony PCR with SoeE and SoeZ primers. *WT* wild type, Δ*x* deletion mutant

6. One Shot TOP10 chemically competent *E. coli* (Invitrogen, Carlsbad, CA, USA).

7. For colony PCR: Prepare 10 % IGEPAL-G630 (Sigma Aldrich) by diluting IGEPAL-G630 with ddH$_2$O, store at 4 °C; BioMix, PCR-grade H$_2$O, pKSV7 multiple cloning site (MCS) primers (pKSV7 fwd 5′-ATGTGCTGCAAGGCGATTA, pKSV7 rev 5′-CCAGGCTTTACACTTTATG).

8. QIAprep spin miniprep kit.

2.2.2 Variant B (Cloning of SoeAD into pORI280)

1. *E. coli* EC101 (RepA$^+$, kindly provided by Jan Kok, University of Groningen, The Netherlands).

2. LB broth and agar complemented with 200 μL/mL erythromycin (Sigma Aldrich, LB-Ery^{200}).

3. Identical to **items 3–8** in Subheading 2.2.1, except for colony PCR use pORI280 MCS primers (fwd 5′-CTCGTTCAT TATAACCCTC, rev 5′-CGCTTCCTTTCCCCCCAT).

2.3 Competent *L. monocytogenes* for Transformation

1. Preparation of 100 mL BHI, 0.5 M sucrose: Dissolve 3.7 g BHI (Oxoid, Cambridge, UK) and 17.5 g sucrose (Sigma Aldrich) in 100 mL ddH$_2$O, autoclave media in 500 mL flask.

2. 500 μL 10 mg/mL penicillin G (Sigma Aldrich), filter sterilized.

3. 150 mL 100 mM HEPES, 0.5 M sucrose: Dissolve 0.03 g HEPES (Sigma Aldrich) and 26.25 g sucrose in 150 mL ddH$_2$O, autoclave, store at 4 °C.

4. For centrifugation: Centrifuge, e.g., Beckman Coulter Avanti J-26XP, sterile 250 mL centrifugation tubes.

2.4 Transformation of *L. monocytogenes* via Electroporation

2.4.1 Variant A: Transformation of pKSV7(SoeAD)

1. Disposable cuvette plus, 2 mm gap size (VWR, Radnor, PA, USA).

2. Electrocell manipulator BTX Harvard apparatus.

3. 10 mL sterile BHI, 0.5 M sucrose.

4. BHI agar plates (Oxoid, Cambridge, UK) supplemented with 10 μg/mL chloramphenicol (BHI-Cm^{10} agar, Sigma Aldrich).

5. Colony-PCR: 10 % IGEPAL-G630, primers (pKSV7 fwd, rev), BioMix, PCR-grade H$_2$O.

2.4.2 Variant B: Cotransformation of pORI280(SoeAD) and pVE6007

1. *L. monocytogenes* pVE6007$^+$ strain (RepA$^+$, kindly provided by Jan Kok, University of Groningen, The Netherlands).

2. For isolation of pVE6007 10 mL BHI, prepare protoplast buffer: 20 mM Tris–HCl (Carl Roth, Karlsruhe, Germany), pH 7.5–8, 10 mg/mL lysozyme, 50 U/mL mutanolysin (both Sigma Aldrich), 0.75 mM sucrose, 5 mM EDTA, filter sterilized and dispensed into 500 μL aliquots, store at –20 °C.

3. Disposable cuvette plus, 2 mm gap size.

4. Electrocell manipulator BTX Harvard apparatus.

5. 1 mL sterile BHI, 0.5 M sucrose.

6. BHI agar plates supplemented with 5 μg/mL erythromycin and 100 μg/mL X-gal (BHI-Ery[5]-X-gal[100], Applichem, Darmstadt, Germany).

7. Colony-PCR: 10 % IGEPAL-G630, primers (pORI280 fwd, rev), BioMix, PCR-grade H_2O.

2.5 Chromosomal Integration of SoeAD

2.5.1 Variant A

1. ~30 tubes of 10 mL BHI-Cm[10] and 50 BHI-Cm[10] agar plates.

2. ~200 tubes of 10 mL BHI and 50 BHI agar plates.

3. Colony-PCR: 10 % IGEPAL-G630, primers (SoeE, SoeZ, and pKSV7 fwd, rev primers), BioMix, PCR-grade H_2O.

2.5.2 Variant B

1. BHI-Ery[7.5]-X-gal[100] and BHI-Cm[10] agar plates.

2. Tubes of 10 mL BHI and BHI-X-gal[100] agar plates.

3. Colony-PCR: 10 % IGEPAL-G630, primers (SoeE, SoeZ, and pORI280 fwd, rev primers), BioMix, PCR-grade H_2O.

3 Methods

3.1 SOE PCR

3.1.1 Primer Design (See **Note 1**)

1. SoeA is a forward primer, which binds upstream of the binding site of primer SoeB and contains a 5′ enzyme restriction site to facilitate cloning into the plasmid of interest.

2. SoeB is a reverse primer that binds to a region flanking the gene targeted for deletion.

3. SoeC is a forward primer that binds to a region flanking the gene targeted for deletion and that contains in addition a 5′ region (~20 bp) that is the reverse complement of primer SoeB.

4. SoeD is a reverse primer, downstream of primer SoeC, which contains a 5′ enzyme restriction site to facilitate cloning into the plasmid of interest.

5. SoeZ is a forward primer that binds upstream of the binding site for SoeA.

6. SoeE is a reverse primer that binds downstream of the binding site for SoeD.

Dilute all primers with PCR-grade H_2O to a concentration of 100 μM (stock solution), store at −20 °C; for PCR reaction dilute primers 1/10 with PCR-grade H_2O.

3.1.2 SoeAB-SoeCD VENT PCR Reaction

All components should be mixed and kept on ice.

1. Prepare the following 50 μL reaction in a PCR tube: 5 μL 10× ThermoPol reaction buffer (1×), 1 μL 10 mM dNTP solution mix (200 μM), 1.5 μL of each 20 μM primer (0.6 μM) either

SoeA and SoeB primers or SoeC and SoeD primers, 0.5 µL VENT DNA. Polymerase (1 Unit), optimal 2 µL 100 mM MgSO$_4$ (6 mM), DNA template. Bring to a volume of 50 µL with PCR-grade H$_2$O.

2. Gently mix the reaction, spin down in a microcentrifuge, and keep on ice.

3. Cycling conditions for PCR: Initial denaturation, 95 °C, 5 min; 20–30 cycles: 95 °C 15–30 s, 48–65 °C 15–30 s, 72 °C 1 min per kb; final extension: 72 °C 5 min; hold: 4–10 °C (*see* **Note 2**).

4. Store PCR products at 4 °C (short term) or –20 °C.

5. Analyze an aliquot of the PCR products by agarose gel electrophoresis.

6. If single bands of correct size are visualized on the agarose gel, purify the respective SoeAB and SoeCD-PCR products using QIAquick PCR purification kit.

3.1.3 SoeAD-VENT PCR

1. Perform a second VENT-PCR using the primers SoeA and SoeD, together with a 1:1 ratio of the purified SoeAB PCR product and the purified SoeCD PCR product as the DNA template, with PCR conditions similar for SoeAB or SoeCD-PCR followed by agarose gel electrophoresis (*see* **Note 3**).

2. If a product of correct size is visualized on the agarose gel, purify the PCR product SoeAD using the QIAquick gel extraction kit and store at –20 °C.

3.2 Cloning of SoeAD into Vector

3.2.1 Variant A: Cloning of SoeAD into pKSV7 (See **Note 4***)*

1. Grow *E. coli* DH5α pKSV7 in LB-Amp[50] overnight at 37 °C with shaking, and isolate pKSV7 using the QIAprep spin miniprep kit.

2. Digest 30 µL of PCR product SoeAD and 30 µL of pKSV7 with the appropriate restriction enzyme according to the manufacturer's instructions, e.g., 2 µL *Pst*I and 2 µL *Eco*RI for 6 h at 37 °C.

3. Purify the digested SoeAD and pKSV7 using the QIAquick PCR purification kit, and perform agarose gel electrophoresis.

4. Ligation: Mix digested SoeAD and pKSV7 at a ratio of 5:1, 3:1, and 1:1 (total volume 16 µL); add 2 µL 10× ligation buffer and 2 µL T4 DNA ligase; and incubate on ice for >16 h.

5. Microdialysis: Put 20 mL sterile ddH$_2$O in a petri dish, place 1 MF-Millipore™ membrane filter with a sterile forceps on the water, pipette the ligation reaction on the filter, incubate for 20 min at RT, pipette the dialyzed product to a sterile microfuge tube, and keep dialyzed ligation on ice.

6. Transformation of *E. coli* with the pKSV7 derivative: Use commercially available electrocompetent or chemically competent *E. coli*, e.g., DH5α *E. coli*. Transform according to

the manufacturer's protocols. Plate the transformants on LB-Amp^{50} plates and incubate for at least 24 h at 37 °C.

7. Use colony PCR to confirm that an insert of correct size has been introduced: for cell lysis resuspend a portion of a colony in 10 μL 10 % IGEPAL-G630, and in parallel streak the rest of the colony on a fresh LB-Amp^{50} agar plate; denature for 10 min at 95 °C; add 40 μL of PCR master mix containing 25 μL BioMix, 1.5 μL of each of the pKSV7 fwd and pKSV7 rev primers (i.e., those situated on either site of the MCS); and perform PCR reaction: 3 min 94 °C and 30 cycles (30 s 94 °C, 30 s 50 °C, 1.5 min 72 °C), 10 °C.

8. Prepare stock solutions of colonies containing pKSV7 derivatives with inserts of correct size, isolate the corresponding plasmid using QIAprep spin miniprep kit, and confirm integrity of SoeAD product by sequencing the plasmid insert.

3.2.2   Variant B: Cloning of SoeAD into pORI280

1. Grow *E. coli* EC101 (or *E. coli* EC10B; both RepA$^+$, Ery^R) containing pORI280 in LB-erythromycin^{200} overnight at 37 °C with shaking. Isolate pORI280 using the QIAprep spin miniprep kit.

2. Digestion, purification of the digested SoeAD and vector, ligation, and microdialysis are preformed as in **steps 2–5** of Subheading 3.2.1.

3. Prepare competent *E. coli* EC101 for either chemical or electrotransformation.

4. Transform pORI280 into competent *E. coli* EC101, and plate transformants on LB-Ery^{200} agar plates; incubate for at least 24 h at 37 °C.

5. Perform colony PCR using the primers pORI280fwd and pORI280rev to determine if the SoeAD amplicon has been successfully ligated into the plasmid.

6. Prepare stock solution of colonies containing pORI280 derivatives with inserts of correct size, isolate the corresponding plasmid using QIAprep spin miniprep kit, and confirm integrity of SoeAD product by sequencing the plasmid insert.

3.3   Competent L. monocytogenes for Transformation

1. Inoculate one colony of *L. monocytogenes* in 5 mL BHI and grow overnight at 37 °C with shaking (*see* **Note 5**).

2. Add 1 mL of overnight culture into 100 mL BHI and 0.5 M sucrose in a 500 mL flask.

3. Grow culture with shaking at 300 rpm for 5–6 h at 37 °C to an OD$_{600}$ of 0.2.

4. Add 100 μL of 10 mg/mL penicillin (final concentration 10 μg/mL) and incubate for 2 h.

5. Keep culture at 4 °C for the rest of the method.

6. Centrifuge cells in 350–500 mL sterile tubes for 10 min at 8,700 ×𝑔 at 4 °C.

7. Remove supernatant, and resuspend pellet in 90 mL sterile 1 mM HEPES and 0.5 M sucrose.

8. Centrifuge for 10 min at 8,700 ×𝑔 at 4 °C.

9. Remove supernatant, resuspend pellet in 45 mL sterile 1 mM HEPES and 0.5 M sucrose, and centrifuge for 10 min at 8,700 ×𝑔 at 4 °C.

10. Remove all supernatant, add 200 µL sterile 1 mM HEPES and 0.5 M sucrose, and gently resuspend the pellet.

11. Dispense into 50 µL aliquots, freeze cells at –80 °C, or continue with electroporation (*see* **Note 6**).

3.4 Transformation of L. monocytogenes via Electroporation

3.4.1 Variant A: Transformation of pKSV7(SoeAD) into L. monocytogenes

1. Thaw competent *L. monocytogenes* on ice.

2. Add 4 µL plasmid (1–2 µg) to 50 µL of competent cells.

3. Electroporate at the following parameters: field strength 10 kV/cm, time constant 5 ms, e.g., for a 0.2 cm cuvette: voltage 2 kV, resistance 400 Ω, capacity for 25 µF.

4. Immediately add 1 mL of BHI and 0.5 M sucrose to the cuvette, and incubate for 1 h without shaking at 30 °C.

5. Plate 5 × 200 µL on BHI-Cm[10] and incubate for at least 48 h at 30 °C.

3.4.2 Variant B: Cotransformation of pORI280(SoeAD) and pVE6007 into L. monocytogenes

1. Isolation of pVE6007: Grow *L. monocytogenes* pVE6007[+] in 10 mL BHI overnight at 37 °C, centrifuge for 10 min at 4,000 ×𝑔, resuspend pellet in 250 µL protoplast buffer, incubate at RT for 30 min, centrifuge at 4,000 ×𝑔 for 10 min, resuspend pellet in buffer P1 of QIAprep spin miniprep kit, and proceed with the QIAprep spin miniprep kit according to the manufacturer's instructions.

2. Thaw competent *L. monocytogenes* on ice.

3. Add 4 µL of each plasmid (1–2 µg) to 50 µL of competent cells (*see* **Note 7**).

4. Electroporate at the identical parameters as in **step 3** of Subheading 3.4.1.

5. Immediately add 1 mL of BHI and 0.5 M sucrose to the cuvette and incubate for 1 h without shaking at 30 °C.

6. Plate 5 × 200 µL on BHI-Ery[5]-X-gal[100] agar plates and incubate at 30 °C for at least 48 h.

3.5 Chromosomal Integration of SoeAD

3.5.1 Variant A: Using pKSV7 Vector (Fig. 2)

1. Confirm pKSV7(SoeAD)[+] transformants by colony PCR using pKSV7-MCS primers.

2. Select three pKSV7(SoeAD)[+] transformants for further passages.

3. Selection for chromosomal integration of pKSV7(SoeAD) via AB or CD crossover: Inoculate one colony in 10 mL pre-warmed

BHI-Cm^{10}, grow culture to stationary phase at 42 °C, dilute 1/1,000 with 10 mL pre-warmed BHI-Cm^{10}, grow to stationary phase at 42 °C, and repeat the serial subculturing 2–3 times. With each subculture, in parallel streak the *L. monocytogenes* cultures on pre-warmed (42 °C) BHI-Cm^{10} agar plates and incubate at 42 °C.

4. Confirmation of chromosomal integration of the plasmid: Plate cultures on BHI, incubate for 24 h at 30 °C, and perform colony PCR using the following primer pairs: SoeE-pKSV7 fwd, SoeE-pKSV7 rev, SoeZ-pKSV7 fwd, and SoeZ-pKSV7 rev (*see* **Note 8**).

5. Plasmid excision and curing by continuous passage: Select three to five colonies with chromosomal integration of the plasmid, inoculate in 10 mL BHI, grow to stationary phase at 30 °C, dilute 1/1,000 in 10 mL BHI, and grow bacteria to stationary phase at 30 °C; repeat this process 10–20 times (*see* **Note 9**).

6. Select for the appropriate deletion event by spreading dilute cultures (1:10^5 dilution) every third or fourth passage onto BHI agar and incubate for 24 h at 30 °C; streak colonies in replica plate onto BHI and BHI-Cm^{10}-agar and incubate at 30 °C. Select colonies that are Cm sensitive, and determine if the deletion event has taken place by colony PCR using the SoeE and SoeZ primers (*see* **Note 10**). Confirm gene deletion by sequencing the SoeEZ PCR product.

3.5.2 Variant B: pORI280/pVE6007 System (Fig. 3)

1. Confirm pORI280(SoeAD)$^+$ transformants by colony PCR using pORI280 MCS primers.

2. Select three pORI280(SoeAD)$^+$ transformants for further passages.

3. Induce loss of pVE6007, select for instances of integration of the pORI280(SoeAD) via AB or CD crossover by performing a complex streak from a single blue colony on BHI-Ery$^{7.5\text{-X-gal100}}$ agar, and incubate plate for 24 h at 37 °C.

4. Select one light and one dark blue colony (usually corresponding to respective AD and CD crossovers), streak colony in parallel on BH-Ery$^{7.5}$ and BHI-Cm^{10} agar, and incubate at 37 °C for 24 h. Colonies negative on the BHI-Cm^{10}-agar indicate loss of pVE6007.

5. Inoculate a single light and dark blue colony exclusively grown on BH-Ery$^{7.5}$ in 10 mL BHI, and grow culture to stationary phase at 37 °C without shaking.

6. Perform plasmid excision and curing by continuous passage: Dilute culture 1/1,000 in 10 mL BHI, grow to stationary phase at 37 °C without shaking, and repeat this process for five sequential passages.

7. Every third or fourth passage dilute cultures $1:10^5$, spread 100 µL of diluted cultures onto BHI-X-gal agar plates, and incubate plates for 24 h at 37 °C.

8. Screen white colonies by colony PCR using SoeE and SoeZ primers (*see* **Note 11**). Test colonies for erythromycin sensitivity to confirm the loss of pORI280(SoeAD).

4 Notes

1. Primer design: The primers should consist of ~20 bp (except the SoeC primer). To ensure that the whole gene is deleted SoeB and SoeC should flank the gene. If this is not possible, the deletion event should be in-frame within the coding region of the gene. The region that is being cloned also needs to be examined for restriction enzyme sites. The PCR products of SoeAB and SoeCD should have the same length (~300 bp).

2. SoeAB and SoeCD PCR: Start with a low annealing temperature, perform a gradient PCR for optimization, perform PCR reaction with and without $MgSO_4$, and use different DNA polymerase, e.g., Platinum-DNA Polymerase (Roche) or TaqPolymerase (Invitrogen).

3. For SoeAD PCR use PCR conditions similar for SoeAB and SoeCD; if the annealing temperatures are different use the lower temperature. Alternatively, perform a gradient PCR for optimization; use different ratios of SoeAB and SoeCD PCR products for DNA template; and use chromosomal DNA as a positive control (larger PCR product).

4. If possible use both vectors (pKSV7 and pORI280) in parallel for cloning of SoeAD, confirm the activity of the restriction enzymes by digesting the vectors with each enzyme alone, and use undigested vectors as a control for the agarose gel electrophoresis.

5. Competent *L. monocytogenes*: If an environmental strain or a strain which has not been made competent before is being used, include in parallel a lab strain, e.g., EGDe as a positive control. Before transformation of pKSV7(SoeAD) or pORI280(SoeAD)/pVE6007 confirm that the selected *L. monocytogenes* strain is competent by transformation of the empty vector (pKSV7 or pVE6007).

6. Alternative protocol for competent *L. monocytogenes*: Dilute an overnight culture of *L. monocytogenes* in 500 mL BHI containing 0.5 M sucrose, grow culture with shaking at 37 °C to an OD_{600} of 0.2–0.25, add ampicillin to a concentration of 10 µg/

mL, incubate for further 2 h, cool cells on ice for 10 min, centrifuge at $5{,}000 \times g$ for 10 min at 4 °C, resuspend pellet in 500 mL ice-cold sucrose glycerol wash buffer (SGW; 10 % glycerol, 500 mM sucrose, pH adjusted to 7.0, filter sterilized), centrifuge at $5{,}000 \times g$ for 10 min at 4 °C cells, resuspend cell pellet in 175 mL SGWB, centrifuge at $5{,}000 \times g$ for 10 min at 4'°C, resuspend in 50 mL SGWB, incubate with lysozyme (final concentration 10–25 µg/mL) for 20 min at 37 °C, centrifuge, resuspend cell pellet in 20 mL SGWB, centrifuge, resuspend pellet in SGWB to a final volume of 2.5 mL, and store 50 µL aliquots at 80 °C.

7. Co-transformation of pORI280/pVE6007: If the co-transformation does not work, transform first pVE6007 into competent *L. monocytogenes*, plate 5×200 µL on BHI-Cm[10] agar plates, make *L. monocytogenes* pVE6007[+] transformants competent, transform pORI280 (SoeAD) into competent *L. monocytogenes* pVE6007[+], plate 5×200 µL on BHI-Ery[5]-X-gal agar plates, and incubate at 30 °C for at least 48 h.

8. Confirmation of chromosomal integration of the plasmid: Use a WT colony as negative control; colonies growing on BHI-Cm[10] at 42 °C are an alternative source of colonies into which the plasmid has integrated. Selecting of the appropriate primer pairs is dependent on the orientation in which the insert is cloned into pKSV7 (i.e., which restriction sites are incorporated into SoeA and SoeD) and the site at which crossover occurs. Alternatively, if all primer pairs are used, two such combinations should produce a band in colonies positive for chromosomal integration.

9. The deletion event using pKSV7 should happen between passages 12 and 15; if a deletion mutant cannot be detected after passage 25, repeat the whole process of chromosomal integration.

10. For the screening of the colonies for the deletion event use the WT strain as a control (the wild-type version of the gene should give the original larger PCR product with the SoeE and SoeZ primers, whereas the strain with the deleted gene should give a smaller product, with the size difference corresponding to the deleted region).

11. Deletion event using the pORI280/pVE6007 system should generally happen between passages 5 and 12.

Acknowledgement

This work was supported by the EU FOODSEG grant.

References

1. Arnaud M, Chastanet A, Debarbouille M (2004) New vector for efficient allelic replacement in naturally nontransformable, low-GC-content, gram-positive bacteria. Appl Environ Microbiol 70:6887–6891

2. Li G, Kathariou S (2003) An improved cloning vector for construction of gene replacements in *Listeria monocytogenes*. Appl Environ Microbiol 69:3020–3023

3. Monk IR, Gahan CG, Hill C (2008) Tools for functional postgenomic analysis of *Listeria monocytogenes*. Appl Environ Microbiol 74: 3921–3934

4. Horton RM, Cai ZL, Ho SN (1990) Gene splicing by overlap extension: tailor-made genes using the polymerase chain reaction. Biotechniques 8:528–535

5. Horton RM (1995) PCR-mediated recombination and mutagenesis. SOEing together tailor-made genes. Mol Biotechnol 3:93–99

6. Cotter PD, Gahan CG, Hill C (2001) A glutamate decarboxylase system protects Listeria monocytogenes in gastric fluid. Mol Microbiol 40:465–475

7. Wiedmann M, Arvik TJ, Hurley RJ (1998) General stress transcription factor sigmaB and its role in acid tolerance and virulence of Listeria monocytogenes. J Bacteriol 180: 3650–3656

8. Smith K, Youngman P (1992) Use of a new integrational vector to investigate compartment-specific expression of the Bacillus subtilis spoIIM gene. Biochimie 74:705–711

9. Law J, Buist G, Haandrikman A (1995) A system to generate chromosomal mutations in *Lactococcus lactis* which allows fast analysis of targeted genes. J Bacteriol 177:7011–7018

10. Maguin E, Duwat P, Hege T (1992) New thermosensitive plasmid for gram-positive bacteria. J Bacteriol 174:5633–5638

Chapter 17

Mutant Construction and Integration Vector-Mediated Gene Complementation in *Listeria monocytogenes*

Reha Onur Azizoglu, Driss Elhanafi, and Sophia Kathariou

Abstract

Genes that play role in stress response mechanisms and other phenotypes of bacteria can be identified by construction and screening of mutant libraries. In this chapter, we describe the construction and screening of mutant libraries of *Listeria monocytogenes* using a plasmid, pMC38, carrying a *mariner*-based transposon system (*TC1/mariner*) and constructed by Cao et al. (Appl Environ Microbiol 73:2758–2761, 2007). Following screening of the mutant library, putative mutants are identified and the transposon is localized, leading to identification of the genes that play possible roles in the phenotype of interest. To confirm the role of the gene in the relevant phenotype, transposon mutants are genetically complemented with the wild type gene using the site-specific temperature-sensitive integration vector pPL2, constructed by Lauer et al. (J Bacteriol 184:4177–4186, 2002).

Key words Transposon, Mutant, *Listeria monocytogenes*, Complementation

1   Introduction

Construction and phenotypic screening of mutant libraries is an effective approach to identify genes responsible for particular stress response mechanisms and other phenotypes of bacteria. Transposons are valuable tools for construction of mutant libraries. In order to determine the genes responsible for specific environmental stress responses and other phenotypes of *Listeria monocytogenes*, mutant libraries of *L. monocytogenes* can be constructed using various transposons, e.g., Tn*1745*, Tn*916*, Tn*917* and its derivatives and a *mariner*-based transposon system, *Tc1/mariner* [1, 3–5]. The *mariner*-based transposition system has a number of advantages. It includes a self-encoded transposase and was earlier shown not to require any other factors for transposition [6]. In addition, this transposition system requires the dinucleotide TA for insertion, which results in highly random transposition in low-GC content organisms, such as *L. monocytogenes*. Comparison of this transposition system with a Tn*917*-based system showed that the *mariner*-based

Kieran Jordan et al. (eds.), *Listeria monocytogenes: Methods and Protocols*, Methods in Molecular Biology, vol. 1157, DOI 10.1007/978-1-4939-0703-8_17, © Springer Science+Business Media New York 2014

transposition system had low rates of plasmid retention and higher efficiency for random transposition [1].

Relevant mutants are identified by screening of the *mariner-based* mutant library for mutants with specific phenotypes (e.g., unable to exhibit a phenotype typical for the wild type parental strain). The number of transposon copies in selected mutants is determined by Southern blot hybridization using a transposon-specific (*ermC*) probe [1]. If multiple mutants are available with the desired phenotype, efforts can be best directed towards further characterization of those harboring single insertions. The location of transposon insertion can be determined by sequencing the DNA fragment amplified by semi-arbitrary PCR, using previously published primers [1].

Following determination of the transposon insertion site in the chromosome of *L. monocytogenes*, further genetic characterization of this gene needs to be conducted to confirm the role of the gene in the phenotype of interest. One widely used method is to genetically complement the mutant with the wild type gene using a plasmid that replicates in *L. monocytogenes* or, preferably, integration vectors; the latter eliminate concerns related to plasmid stability and copy number. Here, we describe a complementation protocol using the site-specific temperature-sensitive integration vector pPL2 [2]. If the phenotype of the mutant is restored to the level of the wild type strain following successful introduction and expression of the wild type gene into the mutant, this will confirm the role of the gene in the phenotype of interest. Even though the discussion here focuses on complementation of *mariner*-based transposon mutants, such genetic complementation is valuable for confirmation of the role of a specific gene that has been identified and/or characterized by other procedures, e.g., via other transposon mutagenesis systems or via construction of targeted deletions.

2 Materials

1. *L. monocytogenes* growth media: Dissolve 37 g of powder Brain Heart Infusion (BHI; Difco, Sparks, MD) in 1 L of purified water. For preparation of BHI agar add 12 g of agar (Difco, Sparks, MD). Autoclave media at 121 °C for 15 min. Store at room temperature. When antibiotic selection is needed, supplement with antibiotics at desired concentrations after cooling the sterilized media to approximately 56 °C.

2. *Escherichia coli* growth media: Dissolve 25 g of powder Luria Broth (LB; Difco, Sparks, MD) in 1 L of purified water. For preparation of LB agar add 12 g of agar (Difco, Sparks, MD). Autoclave media at 121 °C for 15 min. Store at room temperature. When antibiotic selection is needed, supplement with antibiotics

at desired concentrations after cooling the sterilized media to approximately 56 °C.

3. Qiagen Plasmid Mini Kit (Qiagen, Valencia, CA).

4. Electroporation buffer: 1 mM HEPES, 0.5 M sucrose, and pH 7.0.

5. Qiagen DNeasy kit (Qiagen, Valencia, CA).

6. Tris/Borate/EDTA (TBE) buffer: 90 mM Tris, 90 mM borate, 2 mM EDTA, and pH 8.0.

7. 10× Saline-Sodium Citrate (SSC) buffer: dilute from 20× SSC (20× SSC: 3 M NaCl, 300 mM Na–Citrate, pH 7.0).

8. Detection buffer (for dilution of chemiluminescent substrate used in Southern Blots): 0.1 M Tris, 0.1 M NaCl, pH 9.5.

9. QIAquick PCR Purification Kit (Qiagen, Valencia, CA).

3 Methods

3.1 Purification of Plasmid pMC38 Carrying a Mariner-Based Transposition System (TC1/Mariner)

1. Transform the plasmid (pMC38) into *E. coli* DH5α by electroporation and subculture a transformant on LB plates supplemented with erythromycin (5 μg/mL) to start a culture for plasmid purification. Incubate at 37 °C for at most 18 h (*see* **Note 1**).

2. Following incubation, purify the plasmid by using a plasmid extraction kit (Qiagen, Valencia, CA).

3.2 Preparation of Electrocompetent L. monocytogenes and Electroporation

1. Grow the *L. monocytogenes* strain of interest in BHI at 37 °C overnight.

2. Dilute the cultures (1 mL) into 99 mL BHI containing 0.5 M sucrose (1:100 dilution) and incubate at 37 °C with shaking (100 rpm) until OD_{600} is approx. 0.2 (~4 h).

3. Add Penicillin G (10 μg/mL; Sigma-Aldrich, St. Louis, MO) and further incubate the cultures for 2 h at 37 °C (*see* **Note 2**).

4. Pellet the cells by centrifugation (6,000 × g; 10 min; 4 °C), and wash 3 times (4 mL, 3 mL and 3 mL, respectively) with electroporation buffer (1 mM HEPES, 0.5 M sucrose, and pH 7.0).

5. Suspend the final pellets in electroporation buffer (250 μL); if needed, store the aliquots (150 μL) of the suspension at −80 °C.

6. Mix the cell suspension (150 μL) with plasmid DNA (10 μL obtained from steps in Subheading 3.1), and incubate on ice for 1 h.

7. Place the mixture in pre-chilled 1 mm gap cuvettes (Eppendorf, Madison, WI) and electroporate (Eppendorf) at 1.0 kV (10 kV/cm).

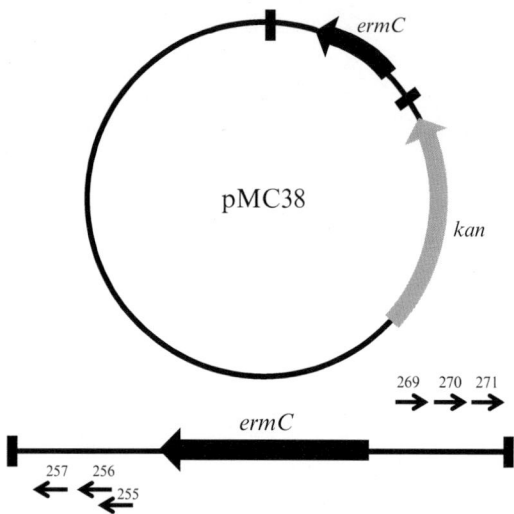

Fig. 1 Schematic representation of the *mariner*-based transposon delivery vector pMC38. *Short vertical bars* indicate the borders of the transposon, which harbors *ermC* and thus confers erythromycin resistance to *Listeria*. *ermC* and *kan*, genes conferring resistance to erythromycin and kanamycin, respectively. *Arrows with numbers* indicate primers used to localize the transposon insertion. Modified from Cao et al. [1]

8. Immediately mix the electroporated suspension with 1 mL BHI with 0.5 M sucrose supplemented with 0.1 µg/mL erythromycin (Em), and incubate at 30 °C for 1 h to induce expression of the transposon-encoded *ermC* gene.

9. Following incubation, plate the culture on BHI supplemented with 5 µg/mL Em and incubate at 30 °C for 3–4 days.

10. Confirm plasmid retention by screening for Em and kanamycin (Km) resistance. Growth of colonies on media with both antibiotics confirms plasmid retention (*see* **Note 3**; Fig. 1).

11. Inoculate a colony resistant to both antibiotics in 3 mL BHI-Em and incubate overnight at 30 °C.

12. Inoculate the overnight culture (10 µL) in 2 mL BHI-Em and incubate at 30 °C with shaking for 1 h; then shift the temperature to 40 °C until $OD_{600} = 0.3$–0.5 (about 5 h) to construct the mutant library (*see* **Note 4**).

13. Plate the culture on BHI-Em and incubate at 30 °C for 2 days.

14. Determine the frequency of transposon mutants (versus cells retaining the plasmid) by streaking individual colonies onto BHI-Em (5 µg/mL) and BHI-Km (10 µg/mL) plates (*see* **Note 5**; Fig. 2a, b). The plasmid retention rate is calculated by dividing the Km-resistant colonies by the total number of Em-resistant colonies.

A   B

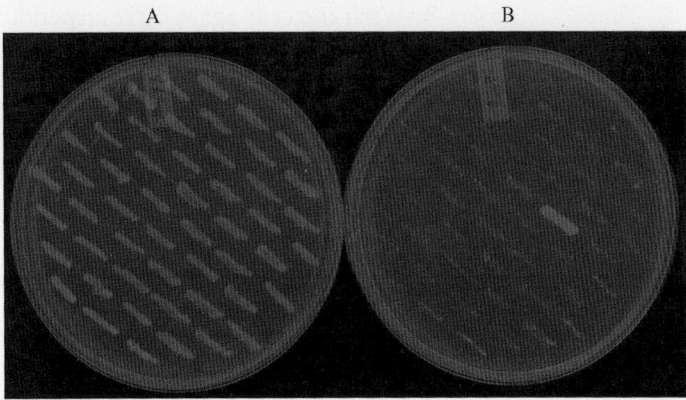

Fig. 2 Screening to determine plasmid retention frequency. Single colonies were tested by streaking a short line in duplicate onto BHI-Em (5 μg/mL) (*plate A*) and BHI-Km (10 μg/mL) (*plate B*). Growth on both plates indicates plasmid retention. In this example, one of the tested mutants retained the plasmid and grew on both media

15. Inoculate transposon mutants into 96-well microtiter plates (Costar, Corning, NY), 250 μL BHI/well, and grow at 37 °C overnight. Transfer the mutants (10 μL) to 96-well microtiter plates with fresh BHI (250 μL/well) using an 8-channel pipette (Eppendorf, Westbury, NY) and incubate at 37 °C overnight. Store the initial 96-well plates at –80 °C without glycerol. The mutants can be screened for specific phenotypes following overnight growth at 37 °C (*see* **Note 6**).

3.3 Preparation of E. coli Electro-competent Cells and Electroporation

1. Streak the *E. coli* strain of interest on LB plates (10 g tryptone, 5 g yeast extract, 10 g NaCl, 15 g agar per liter) and grow overnight at 37 °C.

2. Inoculate a freshly grown single colony into 3–5 mL LB broth and incubate overnight at 37 °C with agitation at 250 rpm.

3. Dilute the overnight culture 1:100 into 100 mL of liquid LB and grow until the culture reaches $OD_{600} = ~0.5$ for about 5–6 h.

4. Place the culture on ice for 10–15 min, and transfer to pre-chilled centrifuge tubes.

5. Centrifuge the cells at $2,300 \times g$ at 4 °C for 5–10 min to harvest the cells.

6. Gently resuspend the pellet in 3 mL of ice-cold sterile 10 % glycerol using a pipette, then add 100 mL of ice-cold sterile 10 % glycerol.

7. Centrifuge the cells at $2,300 \times g$ at 4 °C for 5–10 min to harvest the cells.

8. Wash the cells again by resuspending them in 3 mL of ice-cold sterile 10 % glycerol, then add 50 mL of ice-cold sterile 10 % glycerol.

9. Centrifuge the cells at $2,300 \times g$ at 4 °C for 10–15 min to harvest the cells.

10. Resuspend the cells in 3 mL of ice-cold sterile 10 % glycerol, then add 15 mL of ice-cold sterile 10 % glycerol.

11. Centrifuge the cells at $2,300 \times g$ at 4 °C for 10–15 min to harvest the cells.

12. Resuspend the cells in 250 μL of ice-cold sterile 10 % glycerol.

13. Aliquot the cells into individual pre-chilled microfuge tubes (50 μL per tube). The cells are now ready to use. For long-term storage, freeze the tubes with 50 μL aliquots in –80 °C freezer.

14. Place the mixture of cells (50 μL) and plasmid (or ligation mixture) DNA (1 μL), in pre-chilled 1 mm gap cuvettes (Eppendorf, Madison, WI) and electroporate (Eppendorf) at 1.25 kV (12.5 kV/cm).

15. Select the transformants on LB plates supplemented with the appropriate antibiotic.

3.4 Southern Blots

1. Before determining the insertion site of the transposon, the number of transposons inserted into the chromosome of *L. monocytogenes* needs to be determined. This can be done by Southern blot hybridization. Design the probe for the Southern blot to amplify a fragment within the erythromycin resistance gene (*ermC*) of pMC38, as described by Cao et al. [1] (Fig. 1).

2. Label the PCR products with digoxigenin (Genius kit; Roche, Indianapolis, IN). Labeling is done by adding sterile water to 10 ng to 1 μg DNA to have a final volume of 15 μL. The DNA is denatured by placing in boiling water for 10 min, and placing immediately on ice. Next, mix 2 μL of 10× Hexanucleotide Mix, 2 μL 10× DIG DNA labeling mix and 1 μL Klenow (5 U/μL) with the denatured DNA and incubate at 37 °C overnight. Stop the reaction by the addition of 2 μL of 0.2 M EDTA (pH 8.0). Store the labeled DNA at –20 °C.

3. Isolate genomic DNA of selected mutants by using DNeasy kit (Qiagen, Valencia, CA).

4. Digest the genomic DNAs with the restriction enzyme *Eco*RI (New England Biolabs, Waverly, MA) according to the manufacturer's instructions.

5. Separate the digested genomic DNAs by gel electrophoresis in TBE buffer (90 mM Tris, 90 mM borate, 2 mM EDTA, and pH 8.0) with 0.8 % agar at 85 V for 2.5 h.

Fig. 3 Determination of number of transposon copies into the chromosome of *L. monocytogenes* by Southern blot hybridization. Genomic DNA was digested by *Eco*RI and hybridized with a probe derived from *ermC*, as described [1]. *Lane 1*: Wild type parental strain *L. monocytogenes* F2365. *Lanes 2* and *3*: Two randomly chosen mutants of *L. monocytogenes* F2365, harboring a single copy of the transposon inserted in a different *Eco*RI fragment in each mutant (as suggested by a single band of different size in each mutant)

6. Transfer the DNA fragments onto nylon membrane (Osmonics, Inc. Westborough, MA) in 10× SSC buffer (dilute from 20× SSC,) by capillary action, overnight.

7. Use CSPD (Chemiluminescent substrate, Roche) (dilute 1:100 mL in detection buffer 0.1 M Tris, 0.1 M NaCl, pH 9.5) as the substrate for anti-DIG-alkaline phosphatase. Expose X-ray film (Fuji) to the chemiluminescent light resulting from hybridization of DNA fragments labeled with the probe (Fig. 3).

3.5 Determination of Transposon Insertion Sites

The transposon insertion site is determined by sequencing DNA fragments amplified by semi-arbitrary PCR. In semi-arbitrary PCR, DNA is amplified from one end of the transposon by applying two successive PCRs [1]. In each round of PCR one transposon-specific and one arbitrary primer is used.

1. In the primary PCR, the PCR conditions are 5 min at 95 °C; 30 cycles of 95 °C for 30 s, 42 °C for 45 s, and 72 °C for 1 min; then holding at 72 °C for 5 min [7]. Two primer pairs are used, (1) Marq207 (5′-GGC CAC GCG TCG ACT AGT ACNNNNNNNNNNNGTAAT-3′) and Marq255

(5′-CAG TAC AAT CTGCTC TGA TGC CGC ATA GTT-3′) and (2) Marq207 (5′-GGC CAC GCG TCG ACT AGT ACNNNNNNNNNNGTAAT-3′) and Marq269 (5′-GCT CTG ATA AAT ATG AAC ATG ATG AGT GAT-3′). These amplify the DNA fragments from the left and right end of the transposon, respectively, using the genomic DNA of the mutant as template (Fig. 1).

2. In the secondary PCR, 5 μL of a 4 % dilution from the primary PCR is used as template in a 20-μL reaction. The PCR conditions are 40 cycles of 95 °C for 30 s, 45 °C for 30 s, and 72 °C for 1 min; then holding at 72 °C for 5 min [7]. Again, two primer pairs are used, (1) Marq208 (5′-GGC CAC GCG TCG ACT AGTAC-3′) and Marq256 (5′-TAG TTA AGC CAG CCC CGA CAC CCG CCA ACA-3′) and (2) Marq 208 (5′-GGC CAC GCG TCG ACT AGTAC-3′) and Marq270 (5′-TGT GAA ATA CCG CAC AGA TGC GAA GGG CGA-3′). These amplify the DNA fragments from the left and right ends of the transposon, respectively (Fig. 1).

3. After purification of the PCR products by QIAquick PCR Purification Kit (Qiagen, Valencia, CA), the PCR product is sequenced using Marq257 (5′-CTTACAGACAAGCTG-TGACCGTCT-3′) and Marq271 (5′-GGGAATCATTTGA-AGGTTGGTACT-3′) for the left and right end of the transposon, respectively.

3.6 Genetic Complementation

To confirm the role of the gene in the phenotype of interest, the mutant (in this discussion, a *mariner*-based transposon mutant; however, it can also be a mutant generated via other transposon mutagenesis schemes, spontaneously, chemical mutagenesis, targeted deletion construction, etc.) is complemented with the relevant wild type gene using the site-specific temperature-sensitive integration vector, pPL2 [2]. Other shuttle vectors such as temperature-sensitive shuttle plasmid pKSV7 [8] or pCON-1 [9] can also be used for complementation purposes. However, such vectors can create challenges related to plasmid copy number and stability. The procedures below pertain to genetic complementation using pPL2 [2].

1. Design forward and reverse primers based on the DNA sequence of gene of interest (*see* **Note 7**).

2. Following the amplification, digest the PCR product by suitable restriction enzymes (according to the manufacturer's instructions).

3. The PCR product is purified using PCR purification kit (Qiagen Valencia, CA). Ligate the purified PCR product into the *L. monocytogenes* site-specific integration vector pPL2 [2], which was digested with the same restriction enzymes (*see* **Note 8**).

4. Introduce the recombinant plasmid into *E. coli* SM10 or *E. coli* S17-1 by electroporation. For electroporation, mix 50 μL of electrocompetent cells (*see* above) with 1 μL of plasmid ligation mixture. The mixture is placed in pre-chilled 1 mm gap cuvettes (Eppendorf, Madison, WI) and electroporated (Eppendorf) at 1.25 kV (12.5 kV/cm) (*see* **Note 9**).

5. Select the transformants on LB plates supplemented with chloramphenicol (25 μg/mL). Positive colonies are confirmed by PCR with primers specific to the target genes.

3.7 Bacterial Conjugation

1. Grow *E. coli* SM10 or *E. coli* S17-1 transformants in LB supplemented with chloramphenicol (25 μg/mL) to mid-log phase ($OD_{600} = \sim 0.55$) with shaking (100 rpm) at 30 °C.

2. Grow recipient *L. monocytogenes* (transposon mutant, or other strain of interest) overnight at 37 °C, and incubate at 45 °C for 10 min in water bath.

3. Mix the donor culture (3 mL) and the 45 °C-treated recipient culture (1.5 mL) and filter through 0.45-μm-pore-size HA-type filters (22 mm; Millipore; 10 mL syringe, BD). Wash the filter with BHI (10 mL).

4. Following washing, place the filter onto a freshly prepared BHI plate without antibiotics and incubate overnight at 30 °C.

5. Resuspend the bacteria on the filter in 2.5–4.0 mL of BHI and plate 25 μL or 50 μL of the suspension on BHI plates supplemented with chloramphenicol (6 μg/mL) and nalidixic acid (20 μg/mL) (*see* **Note 10**).

6. Incubate the plates at 30 °C for 2–3 days. In addition, to further enrich the suspension for transconjugants, transfer 50 μL of the suspended bacterial cell suspension into 30–50 mL BHI broth supplemented with chloramphenicol (6 μg/mL) and nalidixic acid (20 μg/mL) and grow at 30 °C overnight. Plate the overnight culture (50 μL) on BHI plates supplemented with chloramphenicol (6 μg/mL) and nalidixic acid (20 μg/mL) and incubate at 30 °C for 2–3 days.

7. Confirm transfer of pPL2 carrying the gene of interest by PCR and test for the restoration of the phenotype of interest.

4   Notes

1. The plasmid, pMC38, is somewhat unstable in *E. coli*; upon extended incubation at 37 °C it may be lost. Therefore, the transformant should be incubated at 37 °C for no longer than 18 h.

2. Supplementation of Penicillin G to the growth medium causes damage of the cell wall of *L. monocytogenes*, and this dramatically increases the transformation efficiency [10].

3. The transposon harbors an erythromycin resistance gene and a kanamycin resistance gene is on the pMC38 plasmid backbone. Therefore, resistance to both erythromycin and kanamycin confirms retention of the plasmid.

4. As pMC38 is a temperature-sensitive plasmid, temperature shift-up leads to the integration of the transposon into the chromosome of *L. monocytogenes* [1].

5. Integration of the transposon into the *L. monocytogenes* chromosome and loss of the plasmid is expected to result in *L. monocytogenes* mutants that are Em-resistant but Km-sensitive.

6. The genome of *L. monocytogenes* has close to 3,000 genes. Therefore, the phenotypic screening of the mutant library could require screening of ~4,000 mutants assuming the transposon is randomly inserted to the genome with minimal insertion in the same gene.

7. Identify the promoter and the transcription termination regions of the gene of interest by using the previously sequenced strains of *L. monocytogenes*. Include the target gene, putative promoter region, and the putative transcription termination site in the PCR fragment. In addition, at the 5′ end of each primer used for the PCR, incorporate suitable restriction enzyme recognition sequences for cloning purposes.

8. Use pPL2 without the gene of interest as negative control.

9. Direct electroporation of the plasmid into electrocompetent *L. monocytogenes* can also be pursued. Here, electroporation of the plasmid into electrocompetent *E. coli* SM10 or *E. coli* S17-1 followed by conjugation of the plasmid from *E. coli* to *L. monocytogenes* is described. This can be used to genetically complement the mutant as well as for heterologous expression of the gene in other *L. monocytogenes* strains, without the need to prepare electrocompetent cells for multiple strains.

10. Nalidixic acid is added to BHI for selection of *L. monocytogenes* (*L. monocytogenes* is naturally resistant to the indicated concentration of nalidixic acid, while *E. coli* is sensitive) and chloramphenicol is added for selection of *L. monocytogenes* with pPL2 (a chloramphenicol resistance determinant is harbored on pPL2). Alternatively, transconjugants can be selected utilizing streptomycin-resistant mutants of the recipient, selected as described [2, 11] and selecting on media with choramphenicol (6 μg/mL) and streptomycin (600 μg/mL).

References

1. Cao M, Bitar AP, Marquis H (2007) A *mariner-base* transposition system for *Listeria monocytogenes*. Appl Environ Microbiol 73: 2758–2761

2. Lauer P, Chow MY, Loessner MJ, Portnoy DA, Calender R (2002) Construction, characterization, and use of two *Listeria monocytogenes* site-specific phage integration vectors. J Bacteriol 184:4177–4186

3. Gaillard JL, Berche P, Sansonetti P (1986) Transposon mutagenesis as a tool to study the role of hemolysin in the virulence of *Listeria monocytogenes*. Infect Immun 52:50–55

4. Kathariou S, Metz P, Hof H, Goebel W (1987) Tn916-induced mutations in the hemolysin determinant affecting virulence of *Listeria monocytogenes*. J Bacteriol 169:1291–1297

5. Camilli A, Portnoy A, Youngman PJ (1990) Insertional mutagenesis of *Listeria monocytogenes* with a novel Tn917 derivative that allows direct cloning of DNA flanking transposon insertions. J Bacteriol 172:3738–3744

6. Lampe DJ, Churcill ME, Robertson HM (1996) A purified *mariner* transposase is sufficient to mediate transposition *in vitro*. EMBO J 15:5470–5479

7. Garsin DA, Urbach J, Huguet-Tapia JC, Peters JE, Ausubel FM (2004) Construction of an *Enterococcus faecalis* Tn917-mediated-gene-disruption library offers insight into Tn917 insertion patterns. J Bacteriol 186: 7280–7289

8. Smith K, Youngman P (1992) Use of a new integrational vector to investigate compartment-specific expression of the *Bacillus subtilis spoIIM* gene. Biochimie 74:705–711

9. Behari J, Youngman P (1998) Regulation of *hly* expression in *Listeria monocytogenes* by carbon sources and pH occurs through separate mechanisms mediated by PrfA. Infect Immun 66:3635–3642

10. Park SF, Stewart GSAB (1990) High efficiency transformation of *Listeria monocytogenes* by electroporation of penicillin-treated cells. Gene 94:129–132

11. Katharios-Lanwermeyer S, Rakic-Martinez M, Elhanafi D, Ratani S, Tiedje JM, Kathariou S (2012) Coselection of cadmium and benzalkonium chloride resistance in conjugative transfers from nonpathogenic *Listeria* spp. to other listeriae. Appl Environ Microbiol 78: 7549–7556

Chapter 18

Absolute and Relative Gene Expression in *Listeria monocytogenes* Using Real-Time PCR

Roberta Mazza and Rina Mazzette

Abstract

Reverse transcription combined with the polymerase chain reaction (RT-PCR) is a viable method widely used to quantify gene expression. There are two ways to quantify gene expression by real-time PCR: relative quantification and absolute quantification. Relative quantification relates the PCR signal of the target gene to a control gene, normally a housekeeping gene. Absolute quantification determines the input copy number, usually by relating the PCR signal to a standard curve. Here we describe both methods from RNA extraction to its quantification by real-time PCR.

Key words Real-time PCR, RNA, Absolute and relative quantification, Standard curve, Plasmid vector

---

1 Introduction

The study of genes and especially their possible expression by the living organism is the basis for knowledge about their function within the same organism. There are two different ways to study DNA and RNA. In qualitative studies, one asks which sequences are present. In quantitative studies, one asks how much of each sequence is present. Gene expression analysis at the mRNA level is one of the key components of functional genomics. Gene expression is the process by which information from a gene is used in the synthesis of a functional gene product. Gene regulation is the basis for cellular differentiation, morphogenesis, and the versatility and adaptability of any organism, it may also assist as a substrate for evolutionary change. In *Listeria monocytogenes*, for example, we can evaluate, through gene expression, which genes may be involved in adaptation. Correlating mRNA expression profiles under different cellular conditions can be used to predict gene functions [1].

Transcriptional profiling can be carried out by several methods. Here, we describe an approach for gene expression analysis in *L. monocytogenes* with real-time PCR. There are two common

Kieran Jordan et al. (eds.), *Listeria monocytogenes: Methods and Protocols*, Methods in Molecular Biology, vol. 1157, DOI 10.1007/978-1-4939-0703-8_18, © Springer Science+Business Media New York 2014

categories of quantification using real-time PCR: Relative and Absolute quantification. (1) Relative quantification is commonly carried out by normalization of the expression levels of the gene of interest with that of the housekeeping (HK) genes. The method of relative quantification allows evaluation of the differences between strains in the levels of expression of a gene. A housekeeping gene, assumed to have equal expression in all strains, is used to normalize expression. Analysis of expression profiles of target genes in *L. monocytogenes* strains is conducted by amplifying the cDNA of the sample for the target gene and the HK gene. Relative quantification, however, only gives a ratio of the gene of interest in comparison to the HK gene and not actual copy numbers in a defined concentration of mRNA population. (2) Absolute quantification relies on a standard plot constructed from the known concentrations of standard template and corresponding levels of real-time PCR data. Absolute quantification therefore would eliminate the ambiguous use of ratios [2].

For absolute quantification, a standard curve is first constructed from a DNA of known concentration. Spectrophotometric measurements at 260 nm can be used to assess the concentration of this DNA, which can then be converted to a copy number value based on the molecular weight of the sample used. This curve is then used as a reference standard for calculating quantitative information for DNA targets of unknown concentrations. cDNA plasmids are the preferred standards for standard curve quantification.

2   Materials

Prepare all the solutions with RNA-free reagents. Before starting all the procedures for RNA extraction, clean all surfaces and equipment with RNA Zap and rinse with RNase-free water. Prepare all the solution at room temperature (unless indicated otherwise) (*see* **Note 1**).

2.1   RNA Extraction

1. Rnasy midi kit (QUIAGEN).

2. RNase-Free DNase Set (50) (QUIAGEN).

3. RNaseZap® RNase Decontamination Solution (Applied Biosystems).

4. TE buffer: add 5 mg/mL of lysozyme to TE Buffer. Use 0.5 mL per sample.

5. RLT buffer: add 20 µL of β-mercaptoethanol to 2 mL of RLT buffer (supplied with the Rnasy midi kit).

6. Ethanol 100 %.

7. Prepare 500 mL of Glucoxyl Buffer 1×, mixing 50 mL of glycoxyl buffer (NorthernMax Gly®, 10× Gelprep/running Buffer (Ambion)).

8. Glycoxal Sample Load Dye (Ambion).

9. Nanodrop (Nanodrop Technologies, Delaware).

2.2 cDNA Preparation

1. SuperScript III First-Strand Synthesis Super Mix Kit (Invitrogen).

2.3 Real-Time PCR

1. LightCycler 480 SYBR Green I master mix (Roche Molecular Diagnostics, Penzburg, Germany).

2. Light Cycler 480 instrument (Roche Molecular Diagnostics, Rotkreuz, Switzerland).

2.4 Cloning in the Vector pCRII-TOPO

1. TOPO® TA Cloning® Kit (Invitrogen).

2. QIAprep Spin Miniprep Kit (QUIAGEN).

3. *Eco*RI (New England Biolabs).

4. Prepare a stock of ampicillin, containing 10 mL of water plus 0.5 mg of ampicillin. Use in 200 mL of LB medium.

5. Prepare LB plates (Luria-Bertani) (Invitrogen) containing 50 µg/mL of ampicillin.

6. Prepare 200 mL of LB (Luria-Bertani) broth medium with 50 µg/mL of ampicillin.

3 Methods

3.1 Total RNA Isolation

1. Inoculate a *L. monocytogenes* colony in 50 mL of BHI broth and incubate at 14 °C (*see* **Note 2**) to stationary phase (*see* **Note 3**).

2. Centrifuge 50 mL culture at 4 °C for 15 min at $5,000 \times g$.

3. Discard the supernatant and resuspend the pellet in 0.5 mL of TE buffer. Leave at room temperature for 10 min.

4. Add 2 mL of RLT Buffer with β-mercaptoethanol and vortex vigorously until complete dissolution of the pellet (*see* **Note 4**).

5. Centrifuge at $5,000 \times g$ for 5 min at room temperature, then gently remove the supernatant, being careful not to remove the pellet, and place it in a new tube.

6. Add 1.4 mL of 100 % ETOH and mix by gently inverting, then transfer the sample to a Rnasy Midi Column and centrifuge for 5 min at room temperature.

7. Put the liquid directly in the membrane of the collection tube (supplied), and centrifuge again for 5 min at room temperature, discard the fluid and keep the tube with the membrane.

8. Add 2 mL of RW1 Buffer (supplied with the kit) and centrifuge at $5,000 \times g$ for 2 min. Discard the fluid and keep the tube.

9. In a separate vial, add 20 µL of DNase I in 140 µL of RDD buffer (supplied in the RNase-Free DNase Set kit) and mix gen-

tly, centrifuge at 3,000 × g briefly to collect all the fluid. Do not vortex. Add the mix directly on the membrane of the column and leave at room temperature for 15 min (*see* **Note 5**).

10. Add 2 mL of RW1 Buffer (supplied with the kit) and leave at room temperature for 5 min. Centrifuge at 5,000 × g for 5 min. Discard the fluid and keep the tube.

11. Add 2.5 mL of RPE Buffer (supplied with the kit) and centrifuge at 5,000 × g for 2 min. Discard the fluid and keep the tube.

12. Add 2.5 mL of RPE Buffer (supplied with the kit) and centrifuge at 5,000 × g for 2 min. Then gently remove the column and transfer the liquid in a new tube.

13. Add 150 μL of RNase-free water directly on the membrane (*see* **Note 6**). Leave at room temperature for 5 min, then centrifuge for 5 min at 5,000 × g (*see* **Note 7**).

14. Recover the 150 μL of fluid and put it back on the membrane, then leave at room temperature for 5 min. Centrifuge at 5,000 × g for 5 min (*see* **Note 7**).

15. Keep an aliquot of 10 μL for the gel electrophoresis and the Nanodrop. Store the rest of the sample in a new RNA-free vial and freeze at –20 °C.

3.1.1 Optical Density Measurements

Optical density measurements can made using a Nanodrop. Good quality RNA will have an OD 260/280 ratio of 1.8–2 and an OD 260/230 of 1.8 or greater. This is because nucleic acid is detected at 260 nm, whereas protein, salt, and solvents are detected at 230 and 280 nm.

1. Open the Nanodrop icon on the computer and select "Nucleic Acid Measurements."

2. Add 1.5 μL of the solvent the RNA has been dissolved in (water).

3. After each and all subsequent measurements clean the pedestal by wiping with a dry lint-free tissue.

4. Add 1.5 μL of water (or the solvent that you used for dilute your RNA) and press "Blank." Repeat the blanking until there is a stable baseline, close to zero.

5. Confirm that the baseline is correct by measuring 1.5 μL of water, as if it were your first sample by pressing "Measure." Each of the measurements is automatically saved by the instrument.

3.1.2 Gel Electrophoresis to Check for Genomic RNA

Before starting be careful to use a clean electrophoresis chambers, leaving Rnase-free water inside the chamber for at least 30 min. Spray Rnase zap and rinse with Rnase-free water.

1. Prepare the electrophoresis gel by mixing 30 mL of Glucoxyl Buffer 1× with 0.45 g of Rnase-free agarose (1.5 % agarose gel).

2. Mix 5 µL of RNA sample and 5 µL of the Glycoxal Sample Loading Dye.

3. Heat at 50 °C for 30 min.

4. Chill on ice for 3 min and centrifuge in a bench-top microfuge for 10 s, prior to loading on a gel.

5. Load 10 µl of the prepared ladder and run the gel at 70 V for 1 h and 30 min.

3.2 cDNA Preparation

1. Prepare the sample by mixing 10 µl of 2× RT reaction mix, 2 µl of RT enzyme mix, RNA 80 µg/µl, and finally DEPC-treated water up to 20 µl (*see* **Note 8**). Vortex the sample briefly to mix and incubate as follows: 10 min at 25 °C, 30 min at 50 °C, 5 min at 85 °C. Chill on ice.

2. Add 1 µl (2U) of *E. coli* RNAse H to the mix and incubate for 20 min at 37 °C. Store the cDNA synthesis reaction at –20 °C or proceed directly to PCR.

3.3 Real-Time PCR

1. In a microtube of 200 µl, mix 2 µl of cDNA template, 0.5 mM of each primer and 2 µl of LightCycler 480 SYBR Green I master mix, made up to the required volume with purified water.

2. The real-time PCR run protocol was: (1) preincubation (10 min at 95 °C); (2) amplification 40 cycles (10 s at 95 °C, 10 s at 53–54 °C, 12 s at 72 °C) (the amplification temperature depends on the primers characteristics); (3) melting curve (65–95 °C at continuous fluorescent measurement); and (4) cooling at 4 °C.

3. Amplification can be monitored in the appropriate LC 480 channel and specific amplification confirmed by single peaks in melting curve analysis.

4. The relative quantification values are calculated using the 2–ΔΔCT method, by expressing the results as fold change, as normalized on the average of the selected group as a control, to which it is given a value of 1 [3].

3.4 Prepare a Standard Curve

3.4.1 Cloning in the Vector pCRII-TOPO

Produce PCR Products

Carry out all the procedures on ice.

1. To produce the PCR product set up the following 50 µL PCR reaction: 10× PCR buffer 5 µL, 50 Mm dNTPs 0.5 µL, PrimerF 5 µL, primerR 5 µL, MgCl 3 µL, H$_2$O 26 µL, Taq 0.5 µL, DNA 5 µL.

2. To produce the control PCR product of 750 bp, set up the following 50 µL PCR: Control DNA Template (100 ng) 1 µL, 10× PCR Buffer 5 µL, dNTP Mix 0.5 µL, Control PCR Primers (0.1 µg/µL each) 1 µL, Water 41.5 µL, Taq Polymerase (1 unit/µL) 1 µL.

3. Amplify using the following cycling parameters: initial denaturation at 94 °C for 2 min, 25 cycles (Denaturation at 94 °C for 1 min, Annealing at 55 °C for 1 min, Extension at 72 °C for 1 min), final extension at 72 °C for 7 min.

4. Remove 10 μL from the reaction and analyze by agarose gel electrophoresis. A discrete 750 bp band should be visible for the control PCR product.

Set Up the TOPO® Cloning Reaction

Once you have produced the desired PCR product, you can clone it in a vector and transform the recombinant vector into competent *E. coli*. It is important to have everything you need set up and ready to use to ensure that you obtain the best possible results.

– For the cloning reaction: mix 0.5–4 μL of PCR product (the amount depends on the freshness of the product), 1 μL of Salt Solution, 1 μL of TOPO vector and add water to a total volume of 5 μL, for a final volume 6 μL. Mix the reaction gently and incubate for 5 min at room temperature. Place the reaction on ice and proceed to transforming Competent Cells or store the TOPO Cloning reaction at –20 °C overnight.

3.4.2 Transfer of the Vector in E. coli Competent Cells

Perform the following before starting the transformation procedure:

– Equilibrate a water bath to 42 °C.

– Equilibrate a shaking incubator at 37 °C and a static incubator at 37 °C.

– Warm the vial of S.O.C. medium (supplied with the kit) and LB medium to room temperature.

– Warm the selective LB plates in a 37 °C incubator for 30 min (use two plates for each transformation).

Then:

1. Thaw, on ice, one vial of One Shot TOP10 chemically competent cells for each transformation.

2. Add 2 μL of the cloning reaction (*see* "Set Up the TOPO® Cloning Reaction") into a vial of One Shot cells and mix gently. Do not mix by pipetting up and down.

3. Incubate the vials on ice for 5 min (up to 30 min).

4. Heat-shock the cells for 30 s at 42 °C without shaking (use a timer).

5. Place them on ice immediately.

6. Aseptically add 250 μL of S.O.C. medium at room temperature to each vial.

7. Cap the vials tightly and shake horizontally at 37 °C for 1 h at 225 rpm in a shaking incubator (*see* **Note 9**).

8. Spread the solution obtained at point 7 in two separate plates, 10 μL plus 20 μL of S.O.C. medium and 50 μL, respectively, and incubate overnight at 37 °C.

9. Select six colonies and transfer them in a new tube containing 5 mL of LB medium plus ampicillin and incubate at 37 °C overnight (one colony for each tube).

10. Isolate the plasmid DNA.

3.4.3 Plasmid DNA Isolation

1. Pellet 1.5 mL bacterial overnight culture by centrifugation at 6,800×g in a conventional, table-top microcentrifuge for 3 min at room temperature. Throw the supernatant.

2. Resuspend pelleted bacterial cells in 250 μl Buffer P1 (supplied with the kit) and transfer to a microcentrifuge tube.

3. Add 250 μl Buffer P2 and mix thoroughly by inverting the tube 4–6 times until it becomes clear (or blue). Do not leave the reaction for more than 5 min.

4. Add 350 μl Buffer N3 and mix immediately and thoroughly by inverting the tube 4–6 times until it again becomes turbid. If using LyseBlue reagent, the solution will turn colorless.

5. Centrifuge for 10 min at 17,900×g in a table-top microcentrifuge.

6. Apply the supernatant from **step 5** to the QIAprep spin column (supplied with the kit).

7. Centrifuge for 60 s and discard the flow-through.

8. Add 500 μL Buffer PB and centrifuge for 60 s and discard the flow-through.

9. Repeat **step 8**.

10. Add 750 μL Buffer PB and centrifuge for 60 s and discard the flow-through.

11. Centrifuge for 1 min to remove residual wash buffer.

12. Place the QIAprep column in a clean 1.5 mL microcentrifuge tube. To elute DNA, add 50 μl Buffer EB (10 mM Tris·Cl, pH 8.5) or water to the center of the QIAprep spin column, let stand for 1 min at room temperature, and centrifuge for 1 min.

13. Once extracted the plasmid DNA, check it with 1 % gel electrophoresis.

3.4.4 DNA Restriction with EcoRI Enzyme

To confirm the presence and the correct orientation of the insert you need to analyze the plasmids by restriction analysis. The correct presence of the insert can be confirmed by sequencing the plasmid or by PCR, amplifying the gene we inserted in the restricted DNA plasmid. The PCR protocol for amplification of the gene of interest used is suitable.

1. Incubate the reaction for 10 min at 94 °C to lyse the cells and inactivate nucleases. Amplify for 20–30 cycles. For the final extension, incubate at 72 °C for 10 min. Store at 4 °C.

Using a restriction enzyme like *Eco*RI it is possible to cut the plasmid once in the vector and once in the insert.

1. In a microtube mix 10 μL of plasmid DNA, 2 μL of Enzyme buffer, 1 μL of *Eco*RI enzyme, and 7 μL of water for a final volume of 20 μL.

2. Incubate from 3 h to overnight at 37 °C.

3. Check all 20 μL by 1 % electrophoresis. Two bands are displayed. One belonging to the plasmid and one to the fragment of the gene of interest.

4. After identifying the correct clone, be sure to prepare a glycerol stock for long-term storage. Store a stock of plasmid DNA at –20 °C.

5. Mix 850 μL of culture (*see* Subheading 3.4.2, **step 9**) with 150 μL of sterile glycerol and transfer to a cryovial. Store at –20 °C.

3.4.5 Construction of the Standard Curve

Standard curve can be constructed using serial dilutions, at known concentrations, of the plasmid DNA obtained from the competent cells after cloning the PCR product of interest in plasmid vectors.

Dilutions are set up by a factor of 10.

Each dilution, for a total of 10, is amplified in triplicate, to have greater reproducibility. The standard curve is constructed by plotting the ordered values of Ct (or rather, the average Ct for each triplicate) and in abscissa the logarithm base 10 of dilutions: it is possible to determine the efficiency reaction considering the slope of the line.

4   Notes

1. It is advisable to keep an area of the bench, glassware, and utensils specifically for working with RNA.

2. We used 14 °C as this is the temperature normally used in the facility where the strain was isolated, and we wanted to study gene expression during growth at that temperature. Any temperature or conditions can be used.

3. It is important to predetermine a growth curve under the conditions used.

4. In the **step 4** of Subheading 3.1, we recommend to vortex for at least 30 s to have in the next step a clear separation of the pellet

from the supernatant. If the pellet is not properly separated from the supernatant you can repeat the previous step.

5. The **step 9** of Subheading 3.1 is not provided in the kit. We recommend it in order to have more certainty with more RNA without any contamination.

6. Avoid touching the membrane.

7. With regard to the **steps 13** and **14** of Subheading 3.1, the incubation time can be increased to gain the amount and the degree of purity. Or repeat one more time **step 14**.

8. Prepare the sample in ice.

9. In Subheading 3.4.2 in **step 7** we used tubes 50 mL vials inside to keep them tight.

References

1. Ding C, Cantor CR (2003) A high-throughput gene expression analysis technique using competitive PCR and matrix-assisted laser desorption ionization time-of-flight. Proc Natl Acad Sci U S A 100:3059–3064

2. Leong DT, Gupta A, Bai HF, Wan G, Yoong LF, Too HP, Chew FT, Hutmacher DW (2007) Absolute quantification of gene expression in biomaterials research using real-time PCR. Biomaterials 28:203–210

3. Livak KJ, Schmittgen TD (2001) Analysis of relative gene expression data using real-time quantitative PCR and 2-ΔΔCT method. Methods 25:402–408

Chapter 19

Genome Sequencing of *Listeria monocytogenes*

Stephan Schmitz-Esser and Martin Wagner

Abstract

Genome sequencing is a key technology in microbiology. A genome sequence is the prerequisite for understanding the molecular basis of a given phenotype; this is of particular importance for pathogens. Particularly for the foodborne pathogen *Listeria monocytogenes*, which is an important model organism in infection biology, genome sequencing has proven to be invaluable in advancing our understanding of its virulence mechanisms and epidemiology. In this chapter, current technologies and software tools for genome sequencing and genome analysis of *L. monocytogenes* are described.

Key words Genome sequencing, *Listeria monocytogenes*

1 Introduction

Listeria monocytogenes is a gram-positive facultative intracellular pathogen responsible for listeriosis, a rare but severe infection in humans (and also animals) with a mortality rate of 20–30 % [1, 2]. Listeriosis is acquired primarily through the consumption of contaminated food. First proof that food can be responsible for listeriosis outbreaks was provided in the early 1980s [3, 4]. Since then *L. monocytogenes* has become increasingly important as a foodborne pathogen and also as a model system for infection biology [5, 6]. *L. monocytogenes* can survive in many habitats such as soil, silage, marine and fresh water, sewage, vegetation, food processing plants, food, domestic and wild animals as well as humans [7, 8]. Due to the ability of *L. monocytogenes* to resist environmental stresses, which normally limit bacterial growth, such as heavy metal ions, high salt concentration, low pH values, low temperatures, and low water activity, this pathogen successfully colonizes food processing environments.

The first *Listeria* genomes (of the type strain *L. monocytogenes* EGDe and of one *L. innocua* strain) were published in 2001 [9]; since then a large number of additional *Listeria* genomes have become available. According to the genomes online database

Kieran Jordan et al. (eds.), *Listeria monocytogenes: Methods and Protocols*, Methods in Molecular Biology, vol. 1157, DOI 10.1007/978-1-4939-0703-8_19, © Springer Science+Business Media New York 2014

(http://www.genomesonline.org/ [10]) and NCBI GenBank, currently, 43 *L. monocytogenes* and 12 other *Listeria* genomes are available and approximately 80 *Listeria* genome projects are ongoing. The published *Listeria* genomes greatly advanced our understanding of *Listeria* pathogenicity, epidemiology of listeriosis outbreaks, and how *L. monocytogenes* replicates in various ecological niches. *Listeria* genomes show a highly conserved genomic structure and only relatively few signs of horizontal gene transfer, particularly from outside the genus *Listeria*. Up to now, *Listeria* genomes have been sequenced using Sanger, Roche/454, Illumina, Life Technologies SOLiD, or a combination of these technologies. A genome sequence is the prerequisite for understanding the molecular basis of a given phenotype (in the case of pathogens, e.g., antibiotic resistance or virulence determinants). With the power of comparative genome analysis and improvements in sequencing technologies, it can easily be imagined that genome sequencing (particularly of multiple strains or species) will become an important experiment in microbiology.

2 Materials

2.1 Disposables (Sterile) and Reagents for DNA Isolation

1. 2 mL Microcentrifuge tubes.
2. 50 mL Polypropylene centrifuge tubes.
3. Pipettes and pipette tips.
4. TE buffer (10 mM Tris, 1 mM EDTA, pH 8.0).
5. 10 % Sodium dodecyl sulfate (SDS).
6. 20 mg/mL Proteinase K (store single-use aliquots at –20 °C).
7. 100 mg/mL Lysozyme.
8. 5 M NaCl.
9. CTAB/NaCl solution [10 % CTAB (hexadecyltrimethyl ammonium bromide)] in 0.7 M NaCl. Dissolve 4.1 g NaCl in 80 mL water and slowly add 10 g CTAB while heating (65 °C) and stirring. It may take up to a few hours to dissolve CTAB. Adjust the final volume to 100 mL and filter sterilize or autoclave.
10. 24:1 Chloroform/isoamyl alcohol.
11. 25:24:1 Phenol/chloroform/isoamyl alcohol.
12. Isopropanol.
13. 70 % Ethanol.

2.2 Equipment

1. Hot plate.
2. Glass beaker.
3. Magnetic stir rod.

4. Thermometer.

5. 65 °C Water bath/heat block.

6. 37 °C Incubator/heat block.

7. 56 °C Water bath/heat block.

8. Fluorometer.

2.3 Computer Hardware Requirements

De novo or reference-based assemblies of bacterial genomes and other calculations during genome analysis can be performed in reasonable time (in minutes or hours) on standard desktop computers with sufficient hard disk space (>1 TB) and memory (16 GB or higher). Most of the software needed runs under Unix/Linux operating systems.

3 Methods

Please note that the software listed below is a non-exhaustive list—the readers are referred to the primary literature for more details and alternative software.

3.1 Bacterial Genomic DNA Isolation Using CTAB

This protocol is based on a protocol for DNA isolation by Wilson (2012) [11] and by the US DOE Joint Genome Institute (JGI) (http://my.jgi.doe.gov/general/protocols/JGI-Bacterial-DNA-isolation-CTAB-Protocol-2012.pdf).

1. Inoculate 5 mL brain–heart infusion (BHI) or tryptic soy broth (TSB) with the bacterial strain of interest. Grow in conditions appropriate for that strain (i.e., appropriate medium, drug selection, temperature) until the culture is saturated.

2. Centrifuge 2 mL of the culture in a microcentrifuge for 2 min at $10,000 \times g$ until a compact pellet forms. Discard the supernatant.

3. Resuspend the cells in 740 µL TE (should correspond to an OD600 of approx. 1.0) (*see* **Note 1**).

4. Add 20 µL lysozyme (100 mg/mL). Mix well (*see* **Note 1**).

5. Incubate for 30 min at 37 °C.

6. Add 40 µL 10 % SDS. Mix well.

7. Add 8 µL Proteinase K (20 mg/mL). Mix well.

8. Incubate for 1–3 h at 56 °C.

9. Add 100 µL 5 M NaCl. Mix well (*see* **Note 2**).

10. Add 100 µL CTAB/NaCl solution (preheated to 65 °C). Mix well (*see* **Note 2**).

11. Incubate at 65 °C for 10 min.

12. Add 500 µL chloroform/isoamyl alcohol (24:1). Mix well.

13. Centrifuge at max speed in a microfuge for 10 min at room temperature (20–25 °C).

14. Transfer the aqueous phase to a clean 2 mL microcentrifuge tube leaving the interface behind.

15. Add 500 μL phenol/chloroform/isoamyl alcohol (25:24:1). Mix well.

16. Centrifuge at maximum speed in a microfuge for 10 min at room temperature (20–25 °C).

17. Transfer the aqueous phase to a clean 2 mL microcentrifuge tube leaving the interface behind.

18. Add 0.6 vol isopropanol (precooled to –20 °C) to precipitate the nucleic acids. Mix well.

19. Incubate at –20 °C for 2 h to overnight.

20. Centrifuge at max speed in a microfuge for 15 min at 4 °C.

21. Wash the pellet with 70 % ethanol (precooled to –20 °C), centrifuge at max speed in a microfuge for 5 min at room temperature (20–25 °C).

22. Discard the supernatant and let pellet dry at room temperature (20–25 °C).

23. Redissolve the pellet in 100 μl TE buffer.

24. Measure the DNA concentration with fluorometer dsDNA assay (e.g., Life Technologies Qubit® Fluorometer).

25. Store the DNA at –80 °C or –20 °C.

3.2 DNA Isolation Using Commercially Available Kits

1. If preferred, DNA isolation can also be performed using commercially available kits (e.g., QIAGEN genomic-tip or others). For protocols for DNA isolation refer to the instructions of the manufacturer.

2. Measure the DNA concentration with a fluorometer dsDNA assay (e.g., Life Technologies Qubit® Fluorometer).

3. Store the DNA at –80 °C or –20 °C.

3.3 Genome Sequencing

1. Due to cost and throughput reasons we suggest to use next generation sequencing technology [e.g., Roche/454, Life Technologies Ion Torrent, Illumina (GAIIx, MiSeq, HiSeq), Pacific Biosciences, Life Techonologies SOLiD]. Briefly, most next generation sequencing technologies are performed in three to four steps: fragmentation of the genomic DNA, tagging with platform-specific adaptors, amplification, and sequencing. After fragmentation of genomic DNA into random overlapping fragments, end repair is performed to generate blunt end fragments to which the platform-specific adaptors can be ligated. After that emulsion PCR and enrichment (in the case of Roche/454 or Ion Torrent technology) or

bridge amplification (Illumina) are performed. Finally, sequencing itself is performed. *See*, e.g., [12–14] for detailed reviews on next generation sequencing technologies.

3.4 Library Preparation

1. Paired-end (DNA fragments are sequenced form each end) or mate-pair (large insert paired-end libraries which are circularized after initial DNA fragmentation and refragmented after circularization) sequencing is essential facilitating assembly in order to resolve repetitive regions or genomic rearrangements (i.e., rRNA operons, transposases).

2. For library preparation, kits are available either from the manufacturer of the respective sequencing platforms or also from other vendors (e.g., New England Biolabs) for a number of different sequencing platforms (*see* **Note 3**).

3. Due to the massive data output of next generation sequencing technologies, it is possible to sequence multiple genomes in a single run. This can be accomplished by including barcodes during library preparation. After sequencing, the reads are sorted according to the barcodes and can be assembled separately. Barcoding (or multiplexing) is particularly interesting for Illumina sequencing technology, which provides much higher number of reads compared to, e.g., Roche/454 sequencing. Usually, a typical bacterial genome (size 2–5 Mb) is sequenced on ¼ picotiter plate of a Roche/454 GS FLX+ instrument (corresponding to approx. 200,000 reads). On Illumina systems due to the much higher throughput, many genomes (depending on the system, 10 or more) can be sequenced on one lane of a flowcell.

4. Independent of the sequencing platform used, it is highly advisable to assess the quality and quantity of the library before sequencing. This can be done using, e.g., the Bioanalyzer from Agilent or Experion from Bio-Rad.

3.5 Sequencing Strategy

1. Perform sequencing at sufficient sequencing depth: coverage of 20–30× using Roche/454 or 50× or higher using Illumina sequencing technology is recommended (*see* **Note 3**).

2. If very high-quality (nearly) complete finished genome sequences are desired, it is advisable to use a combination of two sequencing platforms. Such combinations can be, e.g., Illumina and Roche/454 or Illumina and Pacific Biosciences. Detailed instructions for sequencing and assembly of bacterial genomes can be found in these two recent studies using a combination of Illumina and Pacific Biosciences sequencing technologies and novel assembly algorithms for genome sequencing and assembly [15, 16].

3.6 Genome Assembly

1. For genome assembly, many (mostly freely available) software tools have been designed; *see*, e.g., [12, 17] for reviews. In addition to standalone dedicated assembly software, commercial software suites for assembly and various sequence analyses are available (e.g., DNASTAR Lasergene, CLC genomics workbench).

2. In general, for genome assembly of short shotgun sequence reads into longer contiguous sequences two main options can be used: In the de novo assembly approach, sequence reads are compared to each other, and then overlapped to build longer contiguous sequences (i.e., contigs). In reference-based assembly, the genome is assembled based on mapping each read to a reference genome sequence. The choice of assembly strategy depends mainly on the presence of a good reference sequence and the biological application. If analysis of genetic variation between the genomes of multiple highly related strains is desired, a reference-based assembly is often sufficient. On the other hand, reference-based assemblies are problematic when dealing with reads from repetitive regions or from parts of the genome which the absent in the reference genome (e.g., virulence factors, phages, or plasmids).

3. Visualize, e.g., using the Tablet Viewer (http://bioinf.scri. ac.uk/tablet/ [18]), compare and verify assembly, order contigs according to an alignment to reference genome(s) using, e.g., MAUVE (http://gel.ahabs.wisc.edu/mauve/ [19]) or Mugsy (http://mugsy.sourceforge.net/ [20]). For verification of assembly, many additional parameters can be used such as transcriptome data, genomes of closely related organisms, or the detection of regions (or contigs) with unusually high or low sequencing coverage. Dedicated software tools are available for assembly validation such as AMOSvalidate from the AMOS consortium (http://sourceforge.net/apps/mediawiki/amos/index.php?title=AMOS [21]).

3.7 Genome Annotation and Analysis

1. Load assembly (contigs) into a software of choice for gene prediction and genome annotation such as MaGe/Microsccope (http://www.cns.fr/agc/microscope/mage/index.php [22]), RAST (http://rast.nmpdr.org/ [23]), IMG/ER (http:// img.jgi.doe.gov/er [24]), GenDB (http://www.cebitec.uni-bielefeld.de/groups/brf/software/gendb_info/ [25]).

2. It is often desired to obtain high quality finished genome sequences. Thus, finishing using either software tools as described, e.g., in [26, 27] and/or wet lab gap closing using PCR and Sanger sequencing is suggested. Software tools for automatic primer design for gap closing are freely available, e.g., [28] or as parts of commercial software packages (e.g., DNASTAR Lasergene, CLC genomics workbench).

3.8 Comparative Genomics	1. A wide variety of software tools are available for comparative genomics (e.g., analysis of genes present or absent, detection of chromosomal rearrangements). The reader is referred to primary literature for more details and approaches used for analyses. Often in comparative genomics, the comparison of multiple genomes and its visualization is a first step. For this, a number of tools are available, e.g., the CGView comparison tool (http://stothard.afns.ualberta.ca/downloads/CCT/ [29]), Blast ring image generator (BRIG) (http://source-forge.net/projects/brig/ [30]), Circos (http://circos.ca/ [31]), or the Artemis comparison tool (ACT) (http://www.sanger.ac.uk/resources/software/act/ [32]).

3.9 Submission and Description of Data

1. With the rapid pace that new genomes are sequenced and deposited in public databases, great effort should be made in order to describe genomic investigations as accurately and comprehensively as possible. This is particularly important in the view of (large scale) comparative analyses. It is strongly suggested that researchers comply with the Minimum Information about a (Meta)Genome Sequence (MIGS) specifications (http://gensc.org/gc_wiki/index.php/MIGS/MIMS [33]).

3.10 L. monocytogenes Genome Sequencing Recipe

1. Due to the low number and size of repetitive sequences within *Listeria* genomes and the high degree of similarity between *Listeria* genomes, next generation sequencing technologies are particularly well suited for genome sequencing of *Listeria*. We have sequenced a number of *L. monocytogenes* genomes and according to our experience and other studies 1, genome sequencing of *L. monocytogenes* strains using Illumina sequencing is suggested. Using Illumina sequencing has a number of advantages: many genomes can be sequenced on one lane, the sequencing accuracy is high, the costs are significantly lower compared to Roche/454 and sequencing usually yields a reasonable number of contigs (due to the low amount of repetitive sequences in *Listeria* genomes).

2. Perform genome sequencing using Illumina 100 bp read length and paired-end sequencing (50 bp or 75 bp read length should also work). In our experience this yields 10–50 contigs (size > 500 bp) after de novo assembly covering approx. 97–99 % of the genome. Mostly, the gaps are within repetitive genes such as rRNA operons, transposases, internalin-like genes, and phage genes. If a closed genome is desired, a combination of different sequencing technologies can be applied (*see* above and **Note 4**).

3. Assemble the genomes using de novo assembly algorithms (*see* above).

4. Verify assembly and order contigs to reference genome (*see* above).

5. Of course, completed finished genome sequences are highly desirable, but depending on the application and the scientific question, high-quality draft genome sequence assemblies can also be sufficient. Finishing a genome sequence is a time-consuming (and costly) process consisting of gap closure (by wet lab work or in silico analyses) and rigorous quality control steps. Nearly all genes are represented in most draft genome sequences; however, the main drawback of draft genomes is the lack of order of the contigs or scaffolds which impedes global comparative genome analyses.

6. Perform finishing to desired degree (either in silico and/or wet lab)

4 Notes

1. Lysozyme treatment is important for efficient cell lysis. It is also important not to use too many cells for DNA isolation, DNA will not separate well from protein during extraction.

2. It is important to keep the salt concentration above 0.5 M NaCl, otherwise nucleic acids might coprecipitate with CTAB. In addition, the temperature of all solutions should be kept above 15 °C, as CTAB will precipitate below this temperature.

3. A detailed survey of sequencing technologies is beyond the scope of this chapter. The reader is referred to the literature for details see, e.g., [12, 14, 17]. Briefly, Illumina sequencing technologies provide massive throughput and high accuracy, but relatively short read lengths. Roche/454 provides longer read-lengths, but lower throughput compared to Illumina and has a higher error rate (particularly in homopolymer regions). Third-generation sequencing technologies such as Pacific Biosciences provide extremely long reads with up to many kb in length, but have a markedly lower accuracy. Currently, Illumina sequencing technology is the most cost-effective, particularly for high-throughput sequencing applications (i.e., sequencing multiple genomes). However, at the pace the field is moving it is likely that new technologies will emerge and existing technologies will be improved.

4. A general issue which needs to be considered is whether library preparation is performed by the researcher or by external providers (such as core facilities or commercial vendors). As indicated above, kits for library preparation are available from different providers, but we suggest that library preparation (and sequencing) should be performed by the sequencing provider unless the researchers have the experience and equipment to perform library preparations themselves.

References

1. Allerberger F, Wagner M (2010) Listeriosis: a resurgent foodborne infection. Clin Microbiol Infect 16:16–23

2. Swaminathan B, Gerner-Smidt P (2007) The epidemiology of human listeriosis. Microbes Infect 9:1236–1243

3. Fleming DW, Cochi SL, MacDonald KL, Brondum J, Hayes PS, Plikaytis BD, Holmes MB, Audurier A, Broome CV, Reingold AL (1985) Pasteurized milk as a vehicle of infection in an outbreak of listeriosis. N Engl J Med 312:404–407

4. Schlech WF 3rd, Lavigne PM, Bortolussi RA, Allen AC, Haldane EV, Wort AJ, Hightower AW, Johnson SE, King SH, Nicholls ES, Broome CV (1983) Epidemic listeriosis—evidence for transmission by food. N Engl J Med 308:203–206

5. Cossart P (2011) Illuminating the landscape of host-pathogen interactions with the bacterium *Listeria monocytogenes*. Proc Natl Acad Sci USA 108:19484–19491

6. Hamon M, Bierne H, Cossart P (2006) *Listeria monocytogenes*: a multifaceted model. Nat Rev Microbiol 4:423–434

7. Ivanek R, Grohn YT, Wiedmann M (2006) *Listeria monocytogenes* in multiple habitats and host populations: review of available data for mathematical modeling. Foodborne Pathog Dis 3:319–336

8. Sauders BD, Wiedmann M (2007) Ecology of Listeria species and L. monocytogenes in the natural environment. In: Ryser ET, Marth EH (eds) Listeria, listeriosis and food safety. CRC Press, Boca Raton, pp 21–53

9. Glaser P, Frangeul L, Buchrieser C, Rusniok C, Amend A, Baquero F, Berche P, Bloecker H, Brandt P, Chakraborty T, Charbit A, Chetouani F, Couvé E, de Daruvar A, Dehoux P, Domann E, Domínguez-Bernal G, Duchaud E, Durant L, Dussurget O, Entian KD, Fsihi H, García-del-Portillo F, Garrido P, Gautier L, Goebel W, Gómez-López N, Hain T, Hauf J, Jackson D, Jones LM, Kaerst U, Kreft J, Kuhn M, Kunst F, Kurapkat G, Madueno E, Maitournam A, Vicente JM, Ng E, Nedjari H, Nordsiek G, Novella S, de Pablos B, Pérez-Diaz JC, Purcell R, Remmel B, Rose M, Schlueter T, Simoes N, Tierrez A, Vázquez-Boland JA, Voss H, Wehland J, Cossart P (2001) Comparative genomics of *Listeria* species. Science 294:849–852

10. Pagani I, Liolios K, Jansson J, Chen IMA, Smirnova T, Nosrat B, Markowitz VM, Kyrpides NC (2012) The Genomes OnLine Database (GOLD) v. 4: status of genomic and metagenomic projects and their associated metadata. Nucleic Acids Res 40:D571–D579

11. Wilson K (2001) Preparation of genomic DNA from bacteria. Curr Protoc Mol Biol Chapter 2:Unit 2.4

12. Loman NJ, Constantinidou C, Chan JZ, Halachev M, Sergeant M, Penn CW, Robinson ER, Pallen MJ (2012) High-throughput bacterial genome sequencing: an embarrassment of choice, a world of opportunity. Nat Rev Microbiol 10:599–606

13. MacLean D, Jones JD, Studholme DJ (2009) Application of 'next-generation' sequencing technologies to microbial genetics. Nat Rev Microbiol 7:287–296

14. Metzker ML (2010) Sequencing technologies—the next generation. Nat Rev Genet 11:31–46

15. Bashir A, Klammer AA, Robins WP, Chin CS, Webster D, Paxinos E, Hsu D, Ashby M, Wang S, Peluso P, Sebra R, Sorenson J, Bullard J, Yen J, Valdovino M, Mollova E, Luong K, Lin S, LaMay B, Joshi A, Rowe L, Frace M, Tarr CL, Turnsek M, Davis BM, Kasarskis A, Mekalanos JJ, Waldor MK, Schadt EE (2012) A hybrid approach for the automated finishing of bacterial genomes. Nat Biotechnol 30:701–707

16. Ribeiro FJ, Przybylski D, Yin S, Sharpe T, Gnerre S, Abouelleil A, Berlin AM, Montmayeur A, Shea TP, Walker BJ, Young SK, Russ C, Nusbaum C, MacCallum I, Jaffe DB (2012) Finished bacterial genomes from shotgun sequence data. Genome Res 22:2270–2277

17. Nagarajan N, Pop M (2013) Sequence assembly demystified. Nat Rev Genet 14:157–167

18. Milne I, Bayer M, Cardle L, Shaw P, Stephen G, Wright F, Marshall D (2010) Tablet—next generation sequence assembly visualization. Bioinformatics 26:401–402

19. Darling AE, Mau B, Perna NT (2010) progressiveMauve: multiple genome alignment with gene gain, loss and rearrangement. PLoS One 5:e11147

20. Angiuoli SV, Salzberg SL (2011) Mugsy: fast multiple alignment of closely related whole genomes. Bioinformatics 27:334–342

21. Phillippy AM, Schatz MC, Pop M (2008) Genome assembly forensics: finding the elusive mis-assembly. Genome Biol 9:R55

22. Vallenet D, Engelen S, Mornico D, Cruveiller S, Fleury L, Lajus A, Rouy Z, Roche D, Salvignol G, Scarpelli C, Médigue C (2009) MicroScope: a platform for microbial genome

annotation and comparative genomics. Database (Oxford) 2009:bap021. doi:10.1093/database/bap021

23. Aziz RK, Bartels D, Best AA, DeJongh M, Disz T, Edwards RA, Formsma K, Gerdes S, Glass EM, Kubal M, Meyer F, Olsen GJ, Olson R, Osterman AL, Overbeek RA, McNeil LK, Paarmann D, Paczian T, Parrello B, Pusch GD, Reich C, Stevens R, Vassieva O, Vonstein V, Wilke A, Zagnitko O (2008) The RAST Server: rapid annotations using subsystems technology. BMC Genomics 9:75

24. Markowitz VM, Mavromatis K, Ivanova NN, Chen IMA, Chu KC, Kyrpides NC (2009) IMG ER: a system for microbial genome annotation expert review and curation. Bioinformatics 25:2271–2278

25. Meyer F, Goesmann A, McHardy AC, Bartels D, Bekel T, Clausen J, Kalinowski J, Linke B, Rupp O, Giegerich R, Puhler A (2003) GenDB—an open source genome annotation system for prokaryote genomes. Nucleic Acids Res 31:2187–2195

26. Boetzer M, Pirovano W (2012) Toward almost closed genomes with GapFiller. Genome Biol 13:R56

27. Nagarajan N, Cook C, Di Bonaventura M, Ge H, Richards A, Bishop-Lilly KA, DeSalle R, Read TD, Pop M (2010) Finishing genomes with limited resources: lessons from an ensemble of microbial genomes. BMC Genomics 11:242

28. Galardini M, Biondi EG, Bazzicalupo M, Mengoni A (2011) CONTIGuator: a bacterial genomes finishing tool for structural insights on draft genomes. Source Code Biol Med 6:11

29. Grant JR, Arantes AS, Stothard P (2012) Comparing thousands of circular genomes using the CGView comparison tool. BMC Genomics 13:202

30. Alikhan NF, Petty NK, Ben Zakour NL, Beatson SA (2011) BLAST Ring Image Generator (BRIG): simple prokaryote genome comparisons. BMC Genomics 12:402

31. Krzywinski M, Schlein J, Birol I, Connors J, Gascoyne R, Horsman D, Jones SJ, Marra MA (2009) Circos: an information aesthetic for comparative genomics. Genome Res 19: 1639–1645

32. Carver TJ, Rutherford KM, Berriman M, Rajandream MA, Barrell BG, Parkhill J (2005) ACT: the Artemis comparison tool. Bioinformatics 21:3422–3423

33. Field D, Garrity G, Gray T, Morrison N, Selengut J, Sterk P, Tatusova T, Thomson N, Allen MJ, Angiouli SV, Ashburner M, Axelrod N, Baldauf S, Ballard S, Boore J, Cochrane G, Cole J, Dawyndt P, De Vos P, DePamphilis C, Edwards R, Faruque N, Feldman R, Gilbert J, Gilna P, Glockner FO, Goldstein P, Guralnick R, Haft D, Hancock D, Hermjakob H, Hertz-Fowler C, Hugenholtz P, Joint I, Kegan L, Kane M, Kennedy J, Kowalchuk G, Kottmann R, Kolker E, Kravitz S, Kyrpides N, Leebens-Mack J, Lewis SE, Li K, Lister AL, Lord P, Maltsev N, Markowitz V, Martiny J, Methe B, Mizrachi I, Moxon R, Nelson K, Parkhill J, Proctor L, White O, Sansone SA, Spiers A, Stevens R, Swift P, Taylor C, Tateno Y, Tett A, Turner S, Ussery D, Vaughan B, Ward N, Whetzel T, San Gil I, Wilson G, Wipat A (2008) The minimum information about a genome sequence (MIGS) specification. Nat Biotechnol 26:541–547

Chapter 20

Using Enhanced Green Fluorescent Protein (EGFP) Promoter Fusions to Study Gene Regulation at Single Cell and Population Levels

Marta Utratna and Conor P. O'Byrne

Abstract

Reporter gene fusions based on the enhanced green fluorescent protein (EGFP) are powerful experimental tools that allow real-time changes in gene expression to be monitored both in single cells and in populations. Here we describe the development of a chromosomally integrated transcriptional reporter fusion in *Listeria monocytogenes* that allows real-time measurements of gene expression. To construct a single copy of an EGFP-based fluorescent reporter fused to a promoter of interest (P*x*) in *L. monocytogenes*, a suicide shuttle vector carrying the P*x*::*egfp* gene fusion is first constructed in *Escherichia coli* (as an intermediate host). Then, the vector is transformed into *L. monocytogenes* and integrated into its chromosome by homologous recombination within the selected promoter region. Subsequently, analysis of fluorescence exhibited by cells carrying a single copy reporter can be performed under selected experimental conditions by stringent sample preparation, optimized image acquisition, and processing of the digital data with the image analysis freeware ImageJ. Thus, the methodology described here can be adapted to investigate the activity and regulation of any promoter in *L. monocytogenes* both at the cell and population levels.

Key words EGFP, *Listeria monocytogenes*, Reporter fusion, Gene expression, Promoter activity, Microscopy, Image analysis, ImageJ, Fluorescence intensity, Particle count, Fluorescent population

1 Introduction

Bacteria use complex biological networks to sense physical of chemical signals and to reprogram gene expression in response to environmental changes. Detection of a specific signal affects promoter activities and underpins survival both under long-term and transient changes of environmental conditions. Investigation of these complex responses by development of reporter fusions helps in understanding the microbial stress response and allows engineered microorganisms to carry out specific tasks [1]. The P*lmo2230*::*egfp* fusion is an example of a reporter fusion, based on the enhanced green fluorescent protein (EGFP) gene that we have developed for real-time measurements of gene expression in the

Kieran Jordan et al. (eds.), *Listeria monocytogenes: Methods and Protocols*, Methods in Molecular Biology, vol. 1157, DOI 10.1007/978-1-4939-0703-8_20, © Springer Science+Business Media New York 2014

foodborne pathogen *L. monocytogenes* EGD-e [2]. When introduced into a wild-type strain it allows monitoring of expression from the promoter of *lmo2230* gene (encoding putative arsenate reductase), which is tightly regulated at the transcriptional level by the sigma factor that controls the general stress response, Sigma B (σ^B). Moreover, the fusion was introduced to a $\Delta rsbV$ mutant of *L. monocytogenes* EGD-e lacking one of the σ^B regulatory proteins, giving an opportunity of evaluating some aspects of the current model of posttranslational regulation of σ^B [3]. In this chapter, we describe the development of EGFP-based reporter fusions based on the methodology used to generate the $P_{lmo2230}::egfp$ reporter fusion cloned into pKSV7 [2–4]. The protocol described allows the construction of a chromosomally integrated *egfp* reporter fusion to any promoter of interest (P*x*) in *L. monocytogenes* and describes how the reporter expression can be quantified using fluorescent microscopy and digital image analysis.

A wide variety of imaging software is available for microscopic imaging, many designed for specialist applications and specific microscope setups. In the great diversity of image analyzing tools, ImageJ is unique because it is license-free, can read most of the common formats, runs on every operating system and it is very flexible due to continuous user-driven development [5]. Moreover, it allows automation of repetitive tasks (with Macros and plug-ins) without a requirement for advanced programming skills, making microscopic data analysis less tedious and more time efficient. Thus, the development of genetically encoded fluorescent reporters together with the advances in microscopic imaging described in this chapter have great potential for a wide range of applications including monitoring of gene expression in single cell and population directed studies.

2 Materials

2.1 Media

1. Luria-Bertani (LB) broth was composed of 1 % (w/v) Tryptone-Peptone (BD), 0.5 % (w/v) Yeast Extract (Difco), and 1 % (w/v) NaCl (BDH) in ddH$_2$O. It was prepared in 200 mL volumes and autoclaved (a minimum of 121 °C for a period of 15 min under pressure).

2. LB Agar was prepared as LB broth but supplemented with 1.7 % (w/v) Agar Number 2 (Lab M) before autoclaving. Molten agar was poured into sterile plastic Petri dishes (approximately 25 mL of agar per plate). Plates were set at room temperature and stored at 4 °C until required.

3. Brain Heart Infusion (BHI) broth was prepared as a 3.7 % (w/v) solution of BHI powder (Lab M) in ddH$_2$O.

4. BHI Agar was prepared as a 4.9 % (w/v) solution of BHI Agar powder (Lab M) in 200 mL volumes in ddH$_2$O. Plates were prepared as described for LB Agar.

5. SOC medium; 2 % (w/v) Tryptone, 0.5 % (w/v) Yeast Extract,10 mM NaCl, 2.5 mM KCl, 10 mM MgCl$_2$, 10 mM MgSO$_4$, and 20 mM Glucose, prepared fresh and sterilized by autoclaving.

6. BHI with 0.5 M sucrose, prepared fresh and sterilized by autoclaving.

2.2 Bacterial Strains

1. Chemically competent One Shot TOP10 *Escherichia coli* (Invitrogen).

2. *Listeria monocytogenes*—selected strains EGD-e wild type, Δ*sigB* mutant and Δ*rsbV* mutant.

2.3 Enzymes

1. Restriction endonucleases (Promega)—according to restriction sites (RS) present in the shuttle vector and included in the synthesized element RS1-P*x*::*egfp*-RS2.

2. Shrimp Alkaline Phosphatase (SAP; 1 U/μl, Roche).

3. T4 DNA Ligase (1 U/μl, Roche).

2.4 Buffers and Solutions

1. Reaction buffer (10×) suitable for both restriction endonucleases (Promega).

2. Alkaline phosphatase buffer (10×; Roche).

3. Ligase buffer (10×; Roche).

4. Electroporation buffer; 1 mM HEPES pH 7.4, 0.5 M sucrose.

5. PBS; 137 mM NaCl, 3 mM KCl, 12 mM NaH$_2$PO$_4$, 2 mM KH$_2$PO$_4$, pH 7.4.

6. Fixing solution; 1:1 100 % methanol/100 % ethanol (v/v).

2.5 Antibiotics

1. Penicillin G; stock of 10 mg/mL in ddH$_2$O, stored at −20 °C.

2. Antibiotic for resistance encoded on the commercial vector delivering the P*x*::*egfp* element for selection in *E. coli*. Ampicillin (Sigma-Aldrich) 5 mg/mL stock in dH$_2$O, working concentration 100 μg/mL.

3. Antibiotic for resistance encoded on the shuttle vector for selection in *E. coli*. Chloramphenicol (Sigma-Aldrich) 50 mg/mL stock in 95 % ethanol, working concentration 10 μg/mL for pKSV7.

4. Antibiotic for resistance encoded on the shuttle vector for selection in *L. monocytogenes*. Chloramphenicol (Sigma-Aldrich) 50 mg/mL stock in 95 % ethanol, working concentration 10 μg/mL for pKSV7.

2.6 Equipment

1. Bio-Rad Gene Pulser apparatus.

2. Bio-Rad Gene Pulser 0.2 cm cuvette (Bio-Rad).

3. PCR block and reagents.

4. Agarose gel electrophoresis apparatus, reagents, and visualization system.

5. Qiagen miniprep plasmid purification kit (Qiagen).

6. QIAquick Gel Extraction Kit (Qiagen).

7. NanoDrop spectrophotometer (Thermo Scientific).

8. Non-transparent tubes 15 mL and Eppendorfs.

9. Vortex.

10. Vacuum filtration unit (Millipore).

11. Polycarbonate membrane filters (Millipore) 0.2 μm pores, diameter 10 mm.

12. Paper pads, diameter 12 mm (Whatman).

13. Microscopic slides and cover slips.

14. Immersion oil.

15. Fluorescence microscope—Nikon Eclipse E600 with CCD camera (QICAM 12-bit Mono Fast 1394) and acquisition software (QCapturePro).

3 Methods

3.1 Design of Gene Fusion and Selection of Reporter Vector

To construct an expression vector which could serve as a tool for monitoring the activity of a specific promoter in *L. monocytogenes*, the promoter region must be fused with a gene encoding an easily detectable reporter protein. The gene fusion construct is synthesized (e.g., by Eurofins MWG Operon) and supplied in a standard commercial plasmid or, with additional charge, can be cloned straight into any vector of interest sent by a customer. Here, a cost efficient approach is described for "in house" development of a fluorescent reporter in a suicide shuttle vector, using the expression of EGFP (enhanced green fluorescent protein) from a promoter (Px) of selected gene x. The reporter fusion is designated Px::*egfp* (Fig. 1a).

1. Assuming the genome of the selected strain of *L. monocytogenes* is sequenced (*see* **Note 1**); identify Px promoter region and neighboring sequence. Select the sequence of the promoter region based on published data if available (in *L. monocytogenes* all transcripts have been mapped in a study by Toledo-Arana et al. [6]), or if not it is prudent to select a region that includes 400 nucleotides upstream from the start codon, to ensure that all *cis*-acting regulatory elements are likely to be included.

2. Carefully select the vector where the reporter fusion is going to be cloned into (Fig. 1a). Ideally, it should be a shuttle vector that allows all the cloning steps to be carried out in *E. coli* and can then be transformed into *L. monocytogenes*. If a single copy reporter (i.e., integrated into the host genome) is required, the vector should also allow integration into the chromosome, for example by possessing some means of conditional replication (the temperature sensitive replication origin from pE194 is widely used for this purpose in *L. monocytogenes* [7]; the antibiotic resistance genes on the plasmid can be selected for by growth at the non-permissive temperature, which allows for selection of integrants where the plasmid has homologously recombined with the chromosome). Once the vector is selected, examine its multiple cloning site locus (MCS) for selection of two restriction sites (RS1 and RS2) for two different restriction endonucleases (*RE1* and *RE2*).

3. Combine the sequence for promoter region (P*x*) with a codon optimized *egfp* sequence (accession AY192024.1), including its start codon, and flank this fusion element (designated P*x*::*egfp*) by selected restriction sites (RS1 and RS2) by adding them upstream and downstream of the P*x*::*egfp* sequence in appropriate orientation (Fig. 1b).

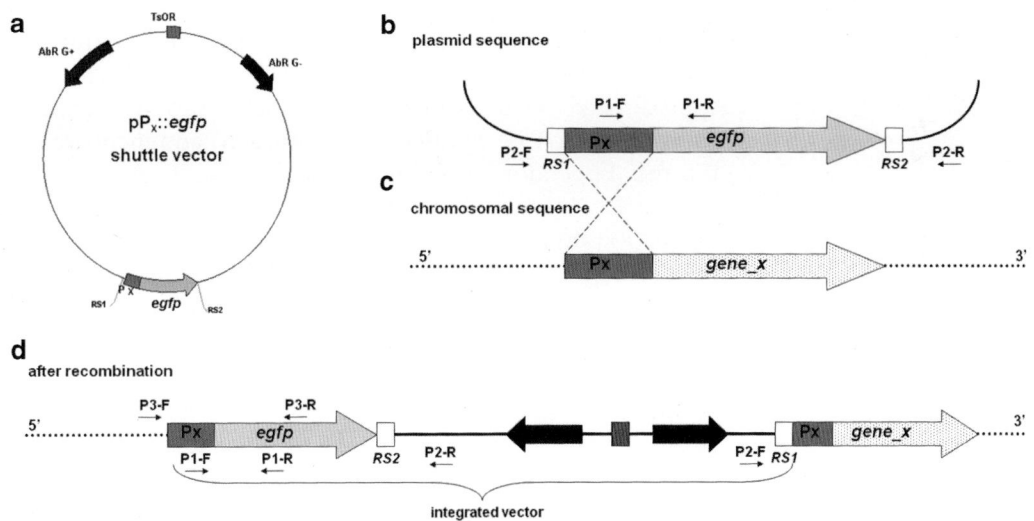

Fig. 1 Schematic illustration of the elements of the pP*x*::*egfp* reporter vector showing the localization of the diagnostic primers used during the construction of the chromosomally integrated reporter fusion. (**a**) Map of suicide shuttle vector carrying genes for antibiotic resistance cassettes for selection in Gram-negative (AbR G−) and Gram-positive (AbR G+) organisms, temperature sensitive origin of replication (TsOR) together with the P*x*::*egfp* fusion cloned into the MCS. (**b**) Detailed illustration of the P*x*::*egfp* element. (**c**) Site of homologous recombination within the chromosome of *L. monocytogenes*. (**d**) Detailed illustration of the gene *x* area within the chromosome of *L. monocytogenes* after recombination and integration of the pP*x*::*egfp* vector

4. Synthesize the fusion element RS1-P*x*::*egfp*-RS2. If another reporter gene is used it will require an additional codon optimization step for the selected organism (*see* **Note 2**).

5. Design and synthesize three pairs of primers (*see* Fig. 1) for amplifying the following products (note that products differ in size and P2 > P3 > P1):

   (a) P1 within the fusion—starting within P*x* region (P1-F) and finished within *egfp* (P1-R); will serve as a control for transformation of ordered fusion into *E. coli* (Fig. 1b)

   (b) P2 outside of the selected restriction sites of the vector (P2-F and P2-R); will serve as a control for successful cloning into the vector (Fig. 1b)

   (c) P3 overlapping chromosomal P*x* region after integration of vector—starting upstream of P*x* (P3-F) and finishing within *egfp* (P3-R) (Fig. 1d). This product can be used to confirm the integration of the vector into the chromosome.

6. Once the fusion is synthesized and supplied in a commercial vector, reconstitute the lyophilized vector in sterile DNAse-free dH$_2$O, then aliquot into smaller volumes and store at −20 °C until required. Check quality control documents showing the sequence of the synthesized product to avoid any mismatches in the fusion element.

3.2 Construction of the Reporter Vector in E. coli

1. The vector should be first transformed into the intermediate host (e.g., One Shot TOP10 *E. coli* from Invitrogen). Add approximately 10 ng of plasmid to a 50 µl aliquot of chemically competent TOP10 *E. coli* cells (Invitrogen) and incubate in a 0.5 mL Eppendorf tube on ice for 30 min. Then move the tube to 42 °C for a 30 s period followed immediately by 2 min incubation on ice (heat shock). Subsequently, supplement the cells with 250 µl of SOC medium and incubate for 30 min with shaking at 37 °C to allow recovery of cells in liquid medium. Plate 100 µl of the transformation mixture onto LB agar plates supplemented with the appropriate antibiotic corresponding to the resistance marker encoded on the commercial vector. Incubate plates overnight at 37 °C. This step can be performed with Electrocompetent One Shot TOP10 *E. coli* cells (Invitrogen) following the manufacturer's protocol and using Bio-Rad Gene Pulser apparatus if higher efficiency of transformation is required.

2. Transfer each of the successful transformants onto fresh LB agar plates with antibiotic selection to allow single colony isolation for individual clones and incubate at 37 °C for an additional 24 h. Next day, prepare single colony suspensions for at least five randomly selected colonies from the restreaked plates.

In each case, use a sterile loop to transfer a single colony into the 0.5 mL Eppendorf tube containing 250 µl of ddH$_2$O and mix by pipetting to prepare cell suspension.

3. Set up a PCR reaction with P1-F/P1-R primers (Fig. 1) using cell suspensions of each of the putative transformants as a template (genomic DNA extraction unnecessary). Include appropriate controls for PCR. Run the PCR products on a 1 % agarose gel for confirmation of the presence of the fusion plasmid in *E. coli* cells after transformation. If the presence of the P*x*::*egfp* element is confirmed in selected transformants, glycerol stocks should be prepared to save clones of the new strain in the lab collection.

4. For plasmid purification, grow one selected transformant overnight at 37 °C with agitation set at 180 rpm, in LB supplemented with the specific antibiotic required for plasmid maintenance. Carry out plasmid purification using the Qiagen miniprep plasmid purification kit (Qiagen) following the manufacturer's protocol. At the final stage, elute plasmid from the column with 50 µl of pre-warmed (50 °C) sterilized ddH$_2$O and store at –20 °C until required.

5. Determine the concentration of purified plasmid by NanoDrop spectrophotometer (Thermo Scientific). Alternatively, use 1 % agarose gel and electrophorese the plasmid suspension alongside a DNA marker of known mass and concentration which allows for estimation of the plasmid concentration.

6. Repeat **steps 4** and **5** for purification of the shuttle vector from *E. coli*.

7. To clone the fusion element into a shuttle vector, perform a restriction digestion reaction using the two restriction endonucleases (*RE1* and *RE2*) whose recognition sequences were incorporated into the synthesized element. The reaction allows a double cut (with *RE1* and *RE2*) of the fusion from the commercial vector and generation of an insert with "sticky" ends corresponding to sites present in the shuttle vector. Mix 10 µl of purified plasmid with 1.25 µl of a 12 U/µl of each restriction endonuclease, 5 µl of reaction buffer suitable for both restriction endonucleases and 0.5 µl of 100 mg/mL acetylated BSA. Add ddH$_2$O up to a final volume of 50 µl. Perform the reaction at 37 °C for a minimum of 2 h, and then inactivate the restriction endonucleases by incubation at 65 °C for 10 min.

8. To prevent self-ligation of multiple copies of the insert carrying the fusion construct, a dephosphorylation of the 5′ phosphate groups on DNA is carried out by adding Shrimp Alkaline Phosphatase (SAP; 1 U/µl, Roche). Mix 10 µl of inactivated restriction digest with 1 µl of SAP, 5 µl of alkaline phosphatase

buffer, and 34 μl of ddH$_2$O. Perform reaction at 37 °C for 2 h and then inactivate the alkaline phosphatase by 10 min incubation at 65 °C.

9. Run the digestion mix in adjacent wells on 1 % agarose gel electrophoresed alongside an appropriate DNA marker to allow identification of the right band corresponding to insert size within the gel. Within each lane of the gel cut out the insert band from the gel under appropriate UV illumination (the time and power of illumination should be minimized to prevent damage to the DNA) and place into an Eppendorf tube. Carry out a purification of the insert from the agarose using the QIAquick Gel Extraction Kit (Qiagen), following the manufacturer's protocols. At the final stage elute the insert from the column with 50 μl of pre-warmed (50 °C) sterilized ddH$_2$O and store the eluate at –20 °C until required.

10. To prepare the shuttle vector for ligation with the purified insert, carry out the procedure described in **steps 7** and **8** with the same restriction endonucleases (*RE*1 and *RE*2).

11. Once insert and the shuttle vector are digested and phosphatase-treated, determine the size and concentration of the products by running of 2 μl of each reaction on a 1 % agarose gel alongside an appropriate DNA molecular weight marker.

12. In order to generate the pP*x*::*egfp* vector by cloning the insert into the shuttle vector both need to be incubated together in predetermined molar ratios of 2:1, 3:1, or 5:1 of insert:vector, respectively. The amount of insert can be calculated using the following formula: [ng insert = ng vector × kb insert/kb vector × (ratio insert:vector)]. Ligation is carried out at 16 °C for a minimum of 2 h by mixing of 1 μl of T4 DNA Ligase (1 U/μl, Roche) 2 μl of ligase buffer, 2 μl of vector at a concentration of 200 ng/mL and insert at the appropriate amount depending on insert:vector ratio. The final volume is adjusted to 20 μl with ddH$_2$O. Ligation reactions were used immediately for transformations of chemically competent TOP10 *E. coli* cells as described in **steps 1** and **2**. Remember to use LB agar plates supplemented with the appropriate selective agent corresponding to the resistance encoded on the shuttle vector that is expressed in *E. coli* (*see* **Note 3**).

13. Set up two PCR reactions with primers P1-F/P1-R and P2-F/P2-R (Fig. 1) using cell suspensions of each of the putative transformants as template in the reactions, as well as appropriate positive and negative controls. Run the PCR products on a 1 % agarose gel for confirmation of the presence of the fusion in *E. coli* cells after transformation. If the presence of the pP*x*::*egfp* vector is confirmed in selected transformants (detection of the expected product for both sets of primers), the corresponding glycerol stocks should be prepared to save clones of the new strain in the lab collection.

14. Perform purification of the pP*x*::*egfp* vector as described in **steps 4** and **5**. Store the plasmid preparation at –20 °C until required.

3.3 Construction of an L. monocytogenes Gene Fusion Strain(s) by Chromosomal Integration of the Reporter Fusion

1. To transform the pP*x*::*egfp* vector into *L. monocytogenes*, electrocompetent cells must be prepared for each of the selected strains used in the study. As a control a strain without the major regulator of the promoter region under investigation (P*x*) should be used to confirm the dependence of the reporter on that regulator. Grow an overnight culture of *L. monocytogenes* in BHI broth at 37 °C with shaking set at 180 rpm. Next morning, prepare 50 mL of sterile BHI broth supplemented with 0.5 M filter sterilized sucrose in a 500 mL Erlenmeyer flask and inoculate this with 0.5 mL of overnight culture. Grow the culture until $OD_{600} = 0.2$ whereupon the culture is supplemented with 50 μl of 5 mg/mL Penicillin G. After 2 h of additional growth at 37 °C with shaking set at 180 rpm, harvest the cells by centrifugation at $9,000 \times g$ for 10 min at 4 °C. Discard the supernatant and immediately move the resulting cell pellet onto ice (*see* **Note 4**). Wash twice with ice-cold electroporation buffer (1 mM HEPES pH 7.4, 0.5 M sucrose) using 45 mL for the first wash and 22.5 mL for the second wash. Resuspend cells in 400 μl of ice-cold electroporation buffer and aliquot into two tubes of 200 μl for immediate use in a subsequent electroporation.

2. Carry out electroporation of competent *L. monocytogenes* cells using a Bio-Rad Gene Pulser and Gene Pulser 0.2 cm cuvette (Bio-Rad) with the following electrical parameters: a voltage of 1,500 V, a capacitance of 25 μF, and a resistance of 200 Ω. First, add 200 μl of freshly prepared competent cells into a pre-chilled (on ice) electroporation cuvette avoiding introduction of air bubbles. Then add 1.5 μg (typical volume 3 μl) of the pP*x*::*egfp* vector into competent cells and place in the Gene Pulser electroporation holder. Immediately after electroporation supplement the cell suspension with 1 mL of sterile and pre-warmed BHI containing 0.5 M sucrose and incubate static at 30 °C for 90 min to allow the recovery of cells.

3. Plate 100 μl aliquots of the transformation mixture onto BHI agar plates supplemented with the appropriate selective agent and incubate the plates for 48 h at 30 °C.

4. Carry out **step 2** from Subheading 3.2 for single colony isolation and preparation of PCR templates using the growth media supplemented with the appropriate antibiotic to select for *L. monocytogenes* cells carrying the shuttle vector.

5. Perform **step 13** from Subheading 3.2 on a *L. monocytogenes* cell suspension to confirm the presence of the pP*x*::*egfp* vector in selected putative transformants. If the presence of the

pP*x*::*egfp* vector is confirmed in selected transformants (expected product for both P1-F/P1-R and P2-F/P2-R sets of primers), they should fluoresce under conditions when P*x* promoter is expected to be expressed. To test this, grow selected transformants under inducing conditions to express *egfp* from P*x* and examine a 5 µl sample of culture by fluorescence microscopy and filter covering EGFP excitation and emission wavelengths (495/509 nm). Permanent glycerol stocks should be prepared to save clones of the new strain in the lab collection.

6. To generate strains bearing a single copy of the P*x*::*egfp* fusion element, integration of the shuttle vector into the *L. monocytogenes* chromosome by homologous recombination within P*x* region is performed (Fig. 1d). Restreak one transformant onto fresh BHI agar with antibiotic selection and incubate plates at 42 °C for an additional 24 h. Restreak colonies from this plate a further three times, incubating each at 42 °C. The suicide shuttle vector is temperature sensitive and unable to replicate autonomously when cells are incubated at 42 °C [4]. Therefore, this process selects for the integration of the vector into the homologous P*x* region on the host chromosome in the presence of selective agent encoded on the vector. Thus, colonies growing at 42 °C represent putative integrants.

7. Prepare single colony templates for at least five randomly selected colonies from the final restreaked plates as described in **step 2** of Subheading 3.2.

8. Set up a PCR with three pairs of primers P1-F/P1-R, P2-F/P2-R, and P3-F/P3-R. Run the PCR products on 1 % (w/v) agarose gel for confirmation of the presence of the integration event (PCR with P1-F/P1-R and P3-F/P3-R pair is expected to produce a product, while product for P2-F/P2-R should not produce a product if integration has occurred). Examine a 5 µl sample of culture, grown under conditions expected to activate P*x*, by fluorescence microscopy. Permanent glycerol stocks should be prepared for clones giving expected outcome and stored as integrants in the laboratory strain collection.

3.4 Microscopic Quantification of EGFP Fluorescence

Once fusion strains of *L. monocytogenes* are developed, then reliable methodology for measurements of fluorescence exhibited by cells expressing EGFP, and effective image processing techniques are required to extract and quantify relevant information concerning P*x* activity. The quantitative fluorescent microscopy involves appropriate (1) sample preparation, (2) stringent image acquisition, and (3) image processing methods. A public domain program, ImageJ (developed at the US National Institutes of Health and available on the internet free of charge at http://rsb.info.nih.gov/ij), is a suitable package for the processing and analysis of the microscopic images obtained [5, 8].

1. In order to stop the physiological processes of bacterial cells and protect fluorescence signals at specific points during growth, cells of *L. monocytogenes* are treated with ice-cold methanol/ethanol fixation. Grow the bacterial culture under the appropriate experimental conditions and measure the OD_{600} at the required point of growth.

2. Prepare a cell suspension equivalent to 1 mL of $OD_{600} = 0.6$ by diluting or concentrating and resuspending in the growth medium, while protecting your samples from light (*see* **Note 5**).

3. Mix the cell suspension immediately with 1 volume of ice-cold 1:1 methanol (100 %)/ethanol (100 %) mixture. Vortex briefly and incubate for 10 min at –20 °C.

4. Centrifuge the samples at $9,000 \times g$ for 10 min at 4 °C, to pellet the fixed cells and remove alcohol/medium residues.

5. Discard the supernatant gently and resuspend the cells in 1 mL of sterile PBS of pH 7.4. Place samples on ice for immediate analysis, or store frozen at –20 °C in 1 mL aliquots for subsequent processing.

6. To obtain reliable images for the quantification of fluorescence in *egfp* expressing bacteria, a flat surface covered with dispersed and non-overlapping cells is needed. Perform a tenfold dilution of the cell suspension ($OD_{600} = 0.6$) in PBS (pH 7.4) and then filter 1 mL of this suspension onto a polycarbonate membrane filter (Millipore) with 0.2 μm pores and 10 mm in diameter. Assemble the vacuum filtration unit (XX1504700; Millipore) to contain the filter placed between two paper pads of 12 mm diameter (Whatman). Apply a vacuum suction to allow a concentration of suspended bacterial cells on the surface of the filter. Remove the filter carefully with tweezers and transfer to a microscope slide containing 3 μl of mineral oil (Sigma-Aldrich) spread as a mounting medium. Place 3 μl of mineral oil in the middle of a cover slip (18 mm × 18 mm) then put it upside down on the top of the filter. Finally, place the cover slip onto the slide and press carefully to dispense the mounting oil equally over the filter.

7. Capture a number (>30) of microscopic fields for each experimental condition for further image analysis with fluorescence microscopy using a filter corresponding to EGFP spectra (495/509 nm), an appropriate neutral density filter. The exposure and focus must be first optimized individually for each microscope as the settings vary depending on the device used and the filters available. Save the images with the highest possible quality and a detailed name corresponding to each experimental setup.

8. Each of the digital images captured is analyzed with ImageJ to determine the pixel intensities corresponding to the fluorescence intensity of captured cells. First install ImageJ software and open the selected image. In order to correct for an unevenly illuminated background and background noise apply the "Process > Background Subtraction" command (Fig. 2a, b). Save this normalized image with a modified name as it will be used in the subsequent processing. Once the background is subtracted (Fig. 2b), obtain the grey-scale histogram and the mean pixel intensity value for the image using the "Analyze > Histogram" command.

9. Export the resulting data to Microsoft Excel with the "File > Save as > .xls file" command and repeat for the other images to compare the relative fluorescence intensities between technical replicates and subsequently between selected experimental conditions (e.g., untreated vs. treated). For presentation of the data, assign a value of 1 to the lowest mean intensity recorded among the selected experimental conditions and recalculate other mean values to obtain relative intensities. Use a Student's t-test to determine the statistical significance of differences between groups.

10. Another approach to comparing the activity of the selected promoter is based on a determination of the number of cells exhibiting detectable fluorescence within a number of randomly selected fields captured by fluorescence microscopy (Particle counting). Open an image with a previously subtracted background in ImageJ. Apply a simple segmentation technique (thresholding algorithm) to distinguish the object from the background using the menu command "Image > Adjust > Threshold" (Fig. 2c step 1). A threshold limits pixels in the image to black (background), while pixels with values outside the defined range are converted to white (objects). To reduce user-dependent bias of manual thresholding, algorithms automatically calculating the threshold ("Default" option) are recommended [9] (Fig. 2c step 2). Once the threshold is defined, create a binary image (black/white) using the menu command "Process > Binary > Make binary" (Fig. 2c step 3). Slightly overlapping elements can be separated in a binary image using a watershedding algorithm, available by application of the "Process > Binary > Watershed" command (Fig. 2c step 4). Then, analyze objects in a segmented image, using the menu command "Analyze > Analyze particles" (Fig. 2c step 5). Additionally, define the minimal size of objects included in particle counting by setting to 5×5 (25 pixels) to minimize the influence of the background noise (Fig. 2c step 6) and use the "Summarise" command to obtain

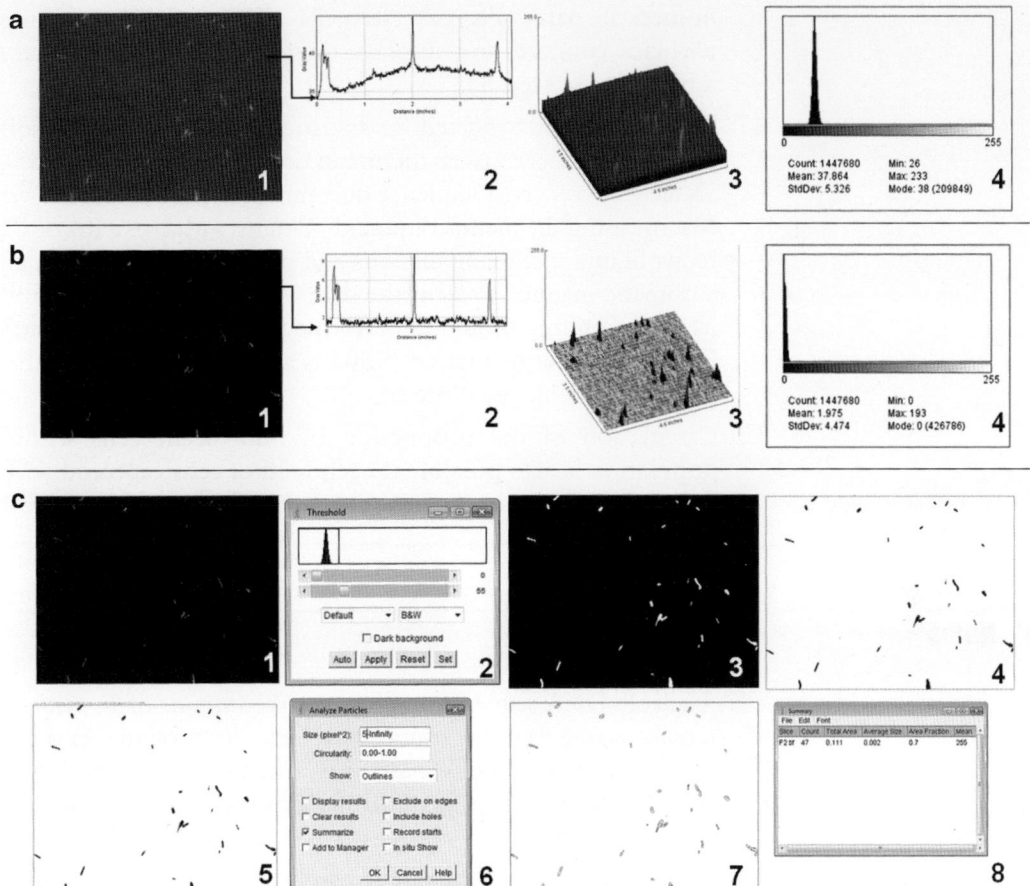

Fig. 2 A schematic illustration of the microscopic image analysis and fluorescence quantification using ImageJ software. (**a**) Pixel intensity measurements for original microscopic image (non-subtracted background): (1) sample image, (2) 2D profile of indicated ROI (*by arrow*), selected from the sample picture and created by a command "Analyse > Plot profile", (3) 3D profile of sample image shown by the "Analyse > Surface Plot" command, (4) A histogram of the image showing the distribution of the pixels over the grey-level scale (0–255; *black–grey–white*). Each pixel is placed in a bin corresponding to the color intensity of that pixel, added up and displayed on a graph. The mean pixel intensity of the sample picture is shown as a numerical value. (**b**) Pixel intensity measurements for microscopic picture with subtracted background using a command, "Process > Background Subtraction" and a "rolling ball" algorithm developed by Stanley Sternberg (Sternberg S, Biomedical image processing, IEEE Computer, Jan 1983). Steps 1–4 as in (**a**). (**c**) Subsequent commands used for particle counting within a sample image obtained with fluorescence microscopy. (1) an original file is opened, (2) thresholding command applied to distinguish between objects and background, (3) the image following thresholding, (4) the image following thresholding and converted into a binary file, (5) separation of overlapping elements by watershedding, (6) settings for automatic counting of particles with the limit of the size as 5×5 pixels to be recognized as a cell, (7) summary data outlining particles counted, and (8) numerical values summarized within a table

numerical data (Fig. 2c step 7). The software outlines particles counted and displays their number in a summary table (Fig. 2c **step 8**).

11. The data can be exported for statistical analysis to a Microsoft Excel spreadsheet, using the menu command "File > Save as > . xls file". Moreover, a sequence of commands can be recorded via Macros using the menu command "Plugins > Macros > Record", to avoid multiple manipulations and perform all the tasks in an automatic manner. Present the data as described in the **step 9**. Manual counting should be performed for three random images from each experimental set (>30) as an evaluation of the automatic processing (*see* **Note 6**).

12. Calculation of the proportion of fluorescent cells within population is also possible if each field of cells captured with fluorescence microscopy is also imaged by phase contrast (*see* **Note 7**).

4　Notes

1. Available DNA sequence data for selected strains of *L. monocytogenes* sequenced in recent years can be found via links below: http://genolist.pasteur.fr/ListiList/; http://www.broadinstitute.org/annotation/genome/listeria_group/GenomesIndex.html.

2. The codon sequence of a gene can be optimized to maximize its expression level by selecting codons that are used frequently within the genome of the selected organism host. This step is routinely availed of by companies offering DNA synthesis services (e.g., GENEius software used by Eurofins MWG). Moreover, a number of freely available servers and software packages can be used, e.g., *OPTIMIZER*, GeneDesign, SyntheticGeneDesigner, and BBOCUS.

3. If ligation does not result in the successful cloning of the insert into the vector the PCR-based approach can be undertaken to increase the insert:vector ratio during the ligation step. Amplify the insert with high fidelity Polymerase (Agilent) and primers overlapping MCS of commercial vector delivering the P*x*::*egfp* element. Digest the PCR product with *RE1*/*RE2* restriction enzymes and purify the amplified insert with a commercial kit to remove short fragments of DNA and primers. Carry out the ligation procedure with an increased the insert:vector ratio. After cloning into the shuttle vector amplify the P*x*::*egfp* element with P2-F/P2-R set of primers and send for sequencing to avoid introduction of any mismatches in the fusion element by this PCR-based approach.

4. Transformation buffer should be pre-chilled on ice for use and all steps should be performed on ice. Tips, disposable pipettes and labeled tubes should also be pre-chilled. Gentle resuspension by swirling the centrifugation pots rather than pipetting up and down is recommended.

5. To minimize fluorescence bleaching, all the steps of sample preparation should be carried out in a preparation room and imaging room where the ambient light is minimized. Nontransparent light protective tubes (black or amber) or wrapping with aluminum foil are recommended for both short- and long-term storage of cell suspensions.

6. A manual count can be performed by either the "Point Picker" or "Cell counter" plugin. This option is also useful for the exclusion of particles that are clearly not objects of interest, arising from nonadjustable equipment limitations.

7. The quantification of particles captured by phase contrast and fluorescence microscopy within the same field would allow an assessment of the proportion of cells (% of fluorescent population) that expressed *egfp* under selected conditions, and were captured by a previously described set of manipulations (sample preparation, filter, exposure time, etc.). The phase contrast image is first inverted using the menu "Edit > Invert" command. Then, the protocol of particle counting (Subheading 3.4, **steps 10** and **11**) is applied for two images of the same microscopic field. The number of particles counted in each image is displayed in a summary table and used for defining the proportion of cells expressing *egfp*. The proportion of fluorescent cells within population can be analyzed automatically by flow cytometry if appropriate device is available.

References

1. van der Meer JR, Belkin S (2010) Where microbiology meets microengineering: design and applications of reporter bacteria. Nat Rev Microbiol 8:511–522. doi:10.1038/nrmicro 2392

2. Utratna M, Cosgrave E, Baustian C, Ceredig R, O'Byrne C (2012) Development and optimization of an EGFP-based reporter for measuring the general stress response in *Listeria monocytogenes*. Bioeng Bugs 3:93–103. doi:10.4161/bbug.19476

3. Utratna M (2012) The development of reporters to measure the general stress response in *Listeria monocytogenes*: population and single cell studies on the activation of σB. PhD thesis, National University of Ireland, Galway

4. Smith K, Youngman P (1992) Use of a new integrational vector to investigate compartment-specific expression of the *Bacillus subtilis* spoIIM gene. Biochimie 74:705–711

5. Schneider CA, Rasband WS, Eliceiri KW (2012) NIH image to ImageJ: 25 years of image analysis. Nat Methods 9:671–675

6. Toledo-Arana A, Dussurget O, Nikitas G, Sesto N, Guet-Revillet H, Balestrino D, Loh E, Gripenland J, Tiensuu T, Vaitkevicius K, Barthelemy M, Vergassola M, Nahori MA, Soubigou G, Regnault B, Coppee JY, Lecuit M, Johansson J, Cossart P (2009) The Listeria transcriptional landscape from saprophytism to virulence. Nature 459:950–956. doi:10.1038/nature08080

7. Villafane R, Bechhofer DH, Narayanan CS, Dubnau D (1987) Replication control genes of plasmid pE194. J Bacteriol 169:4822–4829

8. Collins TJ (2007) ImageJ for microscopy. Biotechniques 43(1 Suppl):25–30

9. Sezgin M, Sankur B (2004) Survey over image thresholding techniques and quantitative performance evaluation. J Electronic Imaging 13:146–165

Part IV

Control Methods

Chapter 21

Control of *Listeria monocytogenes* in the Processing Environment by Understanding Biofilm Formation and Resistance to Sanitizers

Stavros G. Manios and Panagiotis N. Skandamis

Abstract

Listeria monocytogenes can colonize in the food processing environment and thus pose a greater risk of cross-contamination to food. One of the proposed mechanisms that facilitates such colonization is biofilm formation. As part of a biofilm, it is hypothesized that *L. monocytogenes* can survive sanitization procedures. In addition, biofilms are difficult to remove and may require additional physical and chemical mechanisms to reduce their presence and occurrence. The initial stage of biofilm formation is attachment to surfaces, and therefore it is important to be able to determine the ability of *L. monocytogenes* strains to attach to various inert surfaces. In this chapter, methods to study bacterial attachment to surfaces are described. Attachment is commonly induced by bringing planktonic cells into contact with plastic, glass, or stainless steel surfaces with or without food residues ("soil") in batch or continuous (e.g., with constant flow of nutrients) culture. Measurement of biofilm formed is carried out by detaching cells (with various mechanical methods) and measuring the viable counts or by measuring the total attached biomass. Resistance of biofilms to sanitizers is commonly carried out by exposure of the whole model surface bearing the attached cells to a solution of sanitizer, followed by measuring the survivors as described above.

Key words Biofilm, Attachment, Detachment, Antimicrobial resistance, Sanitizer, Bead vortexing, Crystal violet, Food soil

1 Introduction

The presence of spoilage or pathogenic microorganisms in the food industry environment is almost inevitable, as they are usually part of the raw materials used for the preparation of foods. For food quality and safety reasons, proper surface cleaning and disinfection processes constitute prerequisites of a complete Hazard Analysis Critical Control Point (HACCP) system in order to avoid any bacterial proliferation, transmission, and/or cross-contamination of the final products. This risk is multiplied due to the ability of microbes to strongly attach to inadequately sanitized surfaces and create stable microbial aggregations, which are termed biofilms.

Kieran Jordan et al. (eds.), *Listeria monocytogenes: Methods and Protocols*, Methods in Molecular Biology, vol. 1157, DOI 10.1007/978-1-4939-0703-8_21, © Springer Science+Business Media New York 2014

Biofilm is defined as a community of microorganisms, in which bacterial cells adhere to each other and grow on a biotic or an abiotic surface [1]. The complex phenomenon of biofilm formation may be affected by many intrinsic and environmental factors such as the potential of the specific bacterial strain to attach on a surface and form biofilm [2], the properties of the surface material [3], the nutrient availability [4], the pH [5], and the temperature [6].

Considering that biofilm formation requires the presence of small amounts of nutrient residues, it is unlikely such formations to be developed on easy-access, flat surfaces which may be cleaned regularly. In contrast, biofilms are usually present at selective points which may not be reached during typical sanitation processes such as in corners, cracks, screws, joints, drains, and inner parts of pipes [7]. From a food hygiene and safety aspect, the significance of biofilm formation for the food industry is that microbial cells enclosed in the self-produced matrix of extracellular polymeric substances (EPS) may become more resistant to further sanitation processes [8]. During the maturing stage of biofilms, any mechanical forces (i.e., brushing, rubbing, and washing with pressure) may favor the detachment of microbes and their transfer to uncontaminated equipment or final products. Therefore, biofilms may constitute a continuous source of contamination to foods coming in contact with them [8].

The production of a lab-scale microbial biofilm aims to represent the naturally developed biofilms in order to test different experimental objectives. The advantage of these biofilms is that they have constant structure and behavior, contrary to the properties of the natural biofilms which may vary significantly with the environmental conditions. There is a variety of available assays for the in vitro examination of the process of bacterial biofilm formation. These assays may be divided into two main categories: static and chemostat or continuous-flow systems [9]. For the first category, different materials of biotic (e.g., lettuce, cabbage, spinach) [10, 11] or abiotic (e.g., plastic, glass, stainless steel, wood) [12] surfaces are used as substrate for the attachment of the microorganisms and the formation of the biofilm under static conditions. In the next paragraphs, a brief description of the most commonly used static assays for the in vitro formation of biofilms and the techniques which are used for the recovery of the attached cells will be given.

2 Materials

2.1 Microtiter Plate Assay

1. Nonselective nutrient broth medium (i.e., tryptic soy broth, brain–heart infusion broth).

2. Nonselective nutrient agar medium (i.e., tryptone soy agar, brain–heart infusion agar, plate count agar).

3. Isotonic buffer (i.e., phosphate-buffered saline [peptone 10.0 g/L, sodium chloride 5.0 g/L, disodium phosphate 3.5 g/L, mono-potassium phosphate 1.5 g/L], ringer solution).

4. Staining dye (i.e., 0.1 % crystal violet, 0.1–1 % safranin).

5. 95 % ethanol or 30 % acetic acid.

6. Different concentrations of the tested disinfectant/sanitizer/organic acid (*see* **Note 1**).

7. Dey–Engley neutralizing broth (casein enzymatic hydrolase 5.0 g/L, yeast extract 2.5 g/L, dextrose 10 g/L, bromocresol purple 0.02 g/L).

8. 96-Well microtiter plates.

9. Microtiter plate reader.

10. Microtiter plate stirrer (if necessary).

11. Airflow cabinet or dehydrator.

12. Sonicator.

13. Cotton swabs.

14. Vortex.

15. Incubators at 30 or 37 °C.

2.2 Surface/Chip Assay

1. Nonselective nutrient broth medium (i.e., tryptic soy broth, brain–heart infusion broth).

2. Nonselective nutrient agar medium (i.e., tryptone soy agar, brain–heart infusion agar, plate count agar).

3. Isotonic buffer (i.e., phosphate-buffered saline, ringer solution).

4. Acetone.

5. Commercial or industrial disinfectant (use according to the manufacturer's instructions).

6. Phosphoric acid.

7. Sterile water.

8. Alkaline detergent (use according to the manufacturer's instructions).

9. Different concentrations of the tested disinfectant/sanitizer/organic acid (*see* **Note 1**).

10. Abiotic (i.e., stainless steel, plastic, glass) or biotic (i.e., vegetable leaves) coupons of standard dimension.

11. Sonicator/ultrasonic water bath.

12. Stirrer.

13. Plastic/glass tubes of 50 mL or conical flasks of 250 mL.

14. Incubators at 30 or 37 °C.

15. Scrappers, swabs, glass beads.

16. Vortex.

3   Methods

3.1   Microtiter Plate Assay

1. Transfer a single colony (*see* **Note 2**) of each strain to be tested from each slant into 10 mL of a rich nonselective medium such as tryptic soy broth (0.25 % glucose) or brain–heart infusion (0.2 % glucose) and incubate for 24 h at 30 or 37 °C depending on the required temperature of the microorganism.

2. Transfer 100 μl of the activated culture into 10 mL of fresh broth medium and incubate for 18 h at the corresponding temperature in order for the cells to reach the stationary phase.

3. Centrifuge at 3,600 × g for 15 min at 4 °C.

4. Wash the pellet twice in 10 mL phosphate-buffered saline (PBS), and adjust the final concentration to a specific level (*see* **Note 3**).

5. Dilute the activated strains from 1:100 to 1:10,000 in the biofilm assay medium (*see* **Note 4**) in order to reach a final concentration of 10^4 to 10^6 CFU/mL.

6. Transfer 100 or 200 μl of each strain culture to a well of a sterile 96-well microtiter plate and incubate under agitation or static conditions (as required) at the required temperature and for the desired amount of time (4–48 h) (*see* **Note 5**). For quantitative assays, at least 4–8 replicates for each strain and/or treatment should be used.

7. Remove planktonic bacteria from each microtiter dish by aspiration of the medium from the wells.

8. Wash each well by transferring 100 or 200 μl of PBS in each well or by immersing the whole plate in a bath with sterile PBS in order to remove the loosely attached cells. Repeat this step for three to five times until the PBS of the last step is crystal clear (*see* **Note 6**).

9. Turn the microtiter plate upside down in order to remove the excess PBS while gently shaking to assist the procedure. Leave the plates to air-dry in an airflow cabinet for 1 h. For direct quantification of the biofilm formed and determination of antimicrobial resistance follow the steps described in (*see* **Note 7**).

3.2   Evaluation of Antimicrobial Resistance

1. Following biofilm fixation, add 100 μl of the antimicrobial solution into each well of the microtiter plate.

2. Incubate the microtiter plate for the appropriate time and temperature. The selection of these parameters will affect the surviving populations of the biofilm.

3. Add 300 μl of sterile neutralizing broth in order to interrupt the antimicrobial activity of the solution being tested.

4. Remove the supernatant with a pipette or by turning the microtiter plate upside down.

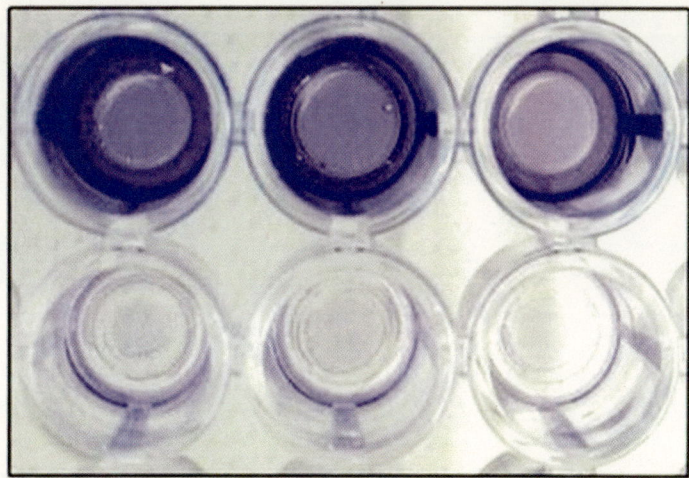

Fig. 1 Biofilm formation of two strains of *L. monocytogenes* (three replicates per strain) on a polystyrene microtiter plate. The biofilm was stained with 0.1 % crystal violet in water. The intensity of *blue color* in each well indicates the potential of each strain to form biofilm under the specific conditions

5. Wash each well with 400 μl of PBS or ringer solution at least twice.

6. Add 125–150 μl of 0.1 % crystal violet stain in each well and incubate for 10–15 min at room temperature (Fig. 1).

7. Rinse off the excess stain from the microplate, and wash thoroughly with tap water two to three times. Invert and vigorously shake the microplate in order to remove any droplets of the remaining water (*see* **Note 6**).

8. Add 200–250 μl of 95 % ethanol or 30 % acetic acid to each well and incubate at room temperature for 10–15 min to solubilize the dyed cells.

9. Transfer 125–150 μl from each well to a clean, sterile microtiter plate, and measure the optical density (O.D.) at 550–595 nm in a microtiter plate reader. Subtract the acquired O.D. values from those obtained from the negative control wells (pure 95 % ethanol or 30 % acetic acid).

3.3 Surface/Chip Assay (See Note 8)

1. Prepare the cultures as described in Subheading 2.1. After the final washing, however, the pellets should be reconstituted in the medium to be tested (e.g., TSB with adjusted pH or addition of NaCl).

2. Prepare the surface to be tested. Prior to use, the surfaces should be properly cleaned and sanitized by soaking the coupons in acetone for at least 30 min followed by washing with a commercial or an industrial detergent solution and rinsing

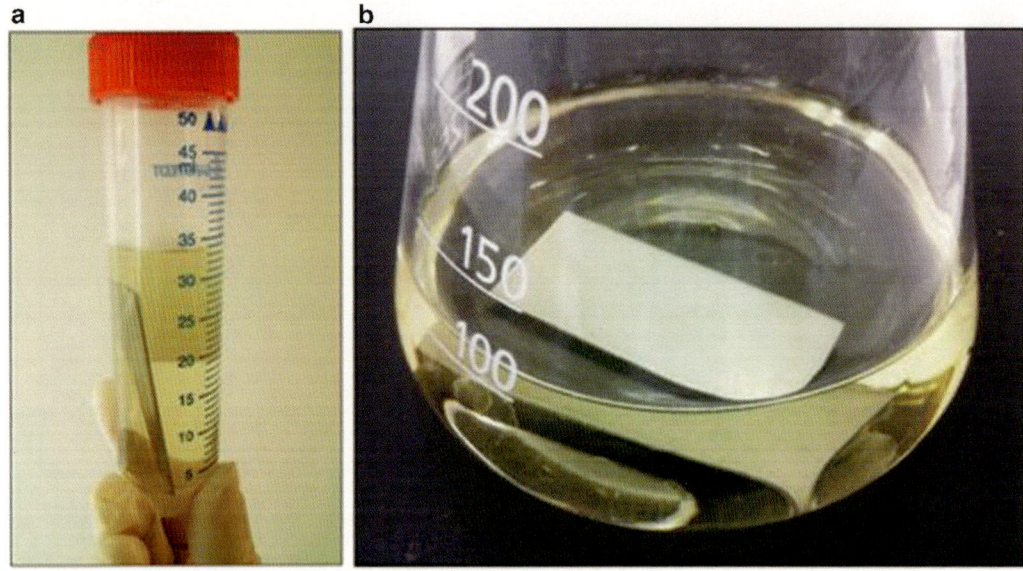

Fig. 2 Vertical (**a**) and horizontal (**b**) positioning of stainless steel coupons dipped in tryptic soy broth

with sterile water followed by air-drying for 1–2 h and further sterilization at 121 °C for 15 min or in hot air oven at 75 °C for 15 min.

3. Transfer 30–40 mL of sterile broth medium (e.g., TSB, TSB + NaCl or meat homogenate) to a 50 mL plastic centrifuge tube or 100 mL to a 250 mL conical flask. The volume of the broth medium should be adjusted in order that the coupons are fully submerged in the medium (Fig. 2).

4. Incubate for the selected duration (20 min to 72 h) and at the storage temperature (4–37 °C) under static conditions or continuous stirring.

5. Following incubation, remove the inoculated coupons from the containers using sterile forceps. Gently rinse the coupons with sterile water or PBS, avoiding direct contact of the stream of fluid with the coupon, as this mechanical force promotes detachment of the formed biofilm (*see* **Note 9**).

6. The surface/chip assay can also be applied in simple continuous-flow systems in order to evaluate the potential of a microorganism to form biofilms without any limitation of nutrients (*see* **Note 10**).

3.4 Evaluation of Antimicrobial Resistance

1. Prepare the antimicrobial solution in a plastic centrifuge tube.

2. Using a sterile pair of forceps, transfer the coupon with the formed biofilm into the plastic tube and incubate at the required temperature and for the appropriate duration. Continuous agitation of the tubes may also be applied to

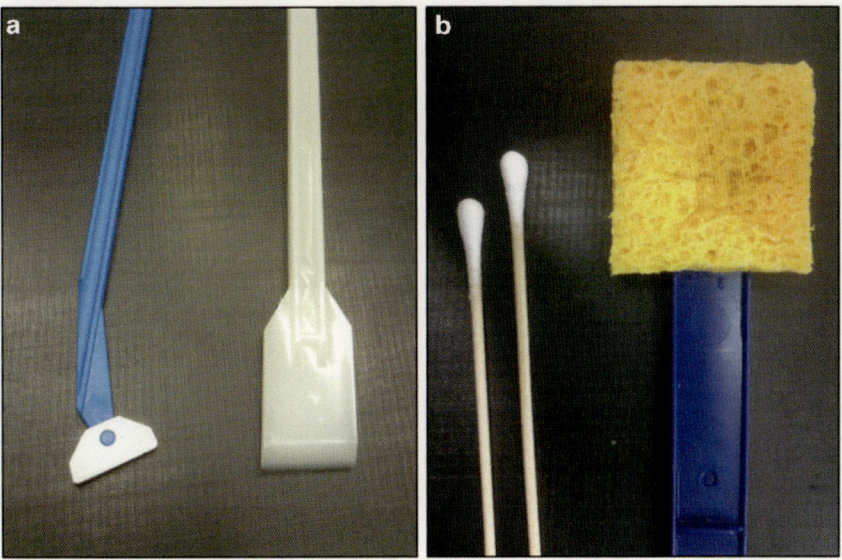

Fig. 3 Two different types of scrapers (**a**) and swabs (**b**) which are commonly used for the detachment of microbial biofilms from surfaces during in vitro studies

ensure that the whole surface is covered by fresh antimicrobial solution. Each coupon should be treated in a different tube.

3. Remove the coupons from the plastic tube and rinse with PBS several times to remove the excess antimicrobial substance.

4. To detach the biofilm cells that survived the antimicrobial treatment use the following: (1) Scrapping, swabbing with a disposable scraper (Fig. 3a), cotton swab, or sponge (Fig. 3b) over the whole surface (both sides) of the coupons eight to ten times applying constant pressure each time for the different experimental cases (*see* **Note 11**): The scrapers or swabs are then placed in containers with an isotonic broth medium (i.e., ringer or buffered peptone water), and after stirring for 2 min the detached populations are enumerated on nutrient media. (2) Vortexing (with or without beads): The coupons are in sterile plastic tubes containing isotonic broth. The tubes are vigorously stirred for 2 min and detached cells enumerated on nutrient media. In order to enhance the cell detachment, ten glass beads (5 mm, Fig. 4) can be added to each plastic tube during bead vortexing. The latter method increases the effectiveness of stirring without affecting the viability of the cells. (3) Ultrasonic method (*see* **Note 12**): The coupons carrying the biofilms are placed into a test tube containing 30–40 mL of the dilution buffer. Following vortexing for 30 s, the tubes are transferred to an ultrasonic bath and are sonicated according to the manufacturer's instructions. Alternatively, the sonication takes place by directly inserting the sonicator probe into the test tube. The detached cells are enumerated on nutrient media.

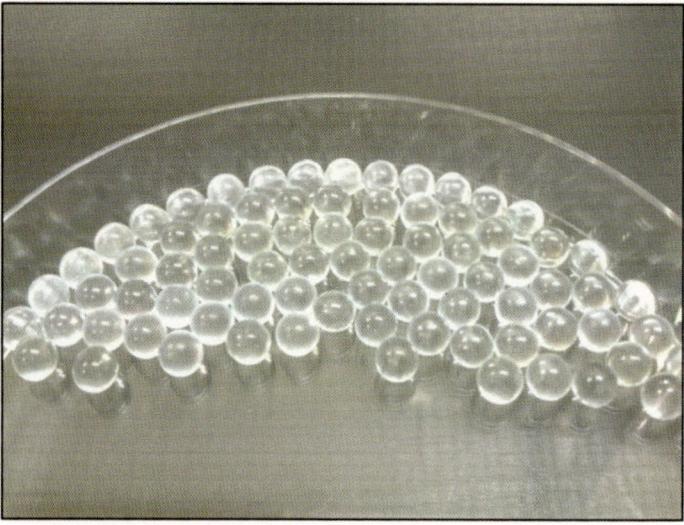

Fig. 4 Glass beads which are used to enhance the detachment of bacterial biofilms from surfaces during vortexing

4 Notes

1. The disinfectants/sanitizers to be tested can vary, as can the concentrations used. For commercial products follow the manufacturer's instructions, while noncommercial detergents/organic acids should be used according to the literature.

2. Use fresh (less than 30 days) slants of nonselective agar media (i.e., tryptic soy agar, brain–heart infusion agar) for the preparation of the cultures.

3. This is crucial, as the inoculum size may significantly influence the amount of the biofilm produced.

4. The properties (e.g., pH, NaCl) of the medium which will be used for the biofilm assay should be adjusted at the desired level before this step.

5. The duration of incubation and the static/dynamic conditions may affect the density of the biofilm formed.

6. It is also recommended that microplates be left for a few hours to dry before continuing. After this stage, the adherent cells on the bottom or the sides of the wells are macroscopically visible as a blue layer (if crystal violet has been used).

7. The quantification of the biofilms formed may also be conducted by using cotton swabs. In particular, a cotton swab is used to scan ten times the walls and the bottom of each well, applying the same pressure all the time. The swabs are further transferred to test tubes with ringer solution and vortexed for

2 min. Following decimal dilutions, the detached populations may be enumerated on nutrient agar media. An alternative approach for the quantification of biofilms, which are formed using the microtiter plate assay, is by detaching the adhered cells with sonication. The whole well is removed from the micro-plate, and 100 μl of PBS is added. The well with the PBS is then transferred to 1.9 mL PBS in a glass tube, and the mixture is sonicated for 8 s at 40 % power. Enumerate the cell concentration in the suspension by plating onto nutrient agar media.

8. This alternative method for the observation and quantification of static biofilms is especially useful when different types of surface materials are to be tested. Thus, the results of this method may be more realistic and representative to the conditions encountered in a food industry, compared with the microtiter plate assay, since more than one type of surface may host the formation of a biofilm. In addition, the potential of biofilm formation on commercial utensils, such as kitchen knives, food containers, and cutting boards, and further study of common decontamination interventions are also feasible with this method. Apart from abiotic surfaces, this methodology allows the study of biofilm formation on food surfaces, such as fresh produce. Briefly, the bacterial cells are adhered on the surface of a glass, metallic, plastic, or wooden coupon with specific dimensions. The quantification of the formed biofilm may be conducted either by classical microbiological methods (microbial counts) or *via* absorbance measurements of the dyed biofilms, as described in microtiter plate assay.

9. Alternatively, rinsing of the coupons may be applied in baths of regularly changed sterile water or PBS. This step enhances the removal of unattached or loosely attached cells from the surfaces.

10. Except for the static biofilm systems, there are also reactors that can be operated under continuous agitation and flow of nutrients, while in parallel, the real-time monitoring of the biofilm formation is also feasible. A characteristic example of such systems is the CDC biofilm reactor (CBR) developed by Donlan et al. [13]. The model system is consisted of a one-liter glass vessel, which is connected with a 10-L container with fresh medium and a second container for the microbial wastes. Eight independent polypropylene rods are vertically placed in the reactor, with each rod holding three round-shaped metal coupons. The glass vessel may be placed in a water bath (if necessary), while continuous agitation is achieved by placing the system on a digital stirrer. Each rod with the three coupons may be easily withdrawn from the top of the vessel at any experimental interval to further quantify the formed biofilm. However, a serial connection of the vessel with an attenuated

total reflectance cell with a similar metallic surface (coupon), which is further placed in an FTIR spectrometer, may enable the real-time monitoring of the differences between the chemical composition of the planktonic cells and the formed biofilms. This type of reactors combines the advantages of the simple surface/chip assay and the continuous-flow systems, while it also provides the opportunity for real-time monitoring of the biofilm formation.

11. The applied pressure may affect significantly the number of the detached cells, and therefore, the scanning approach should be well standardized. If cotton swabs are used, the cotton should be slightly moistened in the dilution buffer and dried at the wall of the tube before scanning of the surface.

12. Sonication is an alternative method for the detachment of bacterial biofilms from surfaces, especially of high porosity. In addition, this method enables the disaggregation of biofilm clumps that may remain intact during vortexing or swabbing. However, since bacteria are sensitive to ultrasounds, attention should be given to the frequency, power output, and duration of sonication. These parameters may vary depending on the type of sonicator used (ultrasonic bath or probe) and the specific instrument.

References

1. Sofos JN, Geornaras I (2010) Overview of current meat hygiene and safety risks and summary of recent studies on biofilms, and control of *Escherichia coli* O157:H7 in nonintact, and *Listeria monocytogenes* in ready-to-eat, meat products. Meat Sci 86:2–14

2. Borucki MK, Peppin JD, White D (2003) Variation in biofilm formation among strains of *Listeria monocytogenes*. Appl Environ Microbiol 69:7336–7342

3. Di Bonaventura G, Piccolomini R, Paludi D, D'Orio V, Vergara A, Conter M, Ianieri A (2008) Influence of temperature on biofilm formation by *Listeria monocytogenes* on various food-contact surfaces: relationship with motility and cell surface hydrophobicity. J Appl Microbiol 104:1552–1561

4. Folsom JP, Siragusa GR, Frank JF (2006) Formation of biofilm at different nutrient levels by various genotypes of *Listeria monocytogenes*. J Food Prot 69:826–834

5. Nilsson RE, Ross T, Bowman JP (2011) Variability in biofilm production by *Listeria monocytogenes* correlated to strain origin and growth conditions. Int J Food Microbiol 150:14–24

6. Dourou D, Beauchamp CS, Yoon Y, Geornaras I, Belk KE, Smith GC, Nychas G-JE, Sofos JN (2011) Attachment and biofilm formation by *Escherichia coli* O157:H7 at different temperatures, on various food-contact surfaces encountered in beef processing. Int J Food Microbiol 149:262–268

7. Berrang ME, Frank JF, Meinersmann RJ (2008) Effect of chemical sanitizers with and without ultrasonication on *Listeria monocytogenes* as a biofilm within polyvinyl chloride drain pipes. J Food Prot 71:66–69

8. Jessen B, Lammert L (2003) Biofilm and disinfection in meat processing plants. Int Biodeter Biodegr 51:265–269

9. Merritt JH, Kadouri DE, Toole GAO (2011). Growing and analyzing static biofilms. Curr Prot Microbiol Unit 1B1.1

10. Patel J, Sharma M (2010) Differences in attachment of Salmonella enterica serovars to cabbage and lettuce leaves. Int J Food Microbiol 139:41–47

11. Choi S, Bang J, Kim H, Beuchat LR, Ryu J-H (2011) Survival and colonization of *Escherichia coli* O157:H7 on spinach leaves as affected by inoculum level and carrier, temperature and relative humidity. J Appl Microbiol 11:1465–1472

12. Belessi C-E, Gounadaki AS, Schvartzman S, Jordan K, Skandamis PN (2011) Evaluation of growth/no growth interface of Listeria monocytogenes growing on stainless steel surfaces, detached from biofilms or in suspension, in response to pH and NaCl. Int J Food Microbiol 145:53–60

13. Donlan RM, Piede JA, Heyes CD, Sanii L, Murga R, Edmonds P, El-Sayed I, El-Sayed MA (2004) Model system for growing and quantifying *Streptococcus pneumoniae* biofilms in situ and in real-time. Appl Environ Mocrobiol 70:4980–4988

Chapter 22

Vaccination Studies: Detection of a *Listeria monocytogenes*-Specific T Cell Immune Response Using the ELISPOT Technique

Mohammed Bahey-El-Din and Cormac G.M. Gahan

Abstract

During systemic infection by *Listeria monocytogenes* the host develops a robust T cell-mediated immune response against the major immunodominant antigens of the pathogen. The enzyme-linked immuno-spot (ELISPOT) test is an accurate and reproducible means of measuring the extent of this T cell response. Here we describe a detailed ELISPOT protocol for measuring an epitope-specific CD8+ T cell-mediated immune response in mice vaccinated with low doses of *L. monocytogenes*. The basic approach can be easily adapted for the analysis of other vaccination regimes and target epitopes.

Key words *Listeria monocytogenes*, ELISPOT, Listeriolysin O, LLO epitope, CD8+ T cell, Cell-mediated immunity

1 Introduction

This chapter provides a step-by-step protocol for the detection of CD8+ cells specific against the major *Listeria monocytogenes* epitope LLO$_{91-99}$ [1]. Listeriolysin O (LLO) is a major virulence factor of *L. monocytogenes* which facilitates its escape from the phagosome to the cytosol of infected cells [2]. This escape allows listerial antigen presentation through the MHC class I presentation pathway with subsequent stimulation of epitope-specific CD8+ cell immune response [3]. The CD8+ immune response develops against the major listerial antigens LLO and P60 with LLO$_{91-99}$ being the epitope with the most prominent immune response [3].

In this protocol, we provide a detailed description of the enzyme-linked immuno-spot (ELISPOT) test for the detection of LLO$_{91-99}$-specific CD8+ cells. This is used as an example of the technique and can be used as a basis for testing specific T cell responses against other epitopes. The ELISPOT test detects epitope-specific IFN-γ-secreting cells, specifically CD8+ cells,

Kieran Jordan et al. (eds.), *Listeria monocytogenes: Methods and Protocols*, Methods in Molecular Biology, vol. 1157, DOI 10.1007/978-1-4939-0703-8_22, © Springer Science+Business Media New York 2014

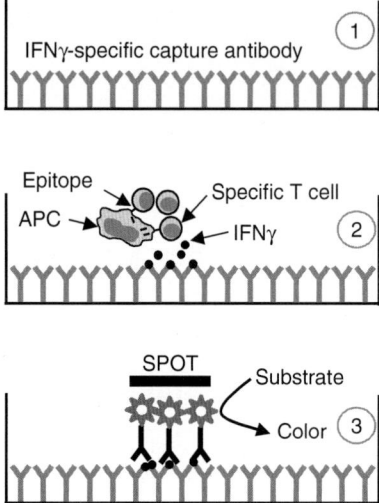

Fig. 1 An overview of the ELISPOT approach. (1) 96-well ELISPOT plates are coated with antibodies specific to interferon gamma (IFNγ). (2) In vitro immune interaction results in the stimulation of T cells specific to the added epitope. Stimulated cells produce IFNγ which binds to the antibodies. (3) Anti-IFN-γ biotinylated antibody is then added to form a sandwich with the IFN-γ captured between both antibodies. The bound IFN-γ is detected by adding peroxidase-labeled streptavidin followed by a peroxidase substrate to form spots. Each spot can be counted using a stereomicroscope to give an index of the number of epitope-specific IFNγ-producing cells

through the use of specialized antigen-presenting cells (APCs) which present the epitope only in the context of the MHC class I complex [4]. The ELISPOT test is a simple, sensitive, and cost-effective technique which is commonly used to detect an antigen-specific CD8+ immune response.

The following protocol describes the detection of LLO$_{91-99}$-specific CD8+ response following sublethal immunization with wild-type *L. monocytogenes* in mice. Nevertheless, the same protocol is applicable to detect CD8+ response against attenuated vaccine strains of *L. monocytogenes* and other live vectors expressing LLO [5–7].

The principle of the ELISPOT test is quite similar to sandwich enzyme-linked immunosorbent assay (ELISA) except that the substrate in ELISPOT produces spots rather than a colored supernatant (*see* Fig. 1 for a general overview of the ELISPOT test). Consequently, the ELISPOT plate wells have an opaque membrane base to allow spot formation. These spots can be easily counted using a stereomicroscope. In the following protocol, 96-well ELISPOT plates are coated with capture antibody for IFN-γ. Isolated splenocytes from vaccinated mice are stimulated by

APCs presenting the LLO$_{91-99}$ epitope so that only epitope-specific CD8+ cells will secrete IFN-γ. The secreted IFN-γ will be captured by the antibody at the base of the well. A second detection anti-IFN-γ biotinylated antibody is then added to form a sandwich with the IFN-γ captured between both antibodies. The bound IFN-γ is detected by adding peroxidase-labeled streptavidin followed by a peroxidase substrate to form spots. Each spot corresponds to an IFN-γ-secreting CD8+ cell.

2 Materials

2.1 Animals, Media, Bacteria, Antigens, and Equipment/Consumables

1. Female Balb/C mice, 6–8 weeks in age. All animal procedures must be reviewed and approved by the corresponding national or institutional ethical assessment committee.

2. *Listeria monocytogenes* EGDe serovar 1/2a.

3. The H^2-Kd-restricted LLO epitope, LLO$_{91-99}$, GYKDGNEYI (synthesized by Peptide Protein Research, UK).

4. Mouse mastocytoma cells P815-1-1 (The European Collection of Cell Cultures; ECACC) (*see* **Note 1**).

5. Luria-Bertani (LB) broth (10 g peptone from casein, 5 g yeast extract, and 5 g sodium chloride per litre). Technical agar (Merck) is added (1.5 % w/v) when solid media are required (*see* **Note 2**).

6. 1- DMEM GlutaMAX® medium (Gibco) (*see* **Note 3**) supplemented with 10 % fetal bovine serum (FBS), 1 mM sodium pyruvate, 0.1 mM nonessential amino acids, 10 mM HEPES, and antibiotic mixture (100 U/mL penicillin G and 100 µg/mL streptomycin). 2- RPMI 1640 medium supplemented with 10 % FBS, 23 mM sodium bicarbonate, 2 mM L-glutamine, 1 mM sodium pyruvate, 0.1 mM nonessential amino acids, 10 mM HEPES, and antibiotic mixture (100 U/mL penicillin G and 100 µg/mL streptomycin). All cell culture media, reagents, and components can be purchased from Gibco company or any other reputable company.

7. Tissue culture incubator adjusted at 37 °C and 5 % CO$_2$ atmosphere.

8. Cooling centrifuge (e.g., Hettich Universal 320R centrifuge).

9. Inverted microscope.

10. Vertical laminar airflow (LAF) cabinet.

11. T75 (75 cm^2) tissue culture flasks (Sarstedt).

12. Sterile microcentrifuge tubes, 1.5 mL.

13. Sterile micropipette tips (10, 200, 1,000 µL).

14. Multichannel micropipette.

15. Adjustable-volume micropipettes (10, 200, and 1,000 μL).

16. Stainless-steel dissecting scissors.

17. Stainless-steel forceps.

18. Sterile plastic pipettes (1 and 10 mL) (Sarstedt).

19. Sterile 50-mL falcon tubes.

2.2 ELISPOT Test

1. ELISPOT plates: MultiScreen® 96-well plates (Millipore corporation, cat. no. MAHA S45 10).

2. Primary capture antibody: Anti-mouse IFN-γ monoclonal antibody (mAb) (clone R4-6A2) (Biolegend, USA).

3. Detection antibody: Biotinylated anti-mouse IFN-γ mAb (Clone XMG1.2) (Biolegend, USA).

4. Peroxidase-labeled streptavidin (Biolegend, USA).

5. Peroxidase substrate: 3,3′-diaminobenzidine tetrahydrochloride (DAB) (Sigma Aldrich).

6. Phosphate-buffered saline (PBS) (Gibco), pH 7.4.

7. Tween 20 (Sigma).

8. Sterile stomacher bags.

9. Tris–HCl buffer, pH 7.5 (*see* **Note 4**).

10. Recombinant mouse interleukin-2 (IL-2) (eBioscience).

11. Glacial acetic acid.

12. Hydrogen peroxide 30 % (Sigma Aldrich).

2.3 Counting APCs, Splenocytes, and Spots on the ELISPOT Plate

1. A hemocytometer.

2. Cover slips.

3. Trypan blue 0.4 % solution (Sigma Aldrich).

4. Stereomicroscope.

3 Methods

See Fig. 2 for an overview of the steps involved in preparing and analyzing the ELISPOT plate. Figure 3 shows the results of a typical ELISPOT assay following immunization of mice with a sublethal dose of *L. monocytogenes*.

3.1 Preparation of Bacterial Inoculum

1. Subculture *Listeria monocytogenes* EGDe in 10 mL LB broth with overnight shaking (200 rpm) at 37 °C.

2. Collect bacterial cells by centrifugation at about $5,000 \times g$ for 5–10 min. Wash the cells once with PBS.

3. Resuspend the cells again in PBS and dilute to obtain a bacterial suspension containing 10^4 cfu/mL (*see* **Note 5**).

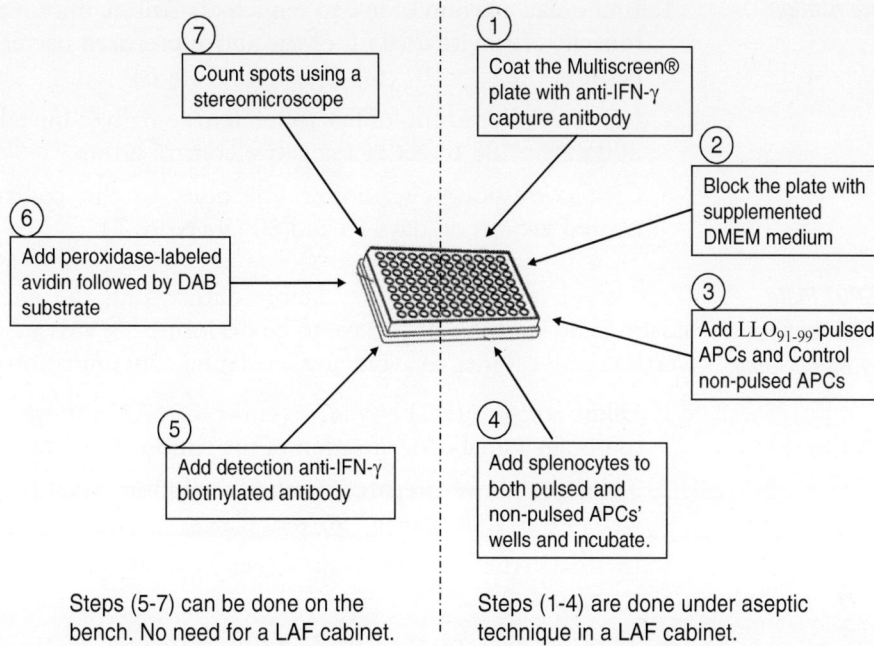

Fig. 2 An outline of the steps for preparing and analyzing the ELISPOT plate for the detection of IFN-γ-secreting cells in response to the listerial LLO$_{91-99}$ peptide epitope. A detailed explanation of each step is outlined in the text

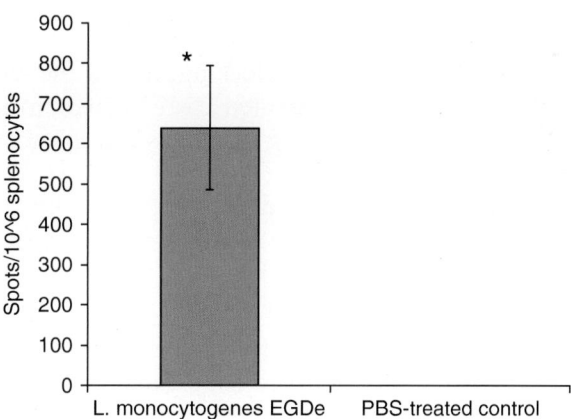

Fig. 3 Typical ELISPOT test results following a three-dose IP immunization regimen. Mice were immunized intraperitoneally with sublethal doses of *L. monocytogenes* EGDe on days 1, 15, and 30 and were euthanized for the ELISPOT test on day 42. A negative control PBS-treated group was included. *$p < 0.05$ as compared to the negative control group. Error bars represent mean ± SEM (standard error of the mean)

3.2 Immunization of Mice

1. Immunize a group of five to ten female Balb/c mice intraperitoneally (IP) with 200 µL of the above-prepared bacterial suspension (i.e., 2×10^3 cfu/dose) (*see* **Note 6**).

2. Inject another group of five to ten female Balb/c mice IP with 200 µL of PBS to act as a negative control group.

3. Give corresponding booster injections to the control and treated groups on days 15 and 30 (*see* **Note 7**).

3.3 ELISPOT Plate Coating with Primary Anti-IFN-γ Monoclonal Antibody

This step is performed 1 day before euthanizing the mice. The coating and washing steps have to be performed in a tissue culture vertical LAF cabinet to avoid any accidental contamination.

1. Dilute anti-mouse IFN-γ mAb (clone R4-6A2) with sterile PBS to obtain a final concentration of the antibody of 10 µg/mL.

2. From the above-prepared antibody solution, take 75 µL per well of the MultiScreen® 96-well plates.

3. Incubate the coated MultiScreen® plates overnight non-shaking at 20–25 °C.

4. Next morning, on the day of mouse euthanization, remove the mAb solution and wash the plates four times with sterile PBS. This washing process can be done by pipetting sterile PBS to fill the wells, then flick the PBS off, and repeat these steps three times more.

5. Block the remaining protein-binding sites in the wells by incubating with 100 µL/well of supplemented DMEM medium for 2 h at least at 37 °C.

6. After blocking, decant the medium, add 100 µL/well of supplemented DMEM medium again, and incubate at 37 °C until adding the APCs and splenocytes (see below).

3.4 Preparation of the Antigen-Presenting Cells P815-1-1

1. 7–10 days prior to the day of mouse euthanization, take one stock vial of the mouse mastocytoma cells P815-1-1 from the liquid nitrogen freezer and immerse immediately in a water bath at 37 °C.

2. After thawing, transfer the whole contents of the vial into pre-warmed supplemented RPMI 1640 medium in a T75 tissue culture flask and examine under an inverted microscope.

3. Incubate in a tissue culture incubator at 37 °C with 5 % CO_2 atmosphere.

4. Examine the cell density daily, and split (passage) the cells when they reach a cell density of $5–10 \times 10^5$ cells per mL.

5. For splitting the cells, take 0.5–1 mL of the P815-1-1 cell suspension, transfer into fresh pre-warmed supplemented RPMI 1640 medium in a T75 tissue culture flask, and incubate at 37 °C and 5 % CO_2 atmosphere.

6. On the day of murine euthanization, transfer the P815-1-1 cell suspension from the T75 tissue culture flask into a sterile 50-mL falcon tube.

7. Wash the cells once with supplemented RPMI 1640 medium (centrifugation at $280 \times g$ for 10 min at 20 °C) (*see* **Note 8**).

8. Resuspend the cells in 1–2 mL of supplemented RPMI 1640 medium.

9. Determine P818-1-1 cell count by taking an aliquot of the cell suspension and diluting it in trypan blue dye 0.4 %.

10. Put a cover slip over the hemocytometer, and using a micropipette fill the hemocytometer chamber under the cover slip with the cells/trypan blue suspension. Avoid overfilling or underfilling.

11. Count the average viable (non-blue) cells, and estimate the number of the cells per mL of culture medium. Viability should be more than 90 %.

12. Adjust the cell count to 10^7 cells/mL using supplemented RPMI 1640 and divide into two equal aliquots in two sterile tubes.

13. Add the LLO_{91-99} peptide epitope to one tube of the above-prepared APC suspension to a final concentration of 10^{-6} M and incubate at 37 °C and 5 % CO_2 atmosphere for 1 h with gentle agitation every 15 min (*see* **Notes 9** and **10**).

14. Incubate the other APC suspension tube without adding any peptide epitope at 37 °C and 5 % CO_2 atmosphere for 1 h with gentle agitation every 15 min. These non-pulsed APCs will serve as a negative control to exclude any nonspecific IFN-γ secretion.

15. After the 1-h incubation, wash the pulsed and non-pulsed APCs separately three times with DMEM GlutaMAX® containing 10 % FBS. Centrifugation should be at $280 \times g$ for 4 min at 4 °C.

16. Resuspend the APCs, pulsed and non-pulsed separately, in supplemented DMEM medium to attain a count of 2×10^6 cells/mL, and put 50 μL per well of this suspension in the MultiScreen® plates. Put the plates in the tissue culture incubator until adding the isolated animal splenocytes.

3.5 Euthanization of Mice and Isolation of Spleen Mononuclear Cells

1. Euthanize mice on day 42 according to institutional ethical guidelines (*see* **Note 11**).

2. Lay the animal on its right side, and sterilize the fur with 70 % ethanol.

3. Using scissors, make a skin incision on the left side of the animal midway between the last rib and the hip joint.

4. Make an incision in the peritoneal wall, and carefully pull out the spleen with a sterile forceps (metal forceps is immersed in alcohol, then flamed, and left enough time to cool down).

5. Put the isolated spleens separately in falcon tubes containing a pre-chilled 5–7 mL DMEM GlutaMAX® medium containing 10 % FBS (*see* **Note 12**).

6. Transfer the spleen and culture medium from one falcon tube to a sterile stomacher bag by carefully pouring the whole contents into the stomacher bag.

7. Using a 10 mL pipette or a long rod, make repeated rolling over the stomacher bag pushing the fluid inside towards the corner of the bag. This process will disintegrate and homogenize the spleen in the culture medium.

8. Once you see the spleen completely homogenized, transfer the entire contents back into the falcon tube and keep it on ice. You may pipette the suspension up and down several times to remove any clumps using a sterile 10 mL pipette (*see* **Note 13**).

9. Repeat the previous homogenization steps for all the isolated spleens one by one keeping the final falcon tubes on ice.

10. Centrifuge the falcon tubes at $280 \times g$ for 5 min at 4 °C, and discard the supernatant.

11. Resuspend the cells in 20 mL DMEM GlutaMAX® containing 10 % FBS and centrifuge again at 4 °C.

12. Discard the supernatant and resuspend in 2–3 mL supplemented DMEM, and put the tubes on ice.

13. Determine cell viability and count by taking an aliquot of the cell suspension and diluting it in trypan blue 0.4 % and acetic acid (2 % final concentration). The acetic acid will lyze red blood cells (RBCs) to facilitate the counting of viable splenocytes (*see* **Note 14**).

14. Determine the cell count and viability using a hemocytometer. Viability should be above 90 %.

15. Make twofold serial dilutions of the cells starting from 10^7 cells/mL. Keep tubes on ice while performing the dilution steps. Put 50 µL of each dilution in wells of the Multiscreen® plate containing either the epitope peptide-pulsed APCs or the control non-pulsed APCs (*see* **Note 15**).

16. Add recombinant mouse IL-2, diluted in supplemented DMEM GlutaMAX®, to each well in the Multiscreen® plate to a final concentration of 30 U/mL. The added volume of IL-2 should be as small as possible and preferably not exceeding 20 µL per well (*see* **Note 16**). All the previous steps for cells and plate manipulations must be done under aseptic conditions in a vertical LAF cabinet to avoid any accidental contamination.

17. Incubate the Multiscreen® plate at 37 °C and 5 % CO_2 atmosphere for 20–24 h.

3.6 Detection of IFN-γ-Secreting CD8+ T Cells

The following steps can be performed on the bench without the need for a LAF cabinet.

1. Wash the Multiscreen® plate after incubation by flicking off the cell suspensions of the wells and washing the wells three times with PBS.

2. Wash the plate an additional three to five times with PBS containing 0.05 % Tween 20 (hereafter called PBST). A 500-mL squeezable bottle can be used for the washing steps.

3. Prepare biotinylated anti-mouse IFN-γ mAb (clone XMG1.2) at a concentration of 2 μg/mL in PBST. Add 100 μL/well of this solution, and incubate the Multiscreen® plate overnight at 4 °C (*see* **Note 17**).

4. Remove the unbound mAb by washing the wells six times with PBST.

5. Add 100 μL per well of peroxidase-labeled streptavidin (diluted 1/800 in PBST). Incubate for 1 h at room temperature.

6. Wash the plate three to five times with PBST and then three times with PBS. Keep the wells wet by leaving PBS in the wells after the last wash until you prepare the peroxidase substrate.

7. Immediately before developing the Multiscreen® plate, prepare fresh DAB substrate 1 mg/mL in 50 mM Tris–HCl buffer pH 7.5. Add 0.5 μL/mL of 30 % hydrogen peroxide to the prepared DAB substrate quickly. Add 100 μL per well of this total solution to the Multiscreen® plate and leave at room temperature for 10–15 min until brown spots become visible. Stop the reaction by discarding the substrate solution and rinsing the plate under running tap water.

8. Leave the plates to dry at 20–25 °C, and count the colored spots corresponding to the number of IFN-γ-secreting CD8+ cells using a stereomicroscope (*see* **Notes 18** and **19**).

9. Subtract the number of spots in the control non-pulsed wells from the number of spots in the corresponding splenocyte dilution of pulsed wells.

10. Calculate the number of IFN-γ-secreting CD8+ cells per 10^6 splenocytes.

A representative example of results from our work is shown in Fig. 3. An overall summary of the steps of the ELISPOT technique described in this protocol is shown in Fig. 2.

4   Notes

1. Mouse mastocytoma cells P815-1-1 present antigens only in the context of MHC class I complex; thus, they provide a good specific tool to detect only CD8+ T cell immune responses against presented epitopes.

2. Some investigators use brain–heart infusion broth and agar for *L. monocytogenes* culture.

3. DMEM GlutaMAX® medium contains L-glutamine in the stabilized form L-alanyl-L-glutamine dipeptide which is more stable and does not degrade into ammonia upon storage or prolonged incubation. If regular DMEM culture medium is used, L-glutamine must be added at a final concentration of 2 mM.

4. Tris–HCl buffer, pH 7.5 can be prepared by dissolving the specified amount (usually for 1 M stock) of Tris base (TRIZMA®, Sigma Aldrich) in distilled water followed by pH adjustment using concentrated HCl and final volume adjustment.

5. An overnight culture of *L. monocytogenes* is about 10^9 cfu/mL. A preliminary optical density at 600 nm (OD600)/viable count correlation experiment should be performed for the bacterial suspension. Final animal inoculum must be confirmed by plate count on LB agar plates.

6. This dose is a sublethal dose of *L. monocytogenes* EGDe (2×10^3 cfu/dose). For attenuated Listeria mutants [7] and *Lactococcus lactis* vaccine strains expressing LLO [5, 6, 8] higher doses and different vaccination regimens can be used.

7. The immunization regimen can vary depending on the tested strains and research objectives. Some investigators give two injections only [7, 9], while others use more injections (five to eight) especially for extremely attenuated strains or non-virulent vectors such as *L. lactis*-expressing LLO [5, 8].

8. Do not use a higher speed of centrifugation as this may damage the cells.

9. Some investigators do not use APCs, and they add the whole antigen, not the epitope, directly to the isolated splenocytes with prolonged incubation for 72 h [10]. However, we found that using the precise epitope with APCs as outlined gave the most consistent results.

10. Other listerial antigen epitopes such as P60$_{217-225}$ [1] can also be examined by the same technique, but the CD8+ cell response would be largely dependent on the presence of functional LLO to facilitate antigen presentation via MHC class I presentation pathway.

11. The date of euthanization is determined according to the objective of the experiment; whether examining short- or long-term immune responses against the LLO antigen of *L. monocytogenes*.

12. Cells must be kept on ice from this step onwards; otherwise, cell viability would be greatly diminished.

13. Some investigators use a screen mesh to filter the cell suspension to remove aggregates and clumps after organ homogenization [11]. However, we did not find a difference in the results when the spleens are homogenized properly as described in this protocol. Minor clumps settle very quickly in the falcon tube, and cell suspension can be easily collected from the supernatant immediately after minor swirling. Also some investigators grind the spleens between two slides [11], but we find our described method the easiest to fulfill quick spleen homogenization without contamination. However, using a screen mesh for tissue filtration is important when examining livers for mononuclear cells due to large tissue aggregates and less homogenization efficiency in case of livers.

14. Acetic acid is added only with trypan blue to lyze RBCs so as to facilitate counting of spleen mononuclear cells. However, RBCs are not lyzed or removed for the ELISPOT test itself as they do not affect the results, and consequently no acetic acid should be added to the ELISPOT plates.

15. You must add each splenocyte dilution to a well containing pulsed APCs and to another control well with non-pulsed APCs to subtract spontaneous nonspecific IFN-γ-secreting cells from the results.

16. Addition of IL-2, although not essential for the test, significantly increases IFN-γ secretion by T lymphocytes and increases the ELISPOT test sensitivity [4, 11].

17. Some investigators suggest incubating the biotinylated antibody in the ELISPOT plate for only 2–4 h at room temperature [11]. However, this short-period incubation was not successful with us and we prefer overnight incubation at 4 °C for proper binding of the antibody.

18. An immunospot image analyzer can be used for automatic counting of the spots in the Multiscreen® plate wells.

19. Excessive background can be a problem when the IFN-γ-secreting cells are too numerous per well. This can be solved through the use of different dilutions of the splenocytes. Also, proper washing steps of the Multiscreen® plate greatly reduce the background.

Acknowledgments

This work was supported by a grant from the Irish Health Research Board (grant number RP/2006/23) and by a grant to the Alimentary Pharmabiotic Centre from Science Foundation Ireland under the Centres for Science Engineering and Technology (CSET) programme (grant numbers O7/CE/B1368 and 12/RC/2273).

References

1. Vijh S, Pamer EG (1997) Immunodominant and subdominant CTL responses to *Listeria monocytogenes* infection. J Immunol 158:3366–3371

2. Vazquez-Boland JA, Kuhn M, Berche P, Chakraborty T, Dominguez-Bernal G, Goebel W, Gonzalez-Zorn B, Wehland J, Kreft J (2001) *Listeria* pathogenesis and molecular virulence determinants. Clin Microbiol Rev 14:584–640

3. Pamer EG (2004) Immune responses to *Listeria monocytogenes*. Nat Rev Immunol 4:812–823

4. Miyahira Y, Murata K, Rodriguez D, Rodriguez JR, Esteban M, Rodrigues MM, Zavala F (1995) Quantification of antigen specific CD8+ T cells using an ELISPOT assay. J Immunol Methods 181:45–54

5. Bahey-El-Din M, Casey PG, Griffin BT, Gahan CG (2008) Lactococcus lactis-expressing listeriolysin O (LLO) provides protection and specific CD8(+) T cells against *Listeria monocytogenes* in the murine infection model. Vaccine 26:5304–5314

6. Bahey-El-Din M, Casey PG, Griffin BT, Gahan CG (2010) Expression of two *Listeria monocytogenes* antigens (P60 and LLO) in *Lactococcus lactis* and examination for use as live vaccine vectors. J Med Microbiol 59:904–912

7. McLaughlin HP, Bahey-El-Din M, Casey PG, Hill C, Gahan CG (2013) A mutant in the *Listeria monocytogenes* Fur-regulated virulence locus (frvA) induces cellular immunity and confers protection against listeriosis in mice. J Med Microbiol 62:185–190

8. Bahey-El-Din M, Casey PG, Griffin BT, Gahan CGM (2010) Efficacy of a *Lactococcus lactis* Δ*pyrG* vaccine delivery platform expressing chromosomally integrated hly from *Listeria monocytogenes*. Bioeng Bugs 1:66–74

9. Geginat G, Schenk S, Skoberne M, Goebel W, Hof H (2001) A novel approach of direct ex vivo epitope mapping identifies dominant and subdominant CD4 and CD8 T cell epitopes from *Listeria monocytogenes*. J Immunol 166:1877–1884

10. Stambas J, Pietersz G, McKenzie I, Nagabhushanam V, Cheers C (2002) Oxidised mannan-listeriolysin O conjugates induce Th1/Th2 cytokine responses after intranasal immunisation. Vaccine 20:1877–1886

11. Carvalho LH, Hafalla JC, Zavala F (2001) ELISPOT assay to measure antigen-specific murine CD8(+) T cell responses. J Immunol Methods 252:207–218

Chapter 23

Sampling the Food Processing Environment: Taking Up the Cudgel for Preventive Quality Management in Food Processing Environments

Martin Wagner and Beatrix Stessl

Abstract

The *Listeria* monitoring program for Austrian cheese factories was established in 1988. The basic idea is to control the introduction of *L. monocytogenes* into the food processing environment, preventing the pathogen from contaminating the food under processing. The Austrian *Listeria* monitoring program comprises four levels of investigation, dealing with routine monitoring of samples and consequences of finding a positive sample. Preventive quality control concepts attempt to detect a foodborne hazard along the food processing chain, prior to food delivery, retailing, and consumption. The implementation of a preventive food safety concept provokes a deepened insight by the manufacturers into problems concerning food safety. The development of preventive quality assurance strategies contributes to the national food safety status and protects public health.

Key words Austrian Listeria monitoring program, Processing environment, Food safety

1 Introduction

Listeria monocytogenes is with Salmonella, the pathogen most tightly regulated by food legislation. In EU-27, food investigation for *L. monocytogenes* follows a legal framework that has been amended after a risk assessment approach in 2005. The relevant EU directive EU (DIR) 2073/2005 requires in Article 5 that food business operators (FBOs) manufacturing ready-to-eat foods that pose a *L. monocytogenes* risk for public health shall sample the processing areas and equipment for *L. monocytogenes* as part of their sampling scheme [1]. Monitoring of food processing environments (FPE), however, is not widely harmonized. To close the gap, a working group under the lead of experts of the European Reference Laboratory (EURL) for listeriosis has currently published a guideline that is especially devoted to this problem [2]. This guideline recommends technical procedures;

Kieran Jordan et al. (eds.), *Listeria monocytogenes: Methods and Protocols*, Methods in Molecular Biology, vol. 1157, DOI 10.1007/978-1-4939-0703-8_23, © Springer Science+Business Media New York 2014

however, choice of sample types, locations, and frequency still remain a responsibility of the FBO quality management.

The Institute of Milk Hygiene, Milk Technology and Food Science (IMMF) launched a *Listeria* monitoring program for Austrian cheese factories in 1988. In the 1980s the national cheese industry had to encounter safety concerns raised by consumers who were alerted by reports on the implication of cheese products in outbreaks of listeriosis. To support the quality management board of important Austrian cheese producers in its endeavour to offer a safe product to the consumers, the Austrian Listeria monitoring program was established. This report comprehensively describes the principles of this monitoring and how it is practically performed.

The basic idea is to control the introduction of *L. monocytogenes* into the FPE from outside, preventing the pathogen from contaminating the food under processing. Unlike many other pathogens, *L. monocytogenes* does not persistently colonize the intestine of animals used for food production. *L. monocytogenes* might be found in raw meat and milk to some extent; however, at least in milk production the organism is reliably killed by the heat treatment schemes employed. The picture may be different in meat and fish processing where contaminated raw materials pose a hazard to violate the safety of finished product. Aside from whether raw material may trigger Listeria contamination, *L. monocytogenes* is in many cases transferred from extramural environmental sources into FPEs. Either through unhygienic design of equipment or by inefficient sanitation of FPEs, *L. monocytogenes* starts to reside in niches from where they might not get eradicated. From these niches, *L. monocytogenes* may spread and contaminate batches of food under processing again and again. Persistent contamination of FPEs has been shown in milk, meat, and fish industries, and clonal types of *L. monocytogenes* can occur for many years [3–5].

Whether the persistence of *L. monocytogenes* in FPEs is merely a fact of insufficient hygiene or due to genetic traits enabling some clones to adapt better is still under debate [6]. It seems to be very unlikely that a single factor will be capable of explaining the phenomenon. It is very likely that inadequate hygiene leads to a selection of those clones that can cope better with the hygiene and disinfection procedures applied. Some studies have shown that the genetic diversity of isolates recovered at the beginning of enforced sanitation is reduced as the standard of hygiene is elevated [7]. This fact clearly shows that not all clones seem to be resistant to the same degree. Another study has demonstrated that smaller food enterprises suffer from longer contamination periods than bigger ones. Limited capabilities and resources in combating contamination obviously lead to a higher probability for persistence [8]. Whatever drives the persistence of *L. monocytogenes*, cross-contamination is certainly a major trigger of accidental occurrence

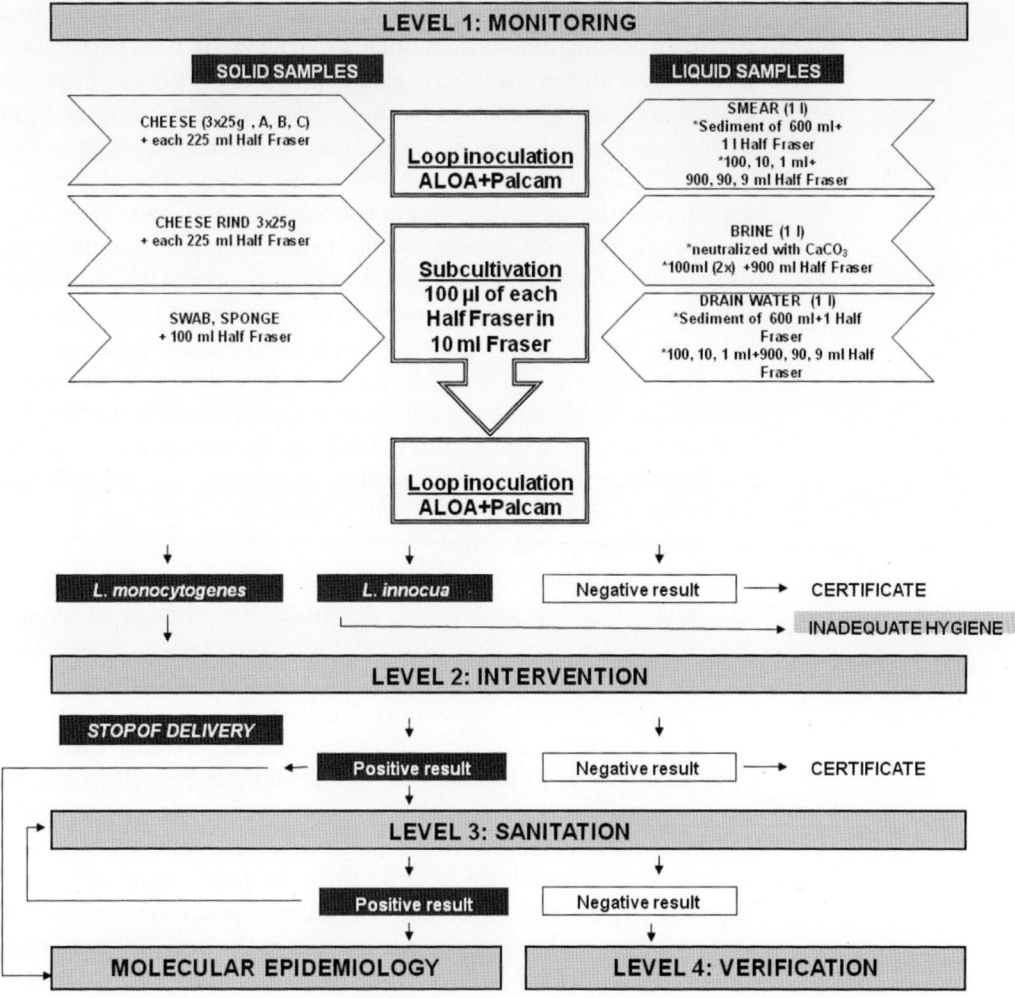

Fig. 1 The concept of the Listeria monitoring program as implemented in Austria since 1988

of *L. monocytogenes* in food products. If *L. monocytogenes* has contaminated a batch of food then it is a matter of growth potential whether such a batch may pose a risk to the health of consumers. This is why EU (DIR) 2073/2005 prescribes in Article 3 that FBOs shall conduct compliance studies to limit during shelf life what on particular applies to ready-to-eat foods that are able to support the growth of *L. monocytogenes* [1].

The Austrian *Listeria* monitoring program comprises four levels of investigation (Fig. 1). Level 1 deals with the routine monitoring of sample matrices supplementary to cheese production such as smear, brine, or wash water. Smear liquid is used to smear cheese lots under production. Monitoring of other cheese types is based on matrices needed for the production of a particular cheese brand. For mould ripened cheeses, since neither smeared nor put into brine for extended periods of time, wash water samples (water used to clean production devices such as trolleys or trays) have been found as an interesting matrix for monitoring purposes. If the processing technology includes neither smear-ripening, other methods of surface ripening, nor brining, drain water from a plant gully is examined. Sampling is performed at least every month with an appropriate device such as smear robot, gully, and brine tank.

Negative results during standard monitoring are certified and used by the company management to document the status of safety. We experienced companies that have been on board the monitoring program for decades, and they have never had a problem with *L. monocytogenes* since. If *L. innocua* or other nonpathogenic Listeria are detected, we document an inappropriate status of hygiene and recommend to reconsider the hygiene management concept applied since it could easily be that pathogenic *L. monocytogenes* are introduced through the same mechanism as a nonpathogenic species obviously was.

If *L. monocytogenes* is detected in a FPE sample, an intervention phase (level 2) is initiated. An increasing number of samples are collected by the factory personnel from sources which have shown to be contaminated and from additional sources (tanks, racks, conveyor belts, ventilation, escalators).

The intervention examination is intended to clarify the extent of the contamination scenario. It should further help the manufacturer to decide whether a risk of cross-contamination to processed food has arisen. Isolates from food contact materials are treated as if those would have been isolated from the food commodity itself. In some papers, the transmission of *L. monocytogenes* within FPEs was followed, and it could be shown that at the end the contamination was transferred from processing line to processing line [9]. In parallel, investigations of cheese samples according to the legal requirements determine whether a FPE contamination has already reached the food batch. If yes, and a test indicates that a food batch does not comply with the legal requirements, then the batch should not be delivered or should be recalled from the market (internal recall).

If the intervention examination confirms the monitoring result, a scrupulous sanitation of the FPE in addition to routine procedures is strongly recommended (level 3). The sanitation usually cannot be performed without advice from external experts. The sanitation should be systematic and include a crucial survey of all factors that might drive the contamination scenario. This in particular includes a critical review of hygiene barriers, internal traffic management, maintenance of buildings and rooms, and cleaning and disinfection procedures applied. Sub-typing of in-house strains supports the sanitation specialist to trace the contamination to hot spots from where Listeriae might recontaminate. It is our experience that a heavily contaminated FPE is hard to sanitize.

In most cases the goal is to repel the contamination to spots from where a food batch contamination can be excluded. This status of a coexistence of FPE contamination with pending food processing is a fragile reality in many food processing enterprises and should be monitored intensively (level 4: verification).

2   Why to Use FPE Samples as Sampling Matrix?

Many FBOs have grown "organically" over decades. They start a business, and if successful they enlarge their premises. In many cases, FBOs are conglomerates of rooms signified by complex trafficking of raw materials and half processed and finished product. Clear one-way trafficking systems might not exist, and contamination hot spots are usually to be found where trafficking routes cross. Monitoring FPEs rather than food commodities is therefore a far more efficient way to draw a "contamination map" and to provide the management with evidence that a risk for food contamination exists.

To facilitate prioritization of (counter) measures, we recommend to clearly define critical control areas (CCAs) where an FPE contamination is not acceptable. It makes a difference whether a *L. monocytogenes* positive gully is located beneath a slicer or if a gully is located in a packaging room. CCA should be clearly marked (e.g., by marks on floors, in construction maps), and hygiene barriers should ensure that CCAs are not visited or trespassed by unqualified personnel. The high hygiene standard that should exist in CCAs can only be monitored by FPE samples.

A further problem is the timely coincidence of sanitation and food processing. FBOs share contracts with retailers and are urged to deliver the food in a timely manner. Management of contamination disturbs the smooth running of business, and the consequences of not addressing contamination might therefore be neglected. In our experience, when an FBO management does not actively react upon recognition of a contamination scenario, then this has a very detrimental effect on both the performance of the quality manager (QM) and the personnel that need to perform the countermeasures. FPE samples support the QM to demonstrate hygiene failures since they provide convincing evidence on how *L. monocytogenes* is transmitted within an FPE. The results are highly educational and can be used for internal training sessions.

In some FBOs, sanitation is done by specialized companies in the evenings after processing, with a loose communication to the QM of the FBO. The efficiency of the sanitation can reliably be shown only when FPE samples are monitored.

A last argument, and in our opinion the most striking, is that FPE monitoring has a preventive character, whereas food testing has a final character (if contaminated that adversity is real).

FPE monitoring provides the company with time to react, and in many FPE contamination scenarios, the food commodity is not going to be contaminated when appropriate countermeasures are taken.

3 The Sensitivity of the Method Decides

The Listeria monitoring program as developed in Austria in the recent decades builds on a semiquantitative approach to monitoring. This approach was chosen since the detection limit of ISO 11290:1 [10] is designed for testing food commodities and might be insufficient to monitor FPE samples. If the sample is liquid, volumes of 1, 10, 100, and 900 ml are tested. This strategy allows estimation of the degree of contamination. Positive results in all four sample volumes indicate a higher number of *L. monocytogenes* present in the matrix, whereas positive results in 100 or 900 ml demonstrate a contamination by a low number of Listeriae. In our observation, approximately 15 % of FPE samples are positive in 900 ml only. This approach, however, could be improved, and for this reason we are working to develop molecular quantitative tools that provide a comparable level of information.

In the mid-1990s, confirmation of presumptive clones was supported by polymerase chain reaction assays targeting the *hly* gene [11] and a multiplex PCR procedure that was developed especially for the monitoring program. This PCR assay amplifies a species-specific *iap* gene sequence from *L. monocytogenes*, *L. innocua*, *L. grayi*, and the group of other species [12]. This PCR allows a discrimination of pathogenic from nonpathogenic *Listeria* species and a partial discrimination among the latter. By swabbing the whole agar plate, we overcome the problem that nonpathogenic Listeria may overgrow *L. monocytogenes* during the two-step enrichment procedure. Despite the development of chromogenic media that have certainly speeded up the confirmation of pathogenic Listeria we find the PCR-based approach to be highly informative since also chromogenic plates might be too unselective [13] and overgrown with commensal species such as *Bacillus* spp. that can produce colony morphologies very similar to *L. monocytogenes* (Fig. 2).

4 Concluding Remarks

The experience gained in recent years confirms our conviction that the developed procedure is superior to standard hygiene inspection techniques such as time-to-time swabbing or contact sliding. An investigation of drain water samples encompasses the contamination status of large plant areas. Using drain investigation is largely devoid of the sampling bias which might occur by swabbing of small areas of the plant interior. Investigation of smear and brine

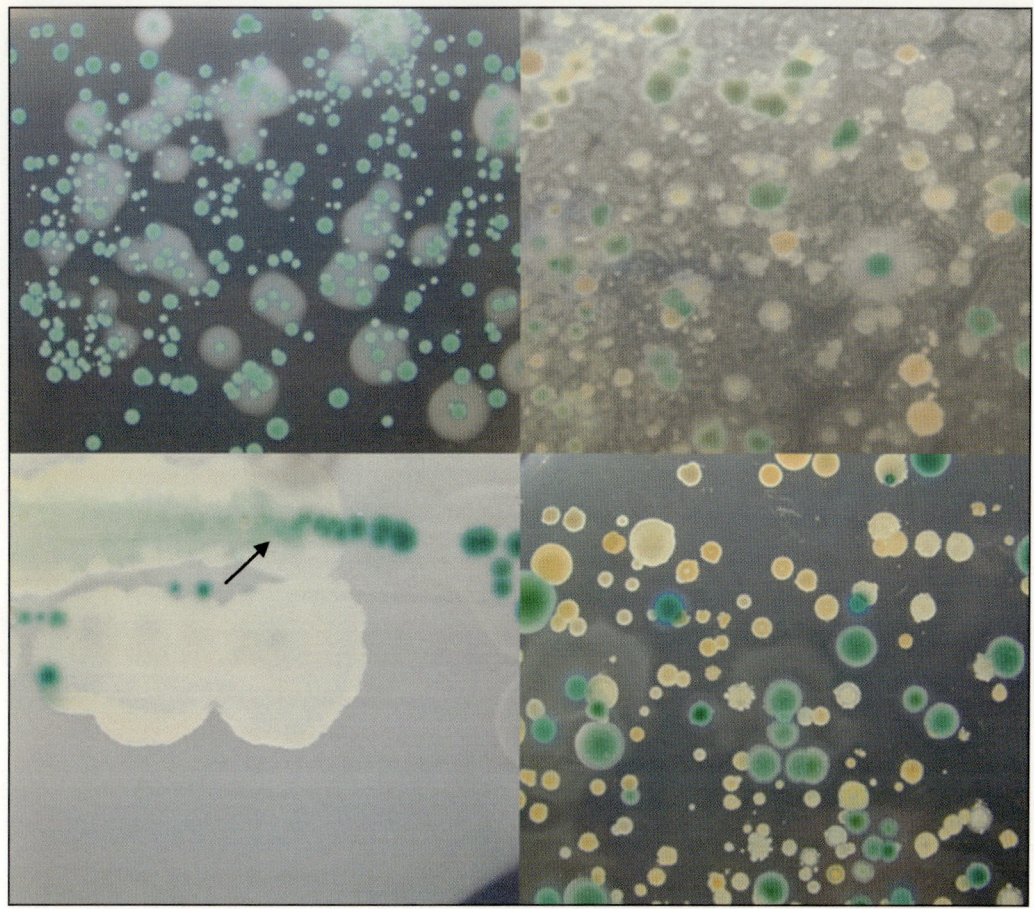

Fig. 2 *L. monocytogenes* confirmation on chromogenic medium ALOA, yes or no? Only a single colony is *L. monocytogenes*; other phenotypes have been produced by Bacilli, Staphylococci, and Corynebacteria

samples is beneficial as these production supplements mimic the contamination status of the whole batch under processing. Although the Listeria monitoring program was initially developed for cheese makers, the national meat industry is adopting the procedures. The most salient difference is the selection of the sampling points. Smear samples are usually replaced by slicer material (Table 1). If a meat product is not going to be sliced then the FPE monitoring is based on drain water samples and contact sliding. However, it must be kept in mind that meat FPEs are more easily contaminated due to transmission of *L. monocytogenes* through contaminated raw materials and the definition of a CCA is usually more challenging.

Preventive quality control concepts attempt to detect a foodborne hazard along the food processing chain, prior to food delivery, retailing, and consumption. The implementation of a preventive food safety concept provokes a deepened insight of the

Table 1
List of procedures and measures taken according to the Listeria monitoring program as implemented in Austria

Sample nature	Method	Sample matrix	Sample volume	If positive action recommended	Legal consequences
FPE[a]	Swab (pad), ISO 11290	Non-FCM[b]	>1 m[c]	Check hygiene barriers[d], training	–
	Swab (pad), ISO 11290	FCM[b]		Intensified testing of FCM, check food batch(es)	Isolate(s) sent to national RLL[c]
	ISO 11290	Liquid (gully)	1–900 ml	Check hygiene barriers, intensified testing, training	–
Food associated	ISO 11290	Liquid (smear, brine)	1–900 ml	Check food batch(es), sanitation, training	Isolate(s) sent to national RLL
	ISO 11290	Solid (slicer material)	5 × 25 g	Check food batch(es), sanitation, training	Isolate(s) sent to RLL
Food batches				Stop of delivery, recall, sanitation, training	Action according to EU (DIR) 2073/32005

[a]FPE Food processing environment
[b]FCM Food contact material
[c]RLL National Reference Laboratory for Listeriosis, acc. to Lebensmittelsicherheits- und Verbraucherschutzgesetz BGBl. I Nr. 13/2006
[d]If gullies are positive that are located in the critical control area, we recommend to intensify testing of FCM

manufacturers into problems concerning food safety. We believe that the development of preventive quality assurance strategies was a contribution to the national food safety status in the recent years. Although our data show a decrease of *L. monocytogenes* prevalence in FPE samples in Austrian cheese factories, we had to experience the biggest multinational outbreak of listeriosis in 2009/2010 [14]. Recalled batches of the contaminated food were heavily contaminated with *L. monocytogenes* [15]. Until June 2009 the food producer involved in this outbreak was taking part in the monitoring program and a positive smear water sample was the first evidence of an ongoing contamination scenario. However, the producer then decided to run a sort of individual intervention procedure and left the monitoring program as offered by the IMMF. Despite this massive backlash, we believe that the Austrian *Listeria* monitoring program is a state-of-the-art concept to control the risk for *L. monocytogenes* transmission in FPE environments. The program is implemented by private national food testing companies, and most cheese and meat manufacturers participate in the program.

References

1. Anonymous (2005) Regulation (EC) No 2073/2005 on microbiological criteria for foodstuffs. http://eur-lex.europa.eu/LexUriServ/LexUriServ.do?uri=CONSLEG:2005R2073:20 060101:EN:PDF

2. Anonymous (2012) Guidelines on sampling the food processing area and equipment for the detection of *Listeria monocytogenes*. Available online: http://www.ansespro.fr/eurl-listeria/Documents/LIS-Cr-201213D1.pdf. Accessed 24 Nov 2013

3. Keto-Timonen R, Tolvanen R, Lundén J, Korkeala H (2007) An 8-year surveillance of the diversity and persistence of *Listeria monocytogenes* in a chilled food processing plant analyzed by amplified fragment length polymorphism. J Food Prot 70:1866–1873

4. Senczek D, Stephan R, Untermann F (2000) Pulsed-field gel electrophoresis (PFGE) typing of *Listeria* strains isolated from a meat processing plant over a 2-year period. Int J Food Microbiol 62:155–159

5. Vogel BF, Jørgensen LV, Ojeniyi B, Huss H, Gram H (2001) Diversity of *Listeria monocytogenes* isolates from cold-smoked salmon produced in different smokehouses as assessed by Random Amplified Polymorphic DNA analyses. Int J Food Microbiol 65:83–92

6. Carpentier B, Cerf O (2011) Persistence of *Listeria monocytogenes* in food industry equipment and premises. Int J Food Microbiol 145:1–8

7. Ortiz S, López V, Villatoro D, López P, Dávila JC, Martínez-Suárez JN (2011) A 3-year surveillance of the genetic diversity and persistence of *Listeria monocytogenes* in an Iberian pig slaughterhouse and processing plant. Foodborne Pathog Dis 7:1177–1184

8. Wagner M, Eliskases-Lechner F, Rieck P, Hein I, Allerberger F (2006) Characterization of *Listeria monocytogenes* isolates from 50 small-scale Austrian cheese factories. J Food Prot 69:1297–1303

9. Pappelbaum K, Grif K, Heller I, Würzner R, Hein I, Ellerbroek L, Wagner M (2008) Monitoring hygiene on- and at-line is critical for controlling *Listeria monocytogenes* during produce processing. J Food Prot 71: 735–741

10. Anonymous (1996) Horizontal method for the detection of Listeria monocytogenes from food stuffs. International Standardisation Organization, Geneva, Switzerland

11. Border PM, Howard JJ, Plastow GS, Siggens KW (1990) Detection of *Listeria* species and *Listeria monocytogenes* using polymerase chain reaction. Lett Appl Microbiol 11:158–162

12. Bubert A, Hein I, Rauch M, Lehner A, Yoon B, Goebbel W, Wagner M (1999) Detection and differentiation of *Listeria* spp. by a single reaction based on multiplex PCR. Appl Environ Microbiol 65:4688–4692

13. Stessl B, Luf W, Wagner M, Schoder D (2009) Performance testing of six chromogenic ALOA-type media for the detection of *Listeria monocytogenes*. J Appl Microbiol 106:651–659

14. Fretz R, Sagel U, Ruppitsch W, Pietzka A, Stoger A, Huhulescu S, Heuberger S, Pichler J, Much P, Pfaff G, Stark K, Prager R, Flieger A, Feenstra O, Allerberger F (2010) Listeriosis outbreak caused by acid cured cheese Quargel, Austria and Germany 2009. Euro Surveill 15(5) pii: 19477

15. Schoder D, Rossmanith P, Glaser K, Wagner M (2012) Fluctuation in contamination dynamics of *L. monocytogenes* in quargel (acid curd cheese) lots recalled during the multinational listeriosis outbreak 2009/2010. Int J Food Microbiol 157:326–331

INDEX

Kieran Jordan et al. (eds.), *Listeria monocytogenes: Methods and Protocols*, Methods in Molecular Biology, vol. 1157,
DOI 10.1007/978-1-4939-0703-8, © Springer Science+Business Media New York 2014

Printed by Printforce, the Netherlands